IMMUNOBIOLOGY OF PROTEINS AND PEPTIDES – III

VIRAL AND BACTERIAL ANTIGENS

ADVANCES IN EXPERIMENTAL MEDICINE AND BIOLOGY

Recent Volumes in this Series

Volume 182
GENE EXPRESSION IN MUSCLE
Edited by Richard C. Strohman and Stewart Wolf

Volume 183
DRUGS AFFECTING LIPID METABOLISM VIII
Edited by David Kritchevsky, William L. Holmes, and Rodolfo Paoletti

Volume 184
MECHANISMS OF CELL-MEDIATED CYTOTOXICITY II
Edited by Pierre Henkart and Eric Martz

Volume 185
IMMUNOBIOLOGY OF PROTEINS AND PEPTIDES III: VIRAL AND BACTERIAL ANTIGENS
Edited by M. Zouhair Atassi and Howard L. Bachrach

Volume 186
MICROENVIRONMENTS IN THE LYMPHOID SYSTEM
Edited by G. G. B. Klaus

Volume 187
AIDS-ASSOCIATED SYNDROMES
Edited by Sudhir Gupta

Volume 188
SOMATOSTATIN
Edited by Yogesh C. Patel and Gloria S. Tannenbaum

IMMUNOBIOLOGY OF PROTEINS AND PEPTIDES—III

VIRAL AND BACTERIAL ANTIGENS

Edited by

M. Zouhair Atassi

Baylor College of Medicine
Houston, Texas

and

Howard L. Bachrach

United States Department of Agriculture
Plum Island, New York

PLENUM PRESS • NEW YORK AND LONDON

Library of Congress Cataloging in Publication Data

International Symposium on the Immunobiology of Proteins and Peptides (3rd: 1984: Tahoe City, Calif.)
Immunobiology of Proteins and peptides III.

(Advances in experimental medicine and biology; v. 185)
"Proceedings of the Third International Symposium on the Immunobiology of Proteins and Peptides, held June 24-27, 1984, in Tahoe City, California"—T.p. verso.
Includes bibliographies and index.
1. Bacterial antigens—Congresses. 2. Viral antigens—Congresses. I. Atassi, M. Z. II. Bachrach, Howard L. III. Title. IV. Title: Viral and bacterial antigens. V. Series. [DNLM: 1. Antigens, Bacterial—immunology—congresses. 2. Antigens, Viral—Immunology—congresses. WL AD559 v.185/QW 573 I61]
QR186.6.B33I58 1984 616.9'2079 85-9440
ISBN 978-1-4684-7976-8 ISBN 978-1-4684-7974-4 (eBook)
DOI 10.1007/978-1-4684-7974-4

Proceedings of the Third International Symposium on the Immunobiology of Proteins and Peptides, held June 24-27, 1984, in Tahoe City, California

Scientific Council of the Symposium

M. Z. Atassi, Chairman
H. L. Bachrach
E. Benjamini
G. R. Dreesman
J. W. Goodman
B. H. Iglewski
J. L. Melnick
J. C. Sadoff
R. G. Webster
N. J. Calvanico, Secretary
C. R. Young, Secretary

Symposium sponsors include

Abbott Laboratories
Beckman Instruments
E.I. Du Pont De Nemours and Co.
Kallestad Laboratories, Inc.
Merck, Sharp and Dohme Research Laboratories
Ortho Diagnostics
Pierce Chemical Company
Syntex Research
Tago, Incorporated
United States Department of Agriculture
United States Army Medical Research and Development Command

Softcover reprint of the hardcover 1st edition 1985
A Division of Plenum Publishing Corporation
233 Spring Street, New York, N.Y. 10013

PREFACE

This symposia series, founded in 1976, is devoted to the advancement and dissemination of knowledge in the field in immunology, particularly as it relates to the immune recognition and responses to protein and peptide antigens. Leading investigators are convened every 2 or 3 years for the purpose of consolidating the research on protein and peptide antigens of defined structure and to focus on these findings in the context of contemporary immunology. Each symposium has focussed on a particular aspect of molecular and cellular immunology of proteins and peptides. It is extremely gratifying that, in the last 2-3 years, the scientific community has shown a heightened interest in the study and understanding of protein and peptide antigens.

The third symposium was devoted to viral and bacterial antigens. Great advances have been made in recent years in the elucidation and synthesis of protein antigenic sites. These, together with advances in cloning, expression and sequencing of protein genes, have offered new avenues for the preparation of synthetic vaccines for viral, bacterial and other antigens. Such vaccines have been the aspiration of immunologists for over 20 years. The meeting has served to integrate and correlate the current knowledge of these systems with developing trends in immunology and to identify the most promising new directions for future investigations.

M.Z. Atassi
H.L. Bachrach

CONTENTS

PERSPECTIVES OF THE IMMUNOLOGY OF PROTEINS+

M. Zouhair Atassi

Verna and Marrs McLean Department of Biochemistry
Baylor College of Medicine
Houston, Texas 77030

When this symposia series was founded in 1976, the major contributions and concepts which constituted the foundations of contemporary immunology of proteins and peptides were quite clear. However, in the last 2-3 years, this field has witnessed an unprecedented activity and has been the focus of a great deal of excitement. As a result, many investigators have entered this area of endeavor from backgrounds other than the molecular immunology or biochemistry of proteins. Therefore, there has been considerable confusion and ambiguity as to the meaning and understanding of various principles and the source of various ideas, concepts and approaches, on which the field is founded. I would like to use the prerogative of these opening remarks, to put a few of these matters in their proper historical perspective.

Origin of Approaches for defining protein antigenic sites

Progress in defining protein antigenic structures has been very slow, and the elucidation of the entire antigenic structure of a protein frustrated many attempts. Early efforts to unravel protein antigenicity can be traced back to the work of Landsteiner (1942) on silk fibroin, La Presle (1955) and Porter (1957) on serum albumin, Brown (1962) on ribonuclease as well as the work of Anderer (1963) on tobacco mosaic virus protein. These attempts underscored the immense difficulties of studying protein antigens and further emphasized the crucial need to know the full recognition profile of a protein by the immune system. Considerable chemical and technical factors were responsible for the slow progress in this field, and these have already been discussed in detail (Atassi, 1975, 1977). Indeed, it is the studies of Porter and La Presle and Brown and Anderer that stimulated my interest, early in 1962, in trying to map out the full antigenic structure of myoglobin.

Analysis of all these previous studies clearly revealed that, because of its complexity, the antigenic structure of a protein cannot be derived

+This text represents the opening and Plenary lecture, edited for publication and references, of the Third International Symposium on Immunobiology of Proteins and Peptides.

Site	Structure and Location	No. of Residues
1	15 16 21 22 (Ala)-Lys-Val-Glu-Ala-Asp-Val-(Ala)	6 (or 7)
2	56 62 Lys-Ala-Ser-Glu-Asp-Leu-Lys	7
3	94 99 Ala-Thr-Lys-His-Lys-Ile	6
4	113 119 His-Val-Leu-His-Ser-Arg-His	7
5	145 146 151 (Lys)-Tyr-Lys-Glu-Leu-Gly-Tyr	6 (or 7)

Fig. 1. Covalent structures of the five antigenic sites of sperm whale Mb. Residues in parentheses are part of the antigenic site only with some antisera. Thus for site 1, the reactive region invariably comprises residues 16-21 and with some antisera Ala-15 is part of the region (which will then correspond to residues 15-21) while with other antisera Ala-22 is an essential part of the region (which will then correspond to residues 16-22). This site occupies either six or seven residues depending on antiserum. A shift in site 2 by two residues to the left was recently found with a monclonal antibody to Mb. For sites 3 and 4 no such 'displacement' or 'shift' has been observed yet (at least with the antisera so far studied). In the case of site 5, Lys-145 can be part of the antigenic region only with some antisera and this site will therefore comprise six or seven residues, depending on the antiserum. (From Atassi, 1975).

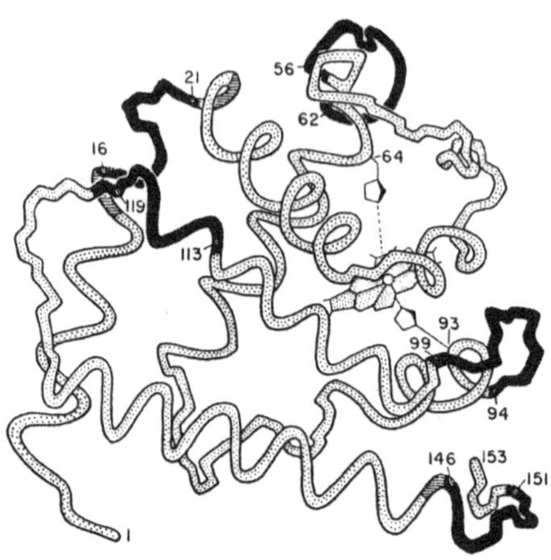

Fig. 2. A schematic diagram showing the mode of the folding of Mb and its antigenic structure. The solid black portions represent segments which have been shown to comprise accurately entire antigenic sites. The striped parts, each corresponding to one amino acid residue only, can be part of the antigenic site with some antisera. The dotted portions represent parts of the molecule which have been shown exhaustively to reside outside antigenic sites. (From Atassi, 1975).

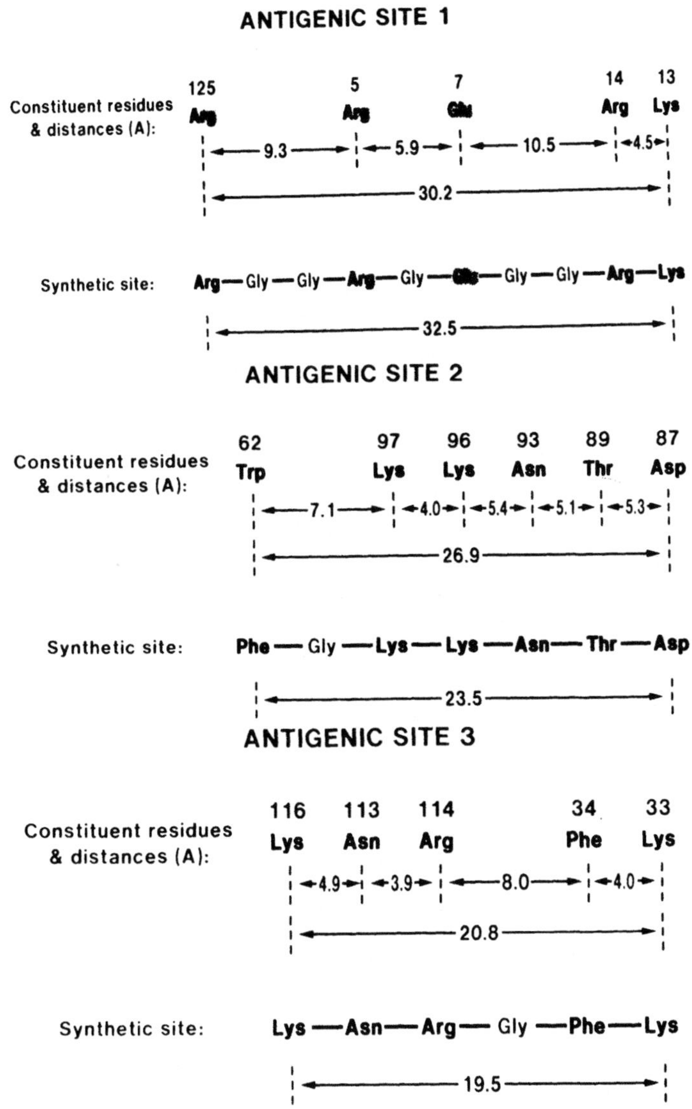

Fig. 3 The three antigenic sites constituting the antigenic structure of lysozyme. The diagram shows the spatially adjacent residue constituting each antigenic site and their numerical positions in the primary structure The distances (in A) separating the consecutive residues and the overall dimension of each site (in its extended form) are given, together with the dimension of each surface-simulation synthetic site. The precise boundary conformational and directional definitions of the sites are described in the text. The three sites account essentially for all (96-100%) the antigenic reactivity of lysozyme. (From Atassi and Lee, 1978).

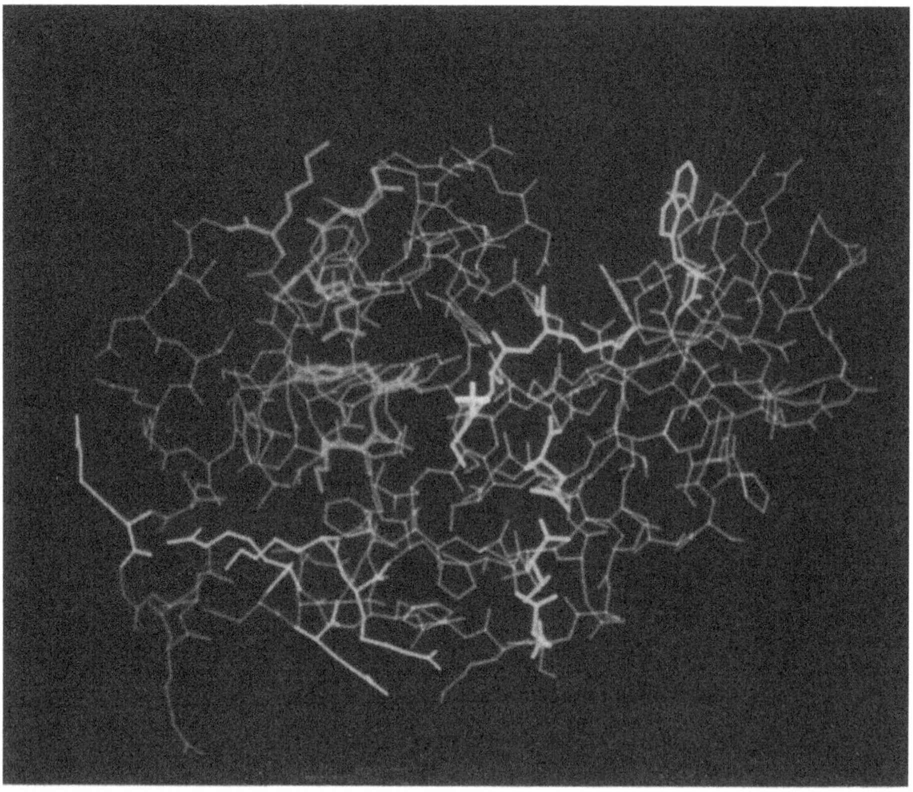

Fig. 4 Photograph of a computer-generated lysozyme model showing the relative positions of the residues constituting the surface-simulation antigenic sites. The preferred 'direction' of site 1 (highlighted in the lower left-hand side of the molecule), at least by surface-simulation synthesis, is Arg-125 to Lys-13. The preferred 'direction' of site 2 (highlighted in the right-hand half of the molecule) with goat antisera is Trp-62 to Asp-87, but no directional preference is observed with rabbit antisera. It must be noted that only Trp-62 should be visible from this perspective; however, with the aid of computer graphics all of the residues of this site are shown. Site 3 (highlighted in the upper left-hand side of the molecule) exhibits a preferred 'direction' of synthesis form Lys-116 to Lys-33 (adapted from Atassi and Lee, 1978).

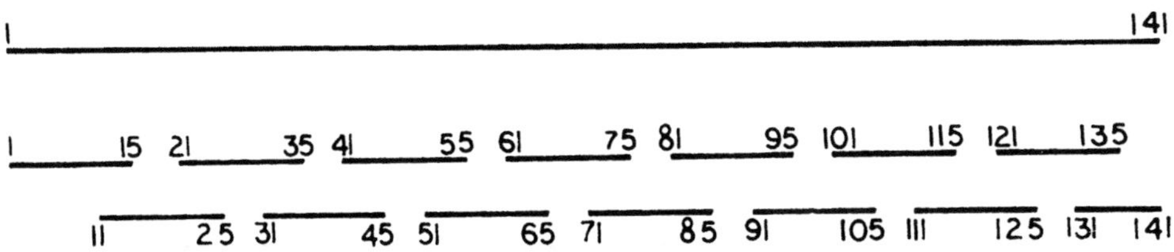

Fig. 5. Synthetic overlapping peptide strategy employed for the delineation of the continuous antigenic sites of the α chain of human adult hemoglobin (Kazim and Atassi, 1980b, 1982). The strategy relied on the synthesis of the entire molecule in 15-residue peptides (except for the C-terminal region 131-141), each overlapping its two adjacent neighbours by five residues on both sides. (From Kazim and Atassi, 1980b)

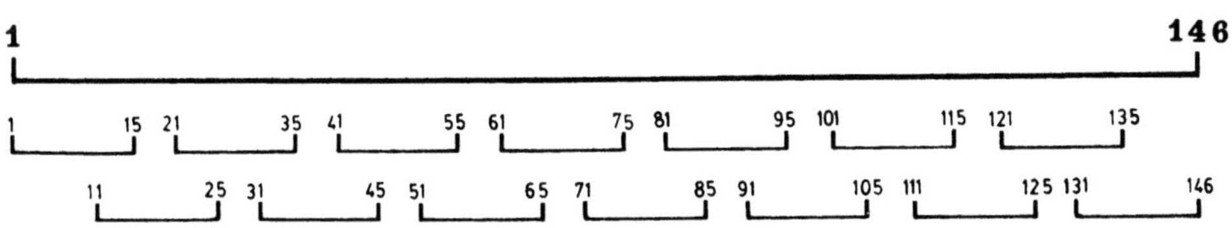

Fig. 6. Synthetic overlapping peptides employed for the delineation of the continuous antigenic sites of β chain of human adult hemoglobin (Yoshioka and Atassi, 1983, 1985). The strategy relied on the synthesis of the entire polypeptide chain in 15-residue peptides, each overlapping its two adjacent neighbours by five residues on both sides. (From Yoshioka and Atassi, 1985)

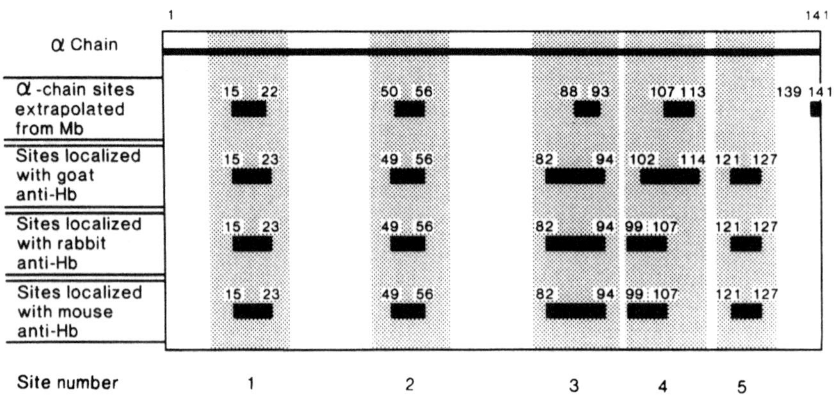

Fig. 7 Schematic diagram of the α chain of Hb, α-chain regions that have been extrapolated from the five antigenic sites of sperm-whale Mb, and the antigenic sites of the α chain that were localized with goat, rabbit and mouse antisera to Hb. The general areas of the sites are highlighted by shading. The numbers refer to the locations of the residues in the α chain. Owing to differences in the sizes of the polypeptide chains of Mb and the α subunit, the antigenic site Mb 145-151 does not really have a full structural counterpart in the α chain. Note that sites 3 and 4 have not yet been narrowed down to their precise boundaries. The regions shown imply that these two sites fall within, but do not necessarily include all of the indicated area. (From Kazim and Atassi and the precise boundary delineation of sites 2 and 5 by McCormick and Atassi, 1983a,b)

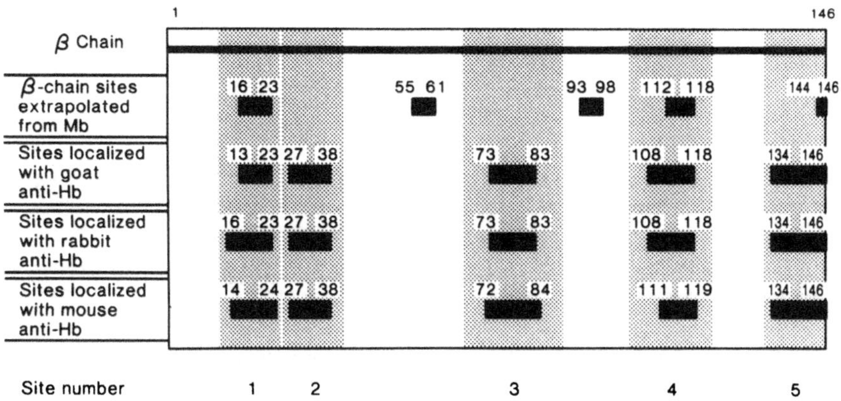

Fig. 8 Schematic diagram showing the β chain of Hb, its regions that have been extrapolated from the five antigenic site of sperm-whale Mb and its antigenic sites that have been localized with goat, rabbit and mouse antisera of Hb. The general areas of the sites are highlighted by shading for clarity. The numbers refer to the locations of the residues in the β chain. It is not implied that the antigenic sites comprise the entire size of the regions shown, but rather that they fall within these regions. (From Yoshioka and Atassi, 1985)

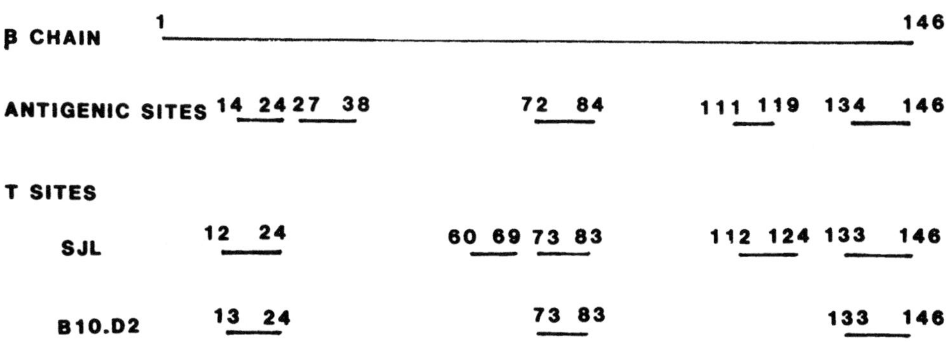

Fig. 9. Schematic diagram showing the full profile of the regions of the β chain of HbA that carry the continuous sites of T-cell recognition (T-sites) in two mouse strains. The T-cell recognition sites are also compared with the sites of mouse antibody (B-cell) recognition (i.e. antigenic sites) which were localized (Figure 8) by Yoshioka and Atassi (1985). It is not implied that the entire regions shown comprise the T site, rather that the sites reside within these regions. Indeed, in the localization by the overlapping peptide strategy, the T sites are localized within intentionally larger regios than their expected size. (From Yoshioka et al., 1985)

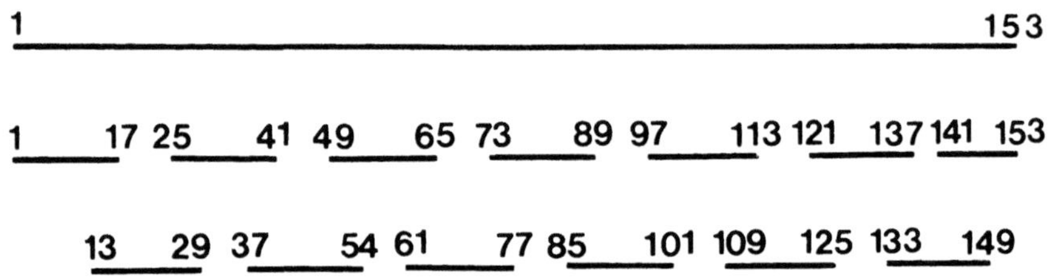

Fig. 10 Synthetic overlapping peptides employed for the delineation of the sites recognized by T-cells on sperm-whale Mb. The strategy depended on the synthesis of the entire polypeptide chain in 17-residue peptides (except for the C-terminal 13-residue peptide 141-153), each overlapping with its adjacent neighbours by five residues on both sides. (From Bixler and Atassi, 1983)

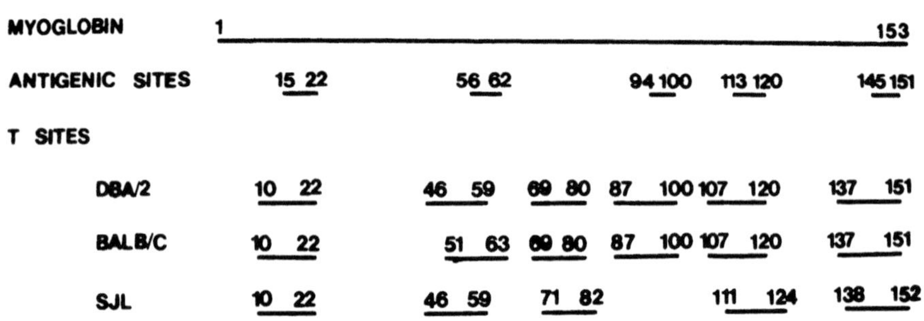

Fig. 11. Schematic diagram showing the full profile of the regions of sperm-whale Mb that carry the continuous sites of T-cell recognition (T sites) in three mouse strains. It is not implied that the entire regions shown comprise the T site, rather that the sites reside within these regions. Indeed, in the localization by the overlapping peptide strategy (Fig. 10), the T sites are localized within intentionally larger regions than their expected size. (From Bixler and Atassi, 1983, 1984b)

by a single chemical approach. A strategy was therefore designed(Atassi, 1972) which relied on five different chemical approaches, none of which by itself was considered capable of defining the full antigenic structure of a protein. These approaches relied on (1) determination of the role of conformation on the immune responses to a protein; (2) immunochemistry and conformation of specific chemical protein derivatives modified at selected amino acid locations; (3) immunochemistry and conformation of several sets of overlapping peptide fragments obtained from the protein; (4) Immunochemistry and conformation of specific chemical derivatives of immunochemically active fragments; (5) Precise definition of the sites, which have been localized through systematic application of the foregoing four approaches, by synthetic peptides spanning various overlaps of each localized region. The results obtained by one approach were always used to confirm or correct those obtained by the others.

This complex chemical strategy which was founded on sound biochemical principles enabled us to define and finally confirm by synthetic peptides the full antigenic structures of myoglobin (Atassi, 1975) and of lysozyme (Atassi, 1978). The elucidation of these structures revealed that in both molecules, the antigenic sites occupy discrete conformationally-sensitive surface areas of the molecule. The sites may comprise residues which are either directly linked by peptide bonds [e.g. the continuous antigenic sites of sperm-whale Mb (Figures 1 and 2) (Atassi, 1975)] or are distant in sequence but come in close spatial proximity by virtue of the folding of the polypeptide chain [exemplified by the discontinuous antigenic sites of lysozyme (Figures 3 and 4); (Atassi, 1978)]. For a protein of an unknown antigenic structure, both architectural alternatives for the sites (i.e. continuous or discontinuous; Atassi and Smith, 1978) should be anticipated.

Because of the complexity of the strategy employed for the above two proteins, a comprehensive synthetic approach was introduced (Kazim and Atassi, 1980b) for the localization of the continuous antigenic sites on a protein. This approach consists of the synthesis and examination of the immunochemical activities of a series of consecutive overlapping peptides that encompass the entire protein chain. The peptides are designed to have a uniform size and each has uniform overlaps with the two adjacent peptides (for examples, see Figures 5 and 6). The approach has proved to be effective and has been employed to localize the full profile of continuous sites, recognized by antibodies on the α chain (Figure 7) (Kazim and Atassi, 1980b, 1982) and β chain (Figure 8) (Yoshioka and Atassi, 1983, 1985) of human adult hemoglobin. In addition, the full T-cell recognition profiles of these two chains (Figure 9) (Yoshioka et al., 1985) and of myoglobin (Figures 10 and 11) (Bixler and Atassi, 1983, 1984b) and lysozyme (Figures 12 and 13) (Bixler et al., 1984a,b) have also been localized by this strategy. Although this approach has been quite effective, it is nevertheless costly and labor-intensive, especially when applied to large molecules. For example, in hemoglobin where the αβ half molecule comprises 287 residues, the synthetic peptides constituted a total of 417 residues because of the synthetic repetition of the peptide segments in the overlap regions.

The determination in this laboratory, however, of several protein antigenic structures has shown that the antigen sites of a protein occupy, without exception surface regions of the molecule (Atassi, 1975, 1978, Kazim and Atassi, 1982; Yoshioka and Atassi, 1983, 1985). Since no interior parts of the molecule are antigenic, any approach that will focus the scanning effort to the surface should shorten the time required

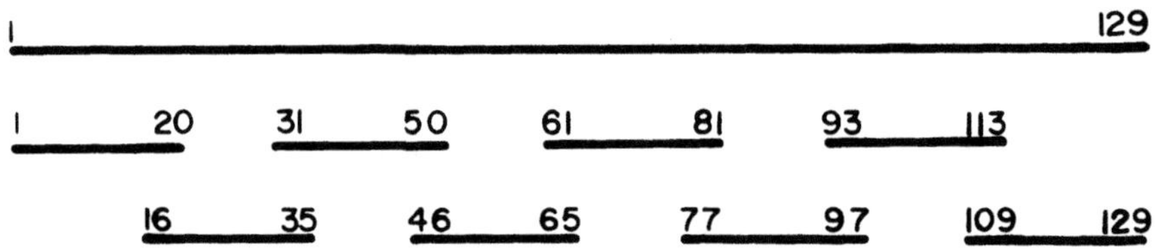

Fig. 12. Synthetic overlapping peptides employed for the delineation of the continuous T-cell recognition sites on hen lysozyme. The strategy relied on the synthesis of the entire polypeptide chain in 20-residue peptides, each overlapping its two adjacent neighbours by five residues on both sides. (From Bixler, Yoshida and Atassi, 1984a)

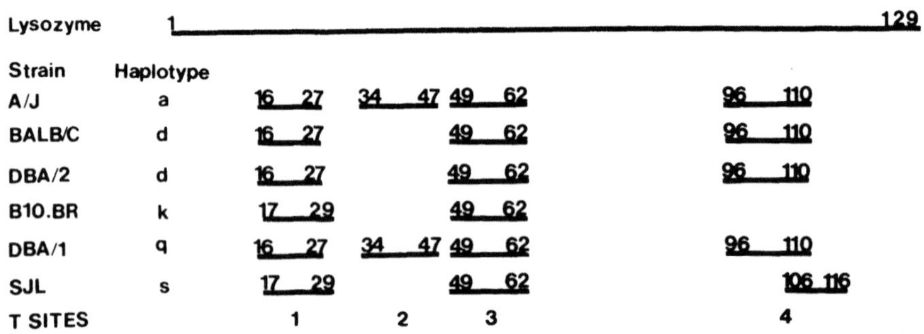

Fig. 13. Schematic diagram showing the regions on hen lysozyme that have been shown to carry the continuous sites of T-cell recognition (T sites) with several mouse strains. The sites were localized by the overlapping peptide strategy shown in Fig. 12. Note that each T site shown falls within but may not necessarily include all the region indicated. In addition to the four T sites shown in this diagram, long-term cultures of lysozyme-primed T-cells were found to mount a proliferative response to the three surface-simulation synthetic sites shown in Fig. 3. (From Bixler, Yoshida and Atassi, 1984a,b)

for the localization and synthesis of protein antigenic sites. "Synthetic surface scanning" has been devised, demonstrated and applied to the localization of sites of T- or B- cell recognition on the hemagglutinin (HA) molecule of influenza A virus (Atassi and Kurisaki, 1984). The three-dimensional structure of influenza A virus HA has been determined by X-ray crystallography at 3 Å resolution (Wilson et al., 1981). This structural information has permitted the systematic scanning of the surface by synthesis of all exposed continuous regions on the molecule and led to the localization of 12 antigenic regions on HA1 and HA2 (Figure 14). It should be emphasized that comprehensive synthetic surface scanning can be applied at this stage only for the localization of the 'continuous' antigenic sites. Discontinuous sites cannot realistically be locaized by this method, because of the large number of alternatives by which surface residues can be linked by surface-simulation synthesis (Atassi, 1978; Atassi et al, 1976b; Lee and Atassi, 1976, 1977a,b). The surface 'contact' residues of a discontinuous site must be identified by chemical or other means before surface-simulation synthesis is attempted.

Other approaches are frequently employed because of their simplicity. Protein mutants are used to implicate, from their immunochemical cross reactions, the involvement of some residues in antigenic sites. This approach ignores charge-transfer, conformational and steric effects induced by amino acid replacements (for more details, see Atassi and Habeeb, 1977; Atassi, 1980, 1984a,b; Atassi and Kazim, 1980). The results derived by this approach have not been synthetically confirmed. Another approach relies on the use of hydrophilicity index to examine the protein sequence for regions of hydrophilicity maxima which are presumed to localize the antigenic sites of the protein (Hopp and Woods, 1981). This approach, which does not take into account three-dimensional considerations of protein structure, is not supported by what is known about protein antigenic sites (for more detail, see Atassi and Webster, 1983; Atassi, 1984a).

Architecture of Protein Antigenic Sites

Immunochemical studies with protein fragments indicated the presence of continuous antigenic sites quite early (Landsteiner, 1942; La Presle, 1955; Porter, 1957; Brown, 1962; Anderer, 1963, Anderer and Schlumberger, 1965; Benjamini et al., 1964, 1965; Crumpton and Wilkinson, 1965; Atassi and Saplin, 1968, 1971; Atassi and Singhal, 1970a; Singhal and Atassi, 1971) and these were the first such sites to be precisely defined by synthesis (Koketsu and Atassi, 1973, 1974a,b). In fact, determination of the first complete antigenic structure of a protein (i.e. myoglobin) revealed that it has five continuous antigenic sites (Atassi, 1975). The existence of continuous antigenic sites has recently been reaffirmed and is also now, ten years later, being proclaimed as a new discovery (Lerner, 1982, 1984; Sutcliffe et al., 1983; Wabuke et al., 1984).

The existence of antigenic sites having a discontinuous architecture was also proposed relatively early (Atassi and Saplin, 1968). However, discontinuous antigenic sites were first discovered in lysozyme (Atassi et al., 1976a) and confirmed unequivocally by surface-simulation synthesis (Atassi et al., 1976b; Lee and Atassi, 1976, 1977a,b). Thus, contrary to recent claims (Berkower et al., 1982; Berzofsky et al., 1982; Benjamin et al., 1984), discontinuous sites (frequently and imprecisely referred to an topographic sites) are not a new discovery. Further, these recent assertions which were based on speculative analysis of the immunological cross-reactions of some protein mutants had no independent

X-31 Hemagglutinin Peptides

HA 1 Peptides

	23	36
Peptide 1	G.T.L.V.K.T.I.T.D.D..Q.I.E.V	
	124	134
Peptide 2	G.F.T.W.T.G.V.T.Q.N.G.	
	138	152
Peptide 3	A.G.K.R.G.P.G.S.G.F.F.S.R.L.N	
	154	167
Peptide 4	L.T.K.S.G.S.T.V.P.V.L.N.V.T.	
	175	188
Peptide 5	D.K.L.Y.I.W.G.V.H.H.P.S.T.N.	
	183	199
Peptide 6	H.H.P.S.T.N.Q.E.Q.T.S.L.Y.V.Q.A.S.	
	201	218
Peptide 7	R.V.T.V.S.T.R.R.S.Q.Q.T.I.I.P.N.I.G.	
	272	288
Peptide 8	A.P.I.D.T.G.I.S.E.G.I.T.P.N.G.S.I	
	300	315
Peptide 9	I.T.G.A.G.P.K.Y.V.K.Q.N.T.L.K	

HA 2 Peptides

	1	1
Peptide 10	G.L.F.G.A.I.A.G.F.I.E	
	56	68
Peptide 11	I.E.K.T.N.E.K.F.H.Q.I.E.K.	
	68	84
Peptide 12	K.E.F.S.E.V.E.G.R.I.Q.D.L.E.K.Y.V	

Fig. 14. Covalent structures of the surface synthetic peptides of influenza virus hemagglutinin that have been shown (Atassi and Webster, 1983; Atassi and Kurisaki, 1984) to have binding activities toward anti-viral antibodies. Also antibodies to the peptides were shown to bind intact virus. In the synthetic peptides, the residues Cys-277, Cys-281 and Cys-305 were replaced by glycine. It is not implied that the antigenic sites constitute the full size of the peptides shown, but rather that the sites reside within the boundaries of these peptides. The single letter notation of the amino acids is: A, alanine; D, aspartic acid; E, glutamic acid; F, phenylalanine; G, glycine, H, histidine; I, isoleucine; K, lysine; L, leucine; N, asparagine; P, proline; Q, glutamine; R, arginine; S, serine; T, threonine; V, valine; W, tryptophan; Y, tyrosine. (From Atassi and Kurisaki, 1984)

experimental verifications by precise chemical and synthetic approaches. In contrast, earlier characterizations of discontinuous protein antigenic sites (Atassi et al., 1976a) were solidly supported by on the concept of surface-simulation synthesis which, although initially devised (Atassi et al., 1976b; Lee and Atassi, 1976, 1977a,b) to provide a synthetic proof for such antigenic sites, was immediately recognized to have wide versatility for mimicking other protein binding sites (Atassi et al., 1976b; Atassi, 1978). The versatility of the concept has been demonstrated, for example, in the synthetic mimicking of antibody combining sites (Atassi and Zablocki, 1977; Twining and Atassi, 1978; Kazim and Atassi, 1980a) and substrate-binding sites of enzymes (Atassi, 1984d, 1985).

Some molecular and functional aspects derived from determination of protein antigenic structures.

A more detailed review of the molecular and functional impact of the determination of protein antigenic structures has recently appeared (Atassi 1984b). The determination of the first complete protein antigenic structures gave the first unique insight into the molecular features responsible for the immune recognition of protein antigens. Thus protein antigenic sites were invariably found to be surface features, limited in number, small in size, and having discrete boundaries. It should be noted that trace antibody responses (1-2 %) to regions interspersed among the immunodominant sites are usually present (for reviews, see Atassi, 1980, 1984b).

The antigenic structures, determined with antisera raised in several host species, of myoglobin (Atassi, 1975, Twining et al., 1980, 1981a), lysozyme (Atassi, 1978), serum albumin (Atassi et al., 1979; Sakata and Atassi, 1980 a,b; Atassi, 1982), human adult hemoglobin (Kazim and Atassi, 1980b, 1982; Yoshioka and Atassi, 1983, 1985), ragweed allergen Ra3 (Atassi and Atassi, 1985) and several of the antigenic sites of influenza virus hemagglutinin (Atassi and Webster, 1983; Atassi and Kurisaki, 1984), have shown that the locations of the antigenic (i.e. antibody binding) sites are independent of the host species. Thus, the same antigenic sites (with minor displacements) are recognized on a given protein regardless of the host species in which the antisera are raised and the time antisera are obtained after the initial immunization (Sakata and Atassi, 1980a; Twining et al., 1981b). Accordingly, it is concluded (Atassi and Kazim, 1978; Kazim and Atassi, 1978) that, contrary to proposals of the others (Reichlin, 1972, 1975; Urbanski and Margoliash, 1977), the immune response is not based on, nor is it directed to, locations of the immunized antigen which differ in sequence from the homologous protein in the immunized host. Rather, the antigenicity of the sites is, to a large extent, determined by the conformational uniqueness of these regions on the molecule [i.e. inherent in their three-dimensional locations (Kazim and Atassi, 1977, 1978; Atassi and Kazim, 1978)]. This led to the discovery of the role of 'structurally inherent' antigenic sites in the autoimmune response to protein antigens (Kazim and Atassi, 1977, 1978, Atassi and Kazim, 1978; Atassi, 1981; Sakata and Atassi, 1981; Yokota et al., 1980). It was concluded (Atassi and Kazim, 1978; Atassi, 1980) that the concept of 'foreignness' is not a necessary prerequisite in immune recognition and responses and that autoimmune T cell clones are not delected (Yokota et al., 1980).

Localization of the full profile of T-cell recognition sites on myoglobin (Bixler and Atassi, 1983, 1984b), lysozyme (Bixler et al., 1984 a,b) ragweed Ra3 (Kurisaki et al., 1985) and hemoglobin (Yoshioka et al.,

1985) and several of those on influenza virus hemagglutinin (Atassi and Kurisaki, 1984) has provided the first opportunity to compare, at the submolecular level, the regions involved in immune recognition by T cells and by B cells (i.e. antibodies). It was found that the sites recognized by B cells are also recognized by T cells (for examples, see Figures 11 and 13). However, the antigen has, in addition, regions that are recognized only by T cells. Like B cell recognition, recognition by T cells is not restricted to a particular antigenic architecture (Bixler and Atassi, 1984a). Determination of the antibody and T-cell recognition profiles for these proteins has shown that the responses to each of the sites recognized on a protein by antibodies and/or by T cells are under separate Ir gene control (Okuda et al., 1979; Twining et al., 1981a; David and Atassi, 1982; Bixler et al., 1984b; Kurisaki et al., 1985). The parameters influencing this genetic control have been reviewed (David and Atassi, 1982).

These investigations gave the first insight into the molecular features responsible for T and B-cell recognition of proten antigens. In addition, they clearly revealed that both T and B cell recognitions are dependent on native conformation. It was concluded that recognition of, and immune responses to, a protein antigen do not follow its internalization and fragmentaton by macrophage (Atassi, 1980). The internalization and proteolytic fragmentation of the antigen by the macrophage is a process which is related to the function of macrophage in clearance and is most likely coincident with, but unrelated to and functionally distinct from, the well established collaborative role of macrophage in antigen presentation.

<u>Distinction between protein antigenic sites and antibody responses to peptides as immunogens.</u>

There seems now to have occured, very frequently in the recent literature (Lerner 1982, 1984; Leach, 1983; Benjamin et al., 1984; Westhof et al., 1984), a great deal of confusion as to what constitutes a protein antigenic site (also often referred to as epitope or determinant). This term has always been used to describe a region of a protein that is recognized (by antibodies and/or T cells) in the immune response when the <u>whole protein</u> is used as an immunogen. However, it has long been known that any peptide from a given protein, if immunized as a conjugate on an appropriate large carrier, will elicit the formation of anti-peptide antibodies. These antibodies can bind to the parent protein if the immunizing peptide represented a surface area of the intact protein. This even applies to peptides representing regions that are not antigenic when the whole protein is used as an immunogen.

To make antibodies to a given peptide and show that they bind to the parent protein was so well known and so routine that, in the late sixties and early seventies, it was hardly an exciting, or even by itself a publishable, observation. Indeed, in a review of the literature, Atassi and Habeeb (1977) previously pointed out: "antibodies obtained by immunization with a peptide (coupled or free) corresponding to a surface region of a protein can recognize that region in the protein, even though that region itself is not part of an antigenic site when the intact protein is used as an immunogen". Thus, the ability of anti-peptide antibodies [i.e. antibodies elicited by immunization with peptide-carrier conjugates or with free peptide (see the following section)] to bind intact protein is not surprising, but should in fact be expected when the peptides represent surface regions on the protein molecule (Atassi and Habeeb, 1977). Contrary to claims in the recent literature

Table 1. Specificity of serum antibodies obtained by immunization with free synthetic antigenic sites of Mb

Serum Antibody[b]	Antibodies bound (cpm) to Mb and peptide-BSA conjugates[a]					
	Site 1	Site 2	Site 3	Site 4	Site 5	Mb
Anti-site 1	16,737 ± 711	0	0	0	0	17,398 ± 1586
Anti-site 2	0	16,859 ± 1457	0	0	0	20,784 ± 2214
Anti-site 3	0	0	17,881 ± 1163	0	0	20,117 ±1973
Anti-site 4	0	0	0	16,437 ± 695	0	21,455 ± 1305
Anti-site 5	0	0	0	0	19, 246 ± 611	20,632 ± 2358
Prebleed[c]	0	0	0	0	0	0

[a] Values represent means (± S.E.M.) of 3-4 replicate analyses (by direct phase RIA) and have been corrected for binding to baseline control antigens. Baseline control antigen for peptide-BSA conjugates was peptide 121-127-BSA conjugate. Baseline control antigen for Mb was BSA. No binding was detected against a control protein hen egg white lysozyme. Values not significantly >0 by Student's T-test ($P<0.05$) reported as 0.

[b] Serum antibodies were obtained by immunization with free peptide in complete Freund's adjuvant.

[c] Prebleed sera was a pooled sera obtained from the spleen cell donor mice before they were injected with any synthetic peptides.

(Table is from Schmitz et al, 1983b).

Table 2. Serum antibodies obtained by immunization with free synthetic peptides representing surface regions that are non-immunogenic in whole Mb.

Peptide/ Protein[b]	Antibodies Bound (CPM)[a]	
	Antisera Against Peptide 1-6	Antisera Against Peptide 121-127
Site 1	0	0
Site 2	0	0
Site 3	0	0
Site 4	0	0
Site 5	0	0
Peptide 1-6	26,050 ± 2,281	0
Peptide 121-127	0	17,082 ± 3,282
Mb	33,058 ± 2,983	29,608 ± 1,542

[a] Values represent Mean (± SEM) of 3-4 replicate (by direct solid phase RIA) and have been corrected for binding to baseline control antigen BSA. Values not significantly >0 by student's T-test (P<0.05) reported as 0. No significant binding was detected using prebled sera preparations obtained from the spleen cell donor mice prior to imunization Representative data shown.

[b] Peptides were conjugated to BSA before use in RIA.

[c] Serum antibodies were obtained by immunizations with free peptide emulsified in complete Freund's adjuvant.

(Table is from Schmitz et al, 1983a)

(Lerner, 1982, 1984; Sutcliffe et al., 1983), it has previously long been known (Arnon et al., 1971; Atassi and Habeeb, 1977; Anderer, 1963; Anderer and Schlumberger, 1965; Fearny et al., 1971; Benjamini et al., 1972) that antibodies to peptides representing surface regions of the molecule bind to the whole protein, regardless of whether or not these regions constitute antigenic sites when the whole molecule is the immunizing antigen (Atassi and Habeeb, 1977, Schmitz et al., 1983a, 1983b). Thus, it is necessary not to confuse this fact with antigenic sites which define the regions that are immunogenic when the whole protein is the immunogen.

Perhaps, it is useful to comment here on the conformational bases of how a peptide is able to bind to anti-protein antibodies and vice versa. A peptide, in solution, exists in conformational equilibrum of a variety of conformational states, the time average of which is random (Atassi and Saplin, 1968). However, these conformational states are not all equally favored and there will be states of energy minima where the peptide spends more time (Atassi and Saplin, 1968). It is apparent that some of these 'favored' conformations, resemble the shape the region has in the native protein and are therefore appropriate for binding to antibody (or to T cells) (Atassi and Saplin, 1968). The probability of attaining these appropriate conformations will improve with peptide concentration (Atassi and Saplin, 1968) and is dependent on conditions of, and other molecules in, the solution (Atassi and Saplin, 1968; Epand and Scheraga, 1968; Singhal and Atassi, 1970; Atassi and Singhal, 1970b). Furthermore, binding to antibody (Crumpton, 1966; Atassi and Saplin, 1968; Atassi, 1975) and to receptors on T cells (Bixler and Atassi, 1983, 1984 a,b, 1985; Bixler et al., 1984a,b) can induce the antigen into appropriate fit. These well established principles (Atassi and Saplin, 1968; Atassi, 1972, 1975, 1977, 1978) have recently been reaffirmed and are now being advanced as a new concept (Lerner, 1984).

Immunogenicity of free peptides

A recent breakthrough in the use of synthetic peptides as immunological reagents has been the discovery that small (6 residues or longer) synthetic peptides can in fact be used in their free form (i.e. without conjugation to a carrier) to elicit antibodies to preselected submolecular regions of a protein (Young and Atassi, 1982a, Schmitz et al., 1983a; Young et al., 1983a,b,c,; Atassi and Webster, 1983). These peptides could represent either the antigenic sites of the protein (Young and Atassi, 1982a; Schmitz et al., 1983a; Young et al., 1983a,b,c,; Atassi and Webster, 1983) (for an example, see Table 1), or surface regions that are non-immuogenic (Schmitz et al., 1983b) (for example, see Table 2) when the intact protein is used as the immunogen. In each case, the antibodies invariably bind specifically to the intact protein, the binding being exclusively at the region used for immunization (e.g., see Tables 1 and 2). Monoclonal antibodies with preselected specificities can also be produced by hybridoma technology from the spleens of the immunized animals (Schmitz et al., 1982, 1983a,b) (e.g. see Tables 3 and 4). Furthermore, we have employed free synthetic peptides to generate T-cell lines and T-cell clones (e.g. see Tables 5, 6 and 7) with specificities to preselected protein immunogenic regions (Yoshioka et al., 1983; Yoshioka and Atassi, 1984) and also to induce tolerance (Young and Atassi, 1983) to such preselected regions. These fascinating breakthroughs indicate that free synthetic peptides may be used in the future as powerful tools in basic investigations and in therapeutic and diagnostic applications [for review of this work, see Atassi (1984c)].

Table 3. Monoclonal antibodies from immunization with free synthetic antigenic sites of Mb

Monoclonal Antibody	Antibody bound (CPM) to Mb and peptide-BSA conjugates					
	Site 1	Site 2	Site 3	Site 4	Site 5	Mb
MB1-4-15 (anti-site 1)	4414 ± 167	0	0	0	0	20,288 ± 613
MB2-5-4 (anti-site 2)	0	9353 ± 745	0	0	0	29,667 ± 933
MB3-5-13 (anti-site 3)	0	0	9735 ± 792	0	0	52,190 ± 2030
MB4-3-22 (anti-site 4)	0	0	0	4357 ± 510	0	17,103 ± 201
MB5-16-1 (anti-site 5)	0	0	0	0	13,392 ± 643	62,763 ± 2010
MB5-14-5 (anti-site 5)	0	0	0	0	27,203 ± 3789	30,570 ± 1410

[a] Values represent means (± S.E.M.) of 4 replicate analyses by direct solid phase RIA and have been corrected for binding to baseline control antigens. Baseline control antigen for peptide-BSA conjugates was peptide 121-127-BSA conjugate. Baseline control antigen for Mb was BSA. No binding was detected against a control protein hen egg white lysozyme. Values not significantly > 0 by Student's t-test ($P < 0.05$) reported as 0. Representative data shown.

(Table is from Schmitz et al., 1983b)

Table 4. Specificity of monoclonal antibodies from immunization with free peptide representing surface regions that are non-antigenic in whole Mb.

Peptide/ Protein[b]	Antibodies Bound (CPM)[a]		
	MB16-6-1	MB16-8-21	MB121-2-5
Site 1	0	0	0
Site 3	0	0	0
Site 4	0	0	0
Site 5	0	0	0
Peptide 1-6	14,231 ± 861	29,665 ± 1060	0
Peptide 121-127	0	0	7,477 ± 374
Mb	18,094 ± 1662	36,390 ± 1095	15,205 ± 1508

[a] Values represent mean (± SEM) of 3-4 replicate analyses by direct solid phase RIA. Results have been corrected for binding to baseline control antigen BSA. Values not significantly >0 by student's T-Test (P<0.05) reported as 0. Representative data shown.

[b] Peptides were conjugated to BSA before use in the RIA.

(Table is from Schmitz et al., 1983a)

Table 5. Proliferative response of an SJL T-cell line obtained by 4 passages with extended site 4.

Challenge Antigen	[^{3}-H]-Thymidine incorporation (Δcpm)[a]		
	2.5 μg/ml	5 μg/ml	10 μg/ml
Mb	92,240	121,760	96,510
Ext. Site 1	0	0	0
Ext. Site 2	0	0	0
Ext. Site 3	0	0	0
Ext. Site 4	73,360	124,760	154,630
Ext. Site 5	0	0	0

[a] The values are given in Δcpm relative to the counts of medium or lysozyme challenged cells. Cells responded to Con A (Δcpm 181,929) but not to LPS or PPD.

(Table is from Yoshioka et al., 1983)

Table 6. Proliferative responses of T-cell clones obtained from SJL long term cultures that were passaged with extended site 4 or extended site 5.

Challenge Antigen[c]	[^{3}-H]-Thymidine incorporated in Δcpm (and S.I.) Clone #4.EIV-1[a] from cell line specific to Ext. Site 4	Clone #5.EV-4[b] from cell line specific to Ext. Site 5
Mb	183,600 (47.6)	5,560 (5.32)
Ext. Site 1	0	0
Ext. Site 2	0	0
Ext. Site 3	0	0
Ext. Site 4	294,450 (75.8)	0
Ext. Site 5	0	5,500 (5.30)

[a] This clone responded to Con A (Δcpm 303,620; S.I. 78.12), but not to PPD, LPS or hen lysozyme (Δcpm 0; S.I., 0.86).

[b] This clone responded to Con A (Δcpm 37,620; S.I. 62.6), but not to PPD, LPS or hen lysozyme (Δcpm 0; S.I. 0.88).

[c] Extended sites were tested in the dose range of 1 to 5 μg/ml. Values represent responses at the optimum dose.

(Table is from Yoshioka et al., 1983)

Table 7. Example of a Balb/c T-cell clone (#8c) from a T-cell line specific to extended site 5.

Challenge Antigen	[^{3}H]-Thymidine incorporation[a] Δcpm	S.I.
Mb	13,500	22.8
Ext. Site 1	0	0.88
Ext. Site 2	0	0.95
Ext. Site 3	0	1.00
Ext. Site 4	0	0.98
Ext. Site 5	13,600	22.9

[a] Cells responded to Con A (Δcpm 19,025) but not to LPS or PPD.

(Data are from Yoshioka et al., 1983).

Origin of synthetic (artificial) antigens as vaccines

The first recorded artificial antigens used for immunological protection against an invasive agent were those reported by Goebel. He showed (Goebel, 1938, 1939) that immunization with a cellobiuronide-conjugate on a protein carrier gave rise to antibodies against pneumococcal type III polysaccharides which agglutinated the pneumococci and gave immunological protection against infection. In 1959, by coupling α-phenyl-colito-pyranoside onto bovine serum albumin, Straub and his co-workers, prepared artificial antigens which elicited agglutinating antibodies against pathogenic gram-negative bacteria (Straub et al., 1959).

These early studies were concerned with polysaccharide antigens which bear no obvious relationship to protein antigens found in other invasive agents such as viruses, bacteria, toxins, allergens, etc. Nevertheless, these studies put forth a promising and reasonable concept that was applied to protein antigens some 20 years later by Anderer (1963) who wrote: "it is of considerable interest to find methods for the preparation of artificial antigens which give immunological protection against natural protein antigens such as viruses with a coat protein". Anderer, drawing from, and unequivocally giving full credit to, the previous studies on artificial polysaccharide antigens, showed (Anderer and Handschuh, 1962) that a hexapeptide of tobacco mosaic virus protein had an inhibitory activity toward antibody binding to the virus. A year later, he showed (Anderer, 1963) that immunization with this hexapeptide conjugated to bovine serum albumin (BSA) elicited antisera that neutralized viral infection. There may be a tendency to dismiss these pioneering studies because of some current erroneous trends that only work with synthetic peptides is relevant. However, in 1965, he reported (Anderer and Schlumberger, 1965) successful protection by immunization with BSA conjugates of four synthetic peptides spanning an antigenic region of the virus. It should be also mentioned that he repeatedly used the terms "artificial antigens" and "synthetic antigens" in his reports.

In spite of this success, Anderer recognized that there must be more than one antigenic site on the viral protein. He pointed out in 1963 that "the search for serologically active sites of globular proteins has not yet been as successful as corresponding investigations carried out with polysaccharide antigens....partly due to the rather complex structure of globular proteins".

The studies of Anderer on tobacco mosaic virus were followed by the very thorough and elegant work of Eli Benjamini on the tobacco mosaic virus protein in the late sixties and early seventies (Fearny et al. 1971; Benjamini et al., 1972). Benjamini and his co-workers have defined the physiochemical attributes of a small peptide constituting an antigenic determinant of the protein and demonstrated that antibodies to the whole protein could be elicited by immunization with the peptide conjugated to a carrier (for review, see Benjamini, 1977). The idea of affording immunological protection by immunization with peptide-carrier conjugates was not further exploited for many years, primarily because of lack of information on protein antigenic structures. The outstanding work of Dr. Howard Bachrach and his co-workers was ongoing during this period on the antigenic recognition of the foot and mouth disease virus. These studies deserve special mention because of the discovery that immunization with a large peptide fragment of VP1 afforded protection against viral infection in guinea pig and cattle (Kleid et al., 1981). Recently, however, there has been renewed interest in investigating the use of peptide-carrier conjugates as synthetic vaccines. From the foregoing brief account it is evident that the idea of using peptides in the hope of designing 'artificial' or 'synthetic' vaccines is not a new concept as has recently been claimed (Lerner, 1982, 1984; Sutcliffe et al., 1983).

Conclusion

As already mentioned, the elegant studies of Anderer and co-workers (1963, 1965) and Benjamini and co-workers (Fearny et al., 1971; Benjamini, 1977) showed that antibodies which neutralize in vitro, and/or bind to, tobacco mosaic virus can be obtained by immunization with conjugates of small viral peptides. Although it is encouraging that anti-peptide antibodies have frequently been reported to bind to an intact virus and even neutralize that virus, it should be emphasized that eliciting in vivo protective immunity in a host by immunization with a small synthetic viral peptide is a different issue. Attempts to develop effective anti-viral vaccines using small synthetic peptides have so far been ineffective. Such studies have relied entirely on the production of anti-peptide antibodies and have been oblivious of the important role played by T cells in the immune recognition and responses to submolecular regions of protein antigens (Okuda et al., 1979; Young and Atassi, 1982b; Bixler and Atassi, 1983, 1984 a,b, 1985; Bixler et al., 1984a,b; Yoshioka et al., 1983) and the regulation of these responses due to critical inter-site regulatory influences exerted by these cells (Okuda et al., 1979; Atassi et al., 1981; Krco et al, 1981a,b,c, 1984). It has been pointed out (Atassi and Kurisaki, 1984) that attempts at protection by peptide immunization must be based not only on knowledge of some of the regions recognized by antibodies but on the full unravelling of the submolecular profiles of both the B- and T- cell responses on the virus and in all likelihood will require good knowledge of the factors involved in the regulation and control of these site-specific responses.

The contributions that have been briefly outlined here have constituted landmarks that have put the progress of contemporary protein immunology in particular and immunology in general on a sound molecular course with the elements of precision brought in by these chemical strategies. It is, of course, reassuring that these principles are now being repeatedly reaffirmed. It is, however, disconserting that they are at the same time being proclaimed as new discoveries. There remains, nevertheless, a school of thought (Benjamin et al., 1984) asserting that antigenic sites are not discrete areas on a protein but rather a continuum over the entire protein molecule and that the very idea of localizing and synthesizing antigenic sites is a meaningless endeavor. The principles, I have briefly reviewed and the work that will be presented by many of the participants throughout this symposium clearly demonstrate that this speculation is incorrect. Your presence here and the data that will be presented in this symposium on several viral and bacterial antigen systems is the best evidence for the soundness and powerful information that can be learnt about the immune system by localizing and synthesizing antigenic sites of protein antigens.

ACKNOWLEDGEMENTS

The work that resulted in the findings outlined in this article was supported by grants AM 33968, AI 21336 from the National Institutes of Health, Q994 from the Welch Foundation, and AHA 81-1012 from the American Heart Association. The author is the holder of the Robert A. Welch Chair of Chemistry at Baylor College of Medicine.

REFERENCES

Anderer, F.A. and Handschuh, D. (1962) Z. Naturforsch., 17b, 536.
Anderer, F.A. (1963) Biochim. Biophys. Acta 71, 246-248.
Anderer, F.A. and Schlumberger, H.D. (1965) Biochim. Biophys. Acta, 97, 503-509.
Arnon, R., Maron, E., Sela, M. and Anfinsen, C.B. (1971) Proc. Natl. Acad. Sci. (USA) 68, 1450-1455.
Atassi, M.Z. (1972) in Specific Receptors of Antibodies, Antigens and Cells, (Pressman, D., Tomasi, T.B., Grossberg, A.L. and Rose, N.R., eds) 3rd Int. Convocation on Immunol. 118-136, S. Karger, Basel.
Atassi, M.Z. (1975) Immunochemistry 12, 423-438.
Atassi, M.Z. (1977) in Immunochemistry of Proteins (Atassi, M.Z., ed.) 2, 77-176, Plenum Press, New York.
Atassi, M.Z. (1978) Immunochemistry 15, 909-936.
Atassi, M.Z. (1980) Mol. Cell. Biochem. 32, 21-44.
Atassi, M.Z. (1981) Molecular Immunol. 18, 1021-1025.
Atassi, M.Z. (1982) Biochim. Biophys. Acta 704, 552-555.
Atassi, M.Z. (1984a) in Molecular Immunology (Atassi, M.Z., van Oss, C.J. and Absolom, D), 15-51, Marcel Dekker, New York.
Atassi, M.Z. (1984b) Europ. J. Biochem. 145, 1-20.
Atassi, M.Z. (1984c) in Conformationally Directed Drug Design (Vida, J.A. and and Gordon, M., Eds.), ACS Symposium Series 251, 29-71.
Atassi, M.Z. (1984d) in Proc. of the 4th Ann. Chem. Def. Bioscience Rev., pp. 355-368.
Atassi, M.Z. (1985) Biochem. J. 226, 477-485.
Atassi, M.Z. and Atassi, H. (1985) Biochem. J. in press.
Atassi, M.Z. and Habeeb, A.F.S.A. (1977) in Immunochemistry of Proteins (Atassi, M.Z., ed.) 2, 177-264, Plenum Press, New York.
Atassi, M.Z. and Kazim, A.L. (1978) in Immunobiology of Proteins and Peptides (Atassi, M.Z. and Stavitsky, A.B., eds.) 1, 19-40, Plenum Press, New York.
Atassi, M.Z. and Kazim, A.L. (1980) Biochem. J. 187, 163-172.
Atassi, M.Z. and Kurisaki, J.I. (1984) Immunol. Commun. (1984), in press.
Atassi, M.Z. and Lee, C.L. (1978) Biochem. J. 171, 429-434.
Atassi, M.Z. and Saplin, B.J. (1968) Biochemistry 7, 688-698.
Atassi, M.Z. and Saplin, B.J. (1971) Biochemistry 10, 4740-4747.
Atassi, M.Z. and Singhal, R.P. (1970a) Biochemistry 9, 3854-3861.
Atassi, M.Z. and Singhal, R.P. (1970b) J. Biol. Chem. 245, 5122-5128.
Atassi, M.Z. and Smith, J.A. (1978) Immunochemistry 15, 609-610.
Atassi, M.Z. and Webster, R. (1983) Proc. Nat. Acad. Sci. (USA) 80, 840-844.
Atassi, M.Z. and Zablocki, W. (1977) J. Biol. Chem. 252, 8784-8787.
Atassi, M.Z., Koketsu, H. and Habeeb, A.F.S.A. (1976a) Biochim. Biophys. Acta 420, 358-375.
Atassi, M.Z. Lee, C.-L. and Pai, R.C. (1976b) Biochim. Biophys. Acta 427, 745-751.
Atassi, M.Z., Sakata, S. and Kazim, A.L. (1979) Biochem. J., 179, 327-331.
Atassi, M.Z., Twining, S.S., Lehmann, H. and David, C.S. (1981) Immunol. Comm. 10, 359-365.
Benjamin, D.C., Berzofsky, J.A., East, I.J., Gurd, F.R.N., Hannum, G., Leach, S.J., Margoliash, E., Michael, J.G., Miller, A., Prager, E.M.,Reichlin, M. Sercarz, E.E., Smith-Gill, S.J., Todd, P.E. and Wilson, A.C. (1984) Ann. Rev. Immunol. 2, 67-101.
Benjamini, E., Michaeli, D. and Young, J.D. (1972) Curr. Top. Microbiol. Immunol. 58, 85-134.

Benjamini, E., Young, J.D., Shimizu, M. and Leung, C.W. (1964) Biochemistry 3, 1115-1120.
Benjamini, E., Young, J.D., Petersen, W.J., Leung, C.W. and Shimizu, M. (1965) Biochemistry 4, 2081-2085.
Benjamini, E. (1977) in Immunochemistry of Proteins (Atassi, M.Z., Ed.) Vol. 2, pp. 265-310, Plenum Press.
Berkower, I., Buckenmeyer, G.K., Gurd, F.R.N. and Berzofsky, J.A. (1982) Proc. Natl. Acad. Sci. (USA) 79, 4723-4727.
Berzofsky, J.A., Buckenmeyer, G.K., Hicks, G., Gurd, F.R.N., Feldmann, R.J. and Minna, J. (1982) J. Biol. Chem. 257, 3189-3198.
Bixler, G.S. and Atassi, M.Z. (1983) Immunol. Comm. 12, 593-603.
Bixler, G.S. and Atassi, M.Z. (1984a) J. Immunogen 11, 245-250.
Bixler, G.S. and Atassi, M.Z. (1984b) J. Immunogen., in press.
Bixler, G.S. and Atassi, M.Z. (1985) Biotechnology 3, 47-54.
Bixler, G.S., Yoshida, T. and Atassi, M.Z. (1984a) Expl. Clin. Immunogen. 1, 99-111.
Bixler, G.S., Yoshida, T. and Atassi, M.Z. (1984b) J. Immunogen., in press.
Brown, R.K. (1962) J. Biol. Chem. 237, 1162-1167.
Crumpton, M.J. (1966) Biochem. J. 100, 223-232.
Crumpton, M.J. and Wilkinson, J.M. (1965) Biochem. J. 94, 545-556.
David, C.S. and Atassi, M.Z. (1982) in: Immunobiology of Proteins and Peptides (Atassi, M.Z., Ed.) 2, 97-125.
Epand, R.M. and Scheraga, H.A. (1968) Biochemistry 7, 2864-2872.
Fearney, F.J., Leung, C.Y., Young, J.D. and Benjamin, E. (1971) Biochim. Acta. 243, 509-514.
Goebel, W.F. (1938) J. Exptl. Med. 68, 469.
Goebel, W.F. (1939) Nature (London) 143, 77.
Hopp, T.P. and Woods, K.R. (1981) Proc. Natl. Acad. Sci. (USA) 78, 3824-3828.
Kazim, A.L. and Atassi, M.Z. (1977) Biochim. Biophys. Acta 494, 277-282.
Kazim, A.L. and Atassi, M.Z. (1978) Immunochemistry 15, 67-70.
Kazim, A.L. and Atassi, M.Z. (1980a) Biochem. J. 187, 661-666.
Kazim, A.L. and Atassi, M.Z. (1980b) Biochem. J. 191, 261-264.
Kazim, A.L. and Atassi, M.Z. (1982) Biochem. J. 203, 201-208.
Kleid, D.G., Yansura, D., Small, B., Dowbenko, B., Moore, D.M., Grubman, M.J., McKercher, P.D., Morgan, D.O., Robertson, B.H. and Bachrach, H.L. (1981) Science 214, 1125-1129.
Koketsu, J. and Atassi, M.Z. (1973) Biochim. Biophys. Acta 328, 289-302.
Koketsu, J. and Atassi, M.Z. (1974a) Biochim. Biophys. Acta 342, 21-29.
Koketsu, J. and Atassi, M.Z. (1974b) Immunochemistry 11, 1-8.
Krco, C.J., Kazim, A.L., Atassi, M.Z. and David, C.S. (1981a) J. Immunogen 8, 315-322.
Krco, C.J., Kazim, A.L., Atassi, M.Z. and David, C.S. (1981b) J. Immunogen. 8, 395-404.
Krco, C.J., Kazim, A.L., Atassi, M.Z., Melvold, R. and David, C.S. (1981c) J. Immunogen. 8, 471-476.
Krco, C.J., Abramson, E.J., Yoshioka, N., Atassi, M.Z. and David, C.S. (1984) J. Immunogen. 11, 33-43.
La Presle, C. (1955) Bull. Soc. Chim. Biol. (Paris) 37, 969-975.
Landsteiner, K. (1942) J. Exp. Med. 75. 269-276.
Leach, S.J. (1983) Biopolymers 22, 425-440.
Lee, C.L. and Atassi, M.Z. (1976) Biochem. J. 159, 89-93.
Lee, C.L. and Atassi, M.Z. (1977a) Biochim. Biophys. Acta 495, 354-368.
Lee, C.L. and Atassi, M.Z. (1977b) Biochem. J. 167, 571-581.
Lerner, R.A. (1982) Nature (London) 299, 592-596.
Lerner, R.A. (1984) Adv. Immunol. 36, 1-44.

McCormick, D.J. and Atassi, M.Z. (1983a) Fed. Proc. 41 Abstr. 2411.
McCormick, D.J. and Atassi, M.Z. (1982b) Fed. Proc. 41, Abstr. 2412.
Okuda, K., Twining, S.S., David, C.S. and Atassi, M.Z. (1979) J. Immunol. 123, 182-188.
Porter, R.K. (1957) Biochem. J. 66, 677-686.
Reichlin, M. (1972) J. Mol. Biol. 64, 485-496.
Reichlin, M. (1975) Adv. Immunol. 20, 71-123.
Sakata, S. and Atassi, M.Z. (1980a) Biochim. Biophys. Acta 625, 159-162.
Sakata, S. and Atassi, M.Z. (1980b) Mol. Immunol. 17, 139-149.
Sakata, S. and Atassi, M.Z. (1981) Mol. Immunol. 18, 961-967.
Schmitz, H.E., Atassi, H. and Atassi, M.Z. (1982) Mol. Immunol. 19, 1699-1702.
Schmitz, H.E., Atassi, H. and Atassi, M.Z. (1983a) Immunol. Commun. 12, 161-175.
Schmitz, H.E., Atassi, H. and Atassi, M.Z. (1983b) Molecular Immunol. 20, 719-726.
Singhal, R.P. and Atassi, M.Z. (1970) Biochemistry 9, 4552-4259.
Singhal, R.P. and Atassi, M.Z. (1971) Biochemistry 10, 1756-1762.
Straub, A.M., Tinelli, R., Luderitz, O. and Westphal, O. (1959) Ann. Inst. Pasteur, 96, 303.
Sutcliffe, J.G., Shinnick, T.M., Green, N. and Lerner, R.A. (1983) Science 219, 660-666.
Twining, S.S. and Atassi, M.Z. (1979) J. Immunol. Methods 30, 139-151.
Twining, S.S., Lehmann, H. and Atassi, M.Z. (1980) Biochem. J. 191, 681-697.
Twining, S.S., David, C.S. and Atassi, M.Z. (1981a) Molecular Immunol. 18, 447-450.
Twining, S.S., Lehmann, H. and Atassi, M.Z. (1981b) Molecular Immunol. 18, 473-479.
Urbanski, G.J. and Margoliash, E. (1977) J. Immunol. 118, 1170-1180.
Wabuke-Bunoti, M.A.N., Taku, A., Fan, D.P., Kent, S. and Webster, R.G. (1984) J. Immunol. 133, 2194-2201.
Westhof, E., Altschuh, D., Moras, D., Bloomer, A.C., Mondragon, A., Klug, A. and Van Regenmortel, M.H.V. (1984) Nature (London) 311, 123-126.
Wilson, I.A., Skehel, J.J. and Wiley, D.C. (1981) Nature (London) 289, 366-373.
Yokota, S., David, C.S. and Atassi, M.Z. (1980) Molecular Immunol. 17, 1079-1082.
Yoshioka, N. and Atassi, M.Z. (1983) 5th Int. Congr. Immunol. Abstr. (232) 0023.
Yoshioka, M. and Atassi, M.Z. (1984) Ret. End. Soc., Abstr., in press.
Yoshioka, N. and Atassi, M.Z. (1985) Biochem. J., in press.
Yoshioka, M., Bixler, G.S. and Atassi, M.Z. (1983) Molecular Immunol. 20, 1133-1137.
Yoshioka, M., Yoshioka, N. and Atassi, M.Z. (1985) Biochem. J., in press.
Young, C.R. and Atassi, M.Z. (1982a) Immunol. Comm. 11, 9-16.
Young, C.R. and Atassi, M.Z. (1982b) Adv. Exp. Med. 150, 79-93.
Young, C.R. and Atassi, M.Z. (1983) J. Immunogen. 10, 161-169.
Young, C.R., Schmitz, H.E. and Atassi, M.Z. (1983a) Molecular Immunol. 20, 567-570.
Young, C.R., Schmitz, H.E. and Atassi, M.Z. (1983b) Immunol. Comm. 12, 419-428.
Young, C.R., Schmitz, H.E. and Atassi, M.Z. (1983c) J. Immunogen. 10. 453-464.

FOOT-AND-MOUTH DISEASE AND ITS ANTIGENS

Howard L. Bachrach

Plum Island Animal Disease Center
U.S. Department of Agriculture, ARS
Greenport, New York 11944

ABSTRACT

Many factors combine to make foot-and-mouth disease (FMD) one of the most damaging and intractable disease of animals. These include its extreme contagion, wide geographic distribution, great multiplicity of both susceptible animal hosts and viral serotypes, a relatively short duration of immunity to a given serotype and a post-recovery carrier state of the virus in many animal species (e.g., cattle, sheep and goats). Nevertheless, import restrictions and other actions of the U.S. Government have kept the United States free of FMD since 1929, even during outbreaks of the disease in Mexico and Canada in the early 1950's. Beginning in the late 1940's, the systematic vaccination of cattle has been practiced in several areas of the world with varying degrees of success. While this procedure has succeeded in Western Europe in greatly reducing the incidence of FMD, the presence of live virus in some batches of vaccine and the escape of virus from vaccine from manufacturing facilities are now responsible for a large proportion of the outbreaks that still occur there. A vaccine is needed that has no possibility of producing the disease. During the last few years, it has been demonstrated that capsid protein VP_1, isolated from type A and C virions or biosynthesized in *E. coli* transformed with the gene for VP_1, can be used to immunize livestock against FMD. Immunization of livestock has also been achieved with a 13 kd fragment (amino acid residues 55 through 179) cleaved with CNBr from the 213 amino acid long VP_1 chain of type A virions. Immunogenic sites on intact virions, 12 S subunit particles and isolated VP_1 chains have been studied by a combination of methods, including: 1) assessment of the immunogenicity of VP_1-specific fragments and synthetic peptides and 2) mapping monoclonal antibodies (Mabs) generated with virus, VP_1 and the 13 kd fragment to virus, 12 S subunits, VP_1, VP_1-specific fragments and a biosynthetic 32mer. Correlation of these results with sites having variant and serotype sequence variability indicates that the 136-179 region of type A_{12} VP_1 possesses four putative neutralization-specific epitopes (ca. 137-143, 146-151, 152-157 and 170-175). Of three neutralization-specific epitopes on type A_{12} virus, two are also present on 12 S subunits and isolated VP_1 chains. Mabs to the three epitopes appear to neutralize virus by different mechanisms:

by viral aggregation, by blocking the site on viral VP_1 that binds to cell receptors or by interfering with a postreceptor attachment step, possibly penetration or uncoating.

Although a peptide synthesized chemically to correspond to amino acid residues 141-160 of type O_1 VP_1 has been shown to produce protective immunity in guinea pigs (Bittle et al., 1982), it has so far been ineffective in cattle. An epitope (residues 146-152) on type O_1 virus and VP_1, co-sequential with the type A_{12} 146-151 epitope, has been detected by the binding of antibody to 208 overlapping hexapeptides covering the entire length of VP_1 (Geysen et al., 1984). Using residue replacement synthetic peptides, the epitope was shown to have an absolute requirement for leucyl residues 148 and 151. The coincidence on VP_1 of this type O_1 epitope and a type A_{12} putative epitope strengthens the evidence that variant and serotype changes in FMD virus can both stem from a common precursor epitope sequence.

INTRODUCTION

Foot-and-mouth disease is reputed to be one of the most dreaded and damaging diseases of domesticated livestock (Bachrach, 1968). Nearly all food- and fiber-producing animals except equines are susceptible (Bachrach, 1978). Infection rarely occurs in humans, and when it does the symptoms are usually very mild. The existence of seven immunotypes of FMD virus (A, O, C; SAT 1, 2, 3; and Asia 1) and some seventy subtypes and strains further complicate the control of the disease. Also, vaccinated animals and animals recovered from active infection are generally not immune to reinfection with the homologous virus for more that 1 to 2 years. The disease is extremely contagious for susceptible animal herds, inducing severe debilitating effects which decrease animal productivity by about one-fourth. While mortality is usually less than 5%, it may reach 50% or more in young animals because of virus-induced myocarditis. Economic losses to exporting countries usually far exceed the direct animal losses because of embargoes that must be placed on the exportation of animals and animal products until the disease has been declared eradicated.

Foot-and-mouth disease appears to have occurred in cattle in Italy as early as 1514 (Fracastorius, 1546). While it has appeared on all the world's continents, Australia has been free of the disease for more than 110 years and North America for over 30 years. Because FMD is a continual problem elsewhere, it is a constant threat to disease-free countries such as the United States which have highly valuable livestock industries comprised of fully susceptible animals.

Foot-and-mouth disease was the first animal disease shown to be caused by a filterable agent (Loeffler and Frosch, 1897). More than 50 years then passed before the first physical parameter of the virus was determined: a sedimentation rate of approximately 140 Svedberg units (Bachrach, 1952). This finding and adaptation of the virus to tissue culture (Bachrach et al., 1955; Bachrach et al., 1957) led to the purification of the virus and its visualization in the electron microscope as a 22 mμ icosahedral particle (Bachrach and Breese, 1958), similar in size and appearance to poliovirus (Bachrach and Schwerdt, 1954). Later the virus was shown to be comprised of 31% RNA and 69% protein (Bachrach et al., 1964), and the isolated RNA was found to be infectious (Mussgay, 1959). Most work in the 1970's involved studies of the viral capsid and noncapsid proteins, the molecular mechanisms of viral replication and mapping of viral-coded proteins on the genome (Bachrach, 1977).

VIRUS AS ANTIGEN

Virus vaccines have been used in Europe since the 1940's to help control FMD. The first effective vaccine for cattle, the Schmidt-Waldmann vaccine, culminated from the work of Schmidt (1936) in Denmark and of Waldmann and Kobe (1938) in Germany. Virus for the vaccine was collected from tongue lesions of cattle, clarified by Seitz filtration, adsorbed onto aluminum hydroxide gel and inactivated with formaldehyde. In 1951, Frenkel substituted this source of virus with virus propagated in strips of bovine tongue epithelium suspended in culture fluid. The Frenkel method produces vaccine of high potency and is still used in some European countries and the U.S.S.R. However, most virus for inactivated vaccines is now produced in baby hamster kidney (BHK) cell lines grown either on glass surfaces or in suspension in fermenters with capacities of up to 10,000 liters or more. It has been determined that the 50% protective dose of inactivated and oil-adjuvanted purified type A_{12} virus is less than 40 ng for cattle and about 160 ng for swine (Morgan et al., 1970; Bachrach and McKercher, 1972). The equivalent dose of type O_1 Brugge for swine is about 3,000 ng. Thus, type O_1, virus appears to be a much weaker immunogen for swine than is type A_{12} virus.

Live virus vaccines attenuated by passages of virus in tissue cultures, small animals or chick embryos have had little success in controlling FMD. When virus has been attenuated properly for cattle, it is generally still too virulent for swine or for cattle under stress, e.g., high yielding milking cows. Consequently, attenuated FMD viral vaccines are forbidden for use in Europe and most other countries.

Systematic vaccination programs for cattle using inactivated vaccines were instituted in the late 1940's in Europe and somewhat later in several countries of South America and Africa. While these programs have markedly reduced outbreaks of FMD in Western Europe, the disease has not been eradicated, and, ironically, vaccines have become a principal cause of the disease in that area. For example, from 1968 to 1981 at least 44% of the outbreaks of FMD in Western Europe were caused by virus in vaccines or by virus that escaped from vaccine factories. Consequently, consideration has been given to replacing vaccination programs there by slaughtering-out procedures that have been used to eradicate FMD in the United States and Great Britain. The use of protein vaccines lacking any possibility of producing FMD would be an alternative. Experiments to perfect such vaccines have been underway since the mid-1970's when it was determined that capsid protein VP_1 from FMD virus induces neutralizing antibodies (LaPorte et al., 1973; Bachrach et al., 1975) and protective immunity in swine (Bachrach et al., 1975).

FMD VIRUS GENOME STRATEGIES

Foot-and-mouth disease virus, a member of the picornavirus family, possesses a single-stranded messenger sense RNA genome approximately 8,000 nucleotides long. Mapping and structural studies have established the gene order shown in Fig. 1. The genome is linked 5' with viral-coded protein VP_g and has the poly C and 3' poly A tracts shown.

The primary translation product of the genome is a poly-protein (ca. 250,000 kd) spanning gene LP_1 through the gene that encodes the viral-coded RNA polymerase (Pol). The entire nucleotide sequence coding for the polyprotein of type O_1 virus has been determined (Carroll et al., 1984). The polyprotein undergoes nascent cleavages by cellular proteases and a cascade of cleavages by the viral-coded protease (Pro) to produce capsid proteins VP_{1-4} and noncapsid proteins (all other proteins shown in Fig. 1). The LP_{1-2} proteins are not components of virions and have no known functions. Three variants of VP_g genes translate to products that link covalently in equi-molar proportions to the 5' termini of viral RNA to form genomes (Forss and Schaller, 1982). Biologically-active Pro has recently been expressed from viral-specific DNA cloned in E. coli (Klump et al., 1984). The viral-coded Pol is required for viral RNA replication. Antibodies to Pol in cattle have been considered diagnostic for previous FMD infection rather than vaccination with inactivated vaccine. However, recent findings of homologies of about 40% between picornavirus Pol's may impair such diagnoses (Kitamura et al., 1981; Robertson et al., 1983b; Palmenberg et al., 1984).

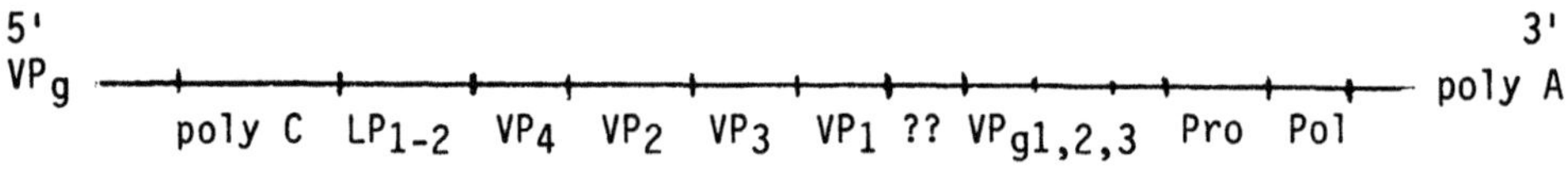

Figure 1. Organization of the RNA genome of FMD virus.

THE VIRION

Foot-and-mouth disease virus is comprised of 60 copies each of proteins VP_{1-4} that encapsidate a single strand of the VP_g-linked RNA genome to form a particle having the form of a regular icosahedron. Actually, two or three of the 60 copies of VP_2 and VP_4 are present in the viral capsid as uncleaved VP_0 precursor molecules. Polyacrylamide gel electrophoresis in SDS/8M urea (PAGE) of disrupted virions separates the capsid proteins as shown in Fig. 2.

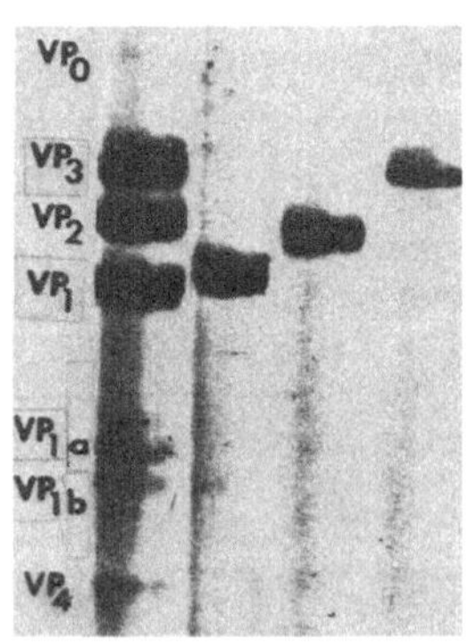

Figure 2. PAGE in SDS/8M urea of FMD virus disrupted by boiling in 8M urea, 2% SDS and 5% mercaptoethanol. Left to right: FMD virus, purified VP_1, VP_2 and VP_3. VP_0 is uncleaved precursor to VP_4 and VP_2. $VP_{1a,1b}$ are tryptic cleavage products of VP_1 in the virion. Adapted from Bachrach et al., 1975.

A number of criteria indicate that the 24 kd VP_1 protein -- named so that genome map loci in all the picornaviruses will be in register (Rueckert and Wimmer, 1984) but previously called VP_3 by this author because it is the third slowest migrating major capsid protein in PAGE -- occupies an external position in the viral capsid: 1) it is cleaved to a 16 kd fragment by treatment of virions with trypsin (Wild et al., 1969), 2) it is the only capsid protein in virions iodinated with ^{125}I, even though VP_2 and VP_3 together have twice as many tyrosyl residues as VP_1 (Bachrach et al., 1978) and iodination is at tyrosyl residue 136 (Robertson et al., 1983a) and 3) some copies of VP_1 are lost from virions on prolonged exposure in CsCl solution (Bernard et al., 1974).

VP_1 AND VP_1-SPECIFIC ANTIGENS

Type A_{12} Capsid Protein VP_1 and its Fragments

Capsid protein VP_1 induces neutralizing antibodies and protective immunity in both swine (Bachrach et al., 1975) and cattle (Bachrach et al., 1982; Kleid et al., 1981) (Table 1). In addition, 15 kd and 16 kd fragments from VP_1 have been shown to elicit similar responses in swine (Table 1), and a 13 kd fragment does so in both swine and cattle. The 13 kd (VP_1 residues 55-179) and 16 kd (residues 1-144) fragments are derived by treatment of virions with CNBr and trypsin, respectively. The 15 kd fragment (residues 1-138) is produced by an obscure protease present in bacterially-contaminated BHK cell cultures.

Cloned Type A_{12} VP_1-Specific Antigens

Based on the immunogenicity of VP_1 and VP_1 fragments in livestock, recombinant DNA (rDNA) technology was used to produce large quantities of VP_1-specific antigen by cloning the gene for VP_1 in _E. coli_ K12 (Kleid et al., 1981). Type A_{12}-specific DNA fragments transcribed from viral RNA were amplified by cloning. Restriction fragments of the DNA were then sequenced until codons were located for amino acid sequences that had previously been determined for the C-terminal region of VP_1 (Bachrach et al., 1973) and for N-terminal regions of VP_1 and its 13 kd and 15 kd fragments (Bachrach et al., 1979, 1982). Plasmid pBR322 containing a _trp_ promoter/operator followed consecutively with the first 190 DNA codons of tryptophan synthetase and DNA corresponding to amino acid residues 8-211 of the 213 amino acid chain of VP_1 was used to transform _E. coli_ and to cause it to express approximately 1-2 million copies per bacterium of a 44 kd fusion protein (LE'-VP_1). After its extraction and purification, the LE'-VP_1 protein was found to elicit neutralizing antibodies and protective immunity in both swine and cattle. United States Secretary of Agriculture John R. Block characterized this achievement "as the first production through gene splicing of an effective vaccine against any disease in animals or humans." Table 1 shows the results of vaccinating swine and cattle with the rDNA-produced 44 kd LE'-VP_1 fusion protein and with other type A_{12} VP_1-specific antigens (i.e., VP_1 isolated from virions; 13 kd, 15 kd and 16 kd fragments from VP_1; and inactivated virus). Protective immunity was achieved with a single dose of the virus vaccine, whereas two doses were required for all the subviral vaccines, includng the LE'-VP_1 vaccine. Challenge of immunity of livestock was carried out one or one and a half months following initial vaccination with the LE'-VP_1 fusion protein.

Table 1. Neutralizing Antibody and Protective Immune Responses in Swine and Cattle Vaccinated with FMD Type A_{12} VP_1 and VP_1-Specific Antigens

Antigen[a]	Animal	No. of Vaccinations	Neut. Ab (-log PD_{50})[b]	Fraction Immune
VP_1	swine	1	ND	1/10
	swine	2	ND	8/10
	swine	3	2.2	2/2
	cattle	2	1.6	0/1
	cattle	2	2.6	5/5
	cattle	3	2.6	2/2
13 kd	swine	2	1.3	3/3
	cattle	2	0.4	2/2
15 kd	swine	2	0.4	2/3
16 kd	swine	2	3.1	3/3
LE'-VP_1	swine	2	1.4,1.5	2/2
	cattle	2	2.1-2.7	6/6[c]
Virus	swine	1	3.1	13/13
	swine	2	4.0	6/6
None	swine	ND	ND	0/30
	cattle	ND	ND	0/7

[a]Virus was inactivated with acetylethylenimine. Incomplete Freund's adjuvant was used with all antigens. The doses of VP_1, VP_1 fragments, LE'-VP_1 and virus were 100, 50, 250, and 10 µg's, respectively. Swine were revaccinated, as required, with VP_1 and VP_1 fragments on days 28 and 60 and were challenge exposed on day 56 or day 80, respectively; cattle were revaccinated, as required, on days 14 and 28 and were challenged on day 72. Revaccinations with LE'-VP_1 were on day 24 for swine and day 28 for cattle; challenges were on days 32 and 42, respectively.
[b]Prechallenge titers were determined as PD_{50} in suckling mice.
[c]One steer had a single small localized foot lesion that healed without any other evidence of disease.
Data are from Bachrach et al., 1975, 1978, 1979, 1982 and Kleid et al., 1981.

Protective immunity has also been observed at 7 and 10.5 months in cattle vaccinated twice (on day 1 and at 3.5 or 7.5 months) with a 26 kd fusion protein containing, in sequence, amino acid residues 1-17 of tryptophan synthetase, a methionyl residue, residues 1-211 of VP_1 and 4 residues from pBR322 (Kleid, 1983; McKercher et al., in press). The challenge of immunity was at 7 months for cattle that received the two lower doses of fusion protein vaccine and at 10.5 months for those that received the two higher dosages (Table 2). Of the cattle vaccinated twice with 10 µg or 50 µg of the 26 kd fusion protein emulsified in an incomplete Freundtype oil adjuvant, 5 of 9 and 7 of 9, respectively, were protected against infection with FMD virus. The two groups of cattle vaccinated twice with either 250 µg or 1250 µg of fusion protein were more fully protected; 8 of 9 and 9 of 9, respectively, resisted the viral challenge. The revaccination schedules for the 26 kd fusion protein were comparable to those used in systematic vaccination programs in Europe using inactivated virus vaccines.

Table 2. Neutralizing Antibody and Protective Immunity in Cattle Vaccinated with Type A_{12} Biosynthetic 26 kd VP_1 Fusion Protein[a]

VP_1 Fusion Protein Dosage	Prechallenge Neut. Ab	Fraction Immune
µg	- log PD_{50}	
10	1.7	5/9
50	2.0	7/9
250	2.4	8/9
1250	2.7	9/9

[a]Conditions were as in footnotes to Table 1, except that revaccinations and challenges were at 3.5 and 7 months, respectively, for the 10 µg and 50 µg doses; and at 7.5 and 10.5 months, respectively, for the 250 µg and 1250 µg doses. The 26 kd VP_1 fusion protein contained a sequence of amino acids corresponding to residues 1-211 of VP_1.
Adapted from McKercher et al., Am. J. Vet. Res., in press.

Cloned Field-Type A and C VP_1-Specific Antigens

While type A_{12} virus is no longer reported to cause outbreaks of FMD, it has recently been demonstrated that VP_1-specific fusion proteins for field type A_{24}, A_{27}, A_{79} and C_3 viruses induce protective immunity in cattle equivalent (McKercher, personal communication) to that reported for the cloned type A_{12} LE'-VP_1 fusion protein (Kleid et al., 1981).

Cloned and Synthetic Type O_1 VP_1-Specific Antigens

Despite the successful immunization of cattle with type A and type C biosynthetic VP_1-specific fusion proteins, similar success has not been achieved with analogous type O_1 products. While guinea pigs and cattle elicited high levels of neutralizing antibody when vaccinated with a biosynthetic fusion protein containing the 125-172 amino acid sequence of VP_1 of type O_1 Campos virus, the cattle were susceptible to challenge with type O_1 virus (Moore et al., 1983).

Similarly, type O_1 VP_1-specific peptides have been synthesized chemically and tested for immunogenic activity in animals. A synthetic 20mer peptide corresponding to amino acid residues 141-160 of type O_1 virus linked to carrier protein and administered with adjuvant was reported to induce neutralizing antibodies and protective immunity in guinea pigs

(Bittle et al., 1982). While the 20mer also induced neutralizing antibodies in cattle, the cattle were not protected against challenge with virus (Wilson, 1984). A synthetic type O_1 VP_1-specific 16mer (residues 144-159) linked to carrier protein has also been reported to elicit neutralizing antibodies in rabbits (Pfaff et al., 1982).

There are several possible explanations why guinea pigs but not cattle have been protected by type O_1 synthetic peptide or fusion protein vaccines: 1) guinea pigs are quite resistant to FMD except by inoculation of virus which has been adapted to them, whereas cattle are highly susceptible natural hosts for FMD, 2) there may have been a strain difference between the type O_1 Campos virus template for the fusion protein containing the 125-172 amino acid sequence and the virus used to challenge the cattle (Moore et al., 1983) -- e.g., a hypervariable region (amino acid residues 147-152) of type A_{12} VP_1 generates strains with marked immunogenic differences (Rowlands et al., 1983), 3) the failure thus far to immunize cattle with type O_1 artificial peptides may be related to an aforementioned finding that type O_1 whole-virus vaccines are intrinsically weaker immunogens for livestock than are type A_{12} virus vaccines (Morgan et al., 1970; Bachrach and McKercher, 1972), 4) immunity in cattle to type O_1 virus may have a more prominent T-cell mediated component than immunity to type A and C viruses and 5) type O_1 VP_1 having two cysteine residues, rather than one as in type A and C VP_1 molecules, may form disulfide-stabilized conformational immunodominant sites in viral capsids that are lacking in isolated type O_1 VP_1 and VP_1-specific antigens. Whatever the cause, its solution is needed before trivalent A, O, C fusion protein vaccines can be formulated for use in Europe and South America where these types are endemic.

Type A_{12} VP_1 Antigenic Variants

A major difficulty in controlling FMD arises from the existence of a multiplicity of viral serotypes (seven) and subtypes (approximately seventy). The complexity is greatest for type A viruses, because at least half of all FMD virus subtypes are of this serotype. Molecular biological approaches have recently shown that type A complexity can be further complicated by the presence of discreet variants within a subtype. Based on reports of the immunogenicity of type A_{12} 44 kd LE'-VP_1 fusion protein (Kleid et al., 1981) and of type O_1 VP_1-specific synthetic peptide 141-160 (Bittle et al., 1982), an analogous peptide 141-160 was synthesized (Rowlands et al., 1983) to correspond to the nucleotide sequence reported by Kleid et al., (1981). Guinea pig antisera raised to this peptide, linked to keyhole limpet haemocyanin and adjuvanted with alum, neutralized the homologous A_{12} virus (called USA virus), but neutralized a stock of British A_{12} virus only poorly. Plaque isolations demonstrated that the British virus consisted of at least three variants of type A_{12} virus termed A, B and C. By sequencing cDNA reverse trancripts from VP_1 regions of the viral genomes, it was determined that variants A, B, C and USA differed at only two positions in the 141-160 amino acid region (positions 147 and 152) and at position 170 (Table 3). The USA variant had been derived years earlier from the British virus stock by plaque purification and repeated passages (about 100) in cell cultures. The USA virus could be a mutant of the B virus. Both have proline residues at position 152, and the difference at positions 147 and 170 could arise by point mutations.

The amino acid substitutions at positions 147, 152, and 170 in the A, B, C and USA variants markedly affect their serological activity as demonstrated in cross neutralization tests with homologous and heterologous antivirion and antipeptide (to the 141-160 VP_1 region) sera elicited in guinea pigs (Table 3). While there is some cross reactivity, marked differences between variants could have practical implications for the choice of viruses for the formulation of conventional whole-virus or protein subunit vaccines. Ideally, antigens used for vaccines should match those present in field isolates.

Table 3. Cross Neutralization of A_{12} Variant Viruses A, B, C and USA by Antivirion and Antipeptide 141-160 Sera[a]

Antiserum[b]		Virus Neutralized[c] A	B	C	USA
		ser	leu	ser	phe at 147
		leu	pro	ser	pro at 152
		val	val	val	glu at 170
A	aV	5.5	3.7	4.1	3.9
	aP	2.5	1.1	1.5	0.9
B	aV	4.5	5.3	3.7	4.3
	aP	nil	3.2	0.5	2.2
C	aV	5.7	5.3	7.3	4.9
	aP	1.5	0.1	2.6	1.5
USA	aV	4.5	3.8	4.1	5.6
	aP	nil	2.0	nil	2.7

[a]Adapted from Rowlands et al., 1983 with permission; actual A_{12} residue numbers are shown.
[b]Antivirion serum (aV) used at 1/2 dilution; antipeptide serum used undiluted.
[c]Log_{10} neutralized by 0.015 ml serum dilution; values are differences between virus plus normal serum and virus plus antiserum.

Hypervariability in the 147-152 region of VP_1 (Table 3) is apparently not restricted to type A_{12} variants, because five amino acid substitutions have been detected in the 147-150 region of VP_1 of the J_1, J_{2-3} and Tub variants of type A_5 virus (Kleid, personal commun.).

ANTIGENIC SITES

The mapping of antigenic sites on FMD viruses and capsid protein VP_1 is being actively investigated, because of its importance to the development of artificial vaccines. Investigative approaches include: 1) predictions of sites based on regions of VP_1 sequence hypervariability between A, O and C types and variants, 2) determination of the immunogenicity of VP_1 fragments and of bio- and chemically-synthesized VP_1--specific peptides and 3) determination of the specificity of monoclonal antibodies generated with virions, VP_1 or VP_1-specific fragments in reactions tions with virions, 12 S viral subunits and VP_1-specific sequences. The most productive of these approaches will be described below.

Immunogenic VP_1-Specific Fragments and Peptides

The loci of immunogenic sites on the 24 kd VP_1 capsid protein of FMD virus have been narrowed to certain regions by examining VP_1-specific fragments and artificial peptides for activity in eliciting neutralizing antibodies and protective immunity in animals. N-terminal 16 kd and 15 kd VP_1 fragments (amino acid residues 1-144 and 1-138, respectively) from proteolyzed type A_{12} virus are reported to have both activities, and an internal 13 kd fragment (residues 55-179) excised with CNBr appears to be as immunogenic as the entire VP_1 molecule. These results (Bachrach et al., 1979; Bachrach et al., 1982) indicated that putative immunogenic sites common to these three fragments of type A_{12} VP_1 were in the region defined by the overlap of the N-terminal 15 kd and internal 13 kd fragments (Fig. 3a), but did not exclude the existence of other putative sites, not common to all three fragments, from being on either side of the overlap region extending from the N-terminus of VP_1 to met 54 or from amino acid residue 138 through met 179 (Fig. 3a). That the latter region contains immunogenic sites was established by the finding that a biosynthetic fusion protein containing the 190-amino acid trp synthetase LE' segment plus a VP_1-specific segment spanning amino acid residues 137-168 in tandem repeat plus 13 vector-specific amino acids (32mer, Fig. 3a) elicits neutralizing antibodies in guinea pigs (Moore, 1983) and both neutralizing antibodies and protective immunity in swine (Morgan et al., 1983). Moreover, part of the VP_1 chain corresponding to the 32mer must be exposed on the surface of the virion, because in virion VP_1 protease-cleavage sites are most likely adjacent to residues 138 and 144, which are also in the 32mer; also the VP_1 tyr 136 available for iodination in the virion immediately precedes the sequence corresponding to the 32mer (Robertson et al., 1983a; Robertson et al., in press).

The location and characterization of antigenic sites on VP_1 of type O_1 Kaufbeuren FMD virus have been investigated similarly using proteolytic and CNBr VP_1 fragments from virions (Strohmaier et al., 1982). The order and immunogenic activities of the fragments along the VP_1 chain were determined. Two protease-sensitive regions of VP_1, 138-154 and 200-213, were established as being on the surface of the virus, and fragments containing these sequences were found to elicit neutralizing antibodies in mice.

Epitopes of VP_1

Experiments have been performed to locate immunodominant epitopes on VP_1 using monoclonal antibodies (Mabs) generated against type A_{12} (large plaque, ab variant) VP_1 and its 13 kd fragment (Robertson et al., in press). Five Mabs were prepared and cloned at least three times: 6AE9, 6HE4, 6HC4 and 6EE2 against VP_1 and 7SF3 against the 13 kd fragment (Fig.

3b). With the exception of 6AE9, all were capable of neutralizing virus in plaque reduction and mouse assays. However, 6AE9 did bind to the 12 S viral subunit (a complex of VP_{1-3} lacking VP_4 and RNA). It also bound to VP_1 and 16 kd and 13 kd fragments, placing its binding epitope in the 55-136 amino acid sequence of VP_1 (Fig. 3b). Mabs 7SF3 and 6HE4 each bound 140 S virions, 12 S subunits, the 13 kd fragment and the biosynthetic 32mer. They did not bind to trypsinized 140 S or 12 S subunits in which VP_1 is cleaved to 16 kd fragments. These reactions indicate that the neutralization epitopes for 7SF3 and 6HE4 Mabs both reside in the 145-168 amino acid region of VP_1 (Fig. 3b). Mabs 6HC4 and 6EE2 bound significant amounts of 140 S virions, 12 S subunits, VP_1 and the 13 kd fragment, but did not bind the 32mer (Fig. 3a). Also, they were not blocked in 12 S competition assays by the 32mer or 16 kd antigens, whereas VP_1, 13 kd and 140 S antigens were effective competitors. Thus, the loci of the epitope(s) for 6HC4 and 6EE2 Mabs appear to be in the 169-179 amino acid sequence of VP_1 (Fig. 3b). The regions of the binding epitope and two distinct neutralization epitopes defined by the five Mabs are all within the 13 kd fragment. While both neutralization epitopes are outside of the 15 kd and 16 kd fragments, other neutralization epitopes not detected with these five Mabs must be present in these fragments, because they are reported to induce neutralizing antibodies and protective immunity in animals (Bachrach et al., 1979, 1982).

Multiple Mechanisms of Viral Neutralization

Using the four neutralization Mabs in Fig. 3b and two other neutralization Mabs, 2PD11 and 2FF11, generated with intact 140 S virions, at least three-neutralizaiton-specific domains have been detected on the surface of type A_{12} virus (Baxt et al., 1984). Two of the domains are retained on isolated VP_1 chains and 12 S protein subunits, and one is a conformation specific site present only on intact 140 S virions. The mechanisms of neutralization by Mabs specific to the three domains appear to be different: Mabs 7SF3 and 6HE4 (reactive in the amino acid 145-168 region of VP_1, Fig. 3b) and 2FF11 (reactive only with 140 S virions and 12 S protein subunits) appear to cause extensive viral aggregation which inhibits virus adsorption to cells, 2) Mabs 6HC4 and 6EE2, which map to the amino acid 169-179 region of VP_1 (Fig. 3b), appear to neutralize viral infectivity by blocking the viral site on VP_1 that attaches to cellular receptors and 3) virion-specific Mab 2PD11, which has about 100-fold greater plaque-reduction titer than most of the other Mabs, may neutralize by inhibiting viral penetration or uncoating because there is little aggregation of virus or interference with virus adsorption to cells.

Specificity of Type A Viral Epitopes

Three Mabs, prepared against type A_{12} (large plaque, ab variant) virus and cloned at least three times, have been used to detect and classify epitopes on five type A virus field isolates (Morgan et al., 1984) (Table 4). While Mab no. 3 reacts with all the field isolates, all except A_{24} Brazil '76 lack epitope no. 1 and have epitope no. 4. The three epitopes appear to be specific to intact virions, because none of the Mabs react with any subviral moieties.

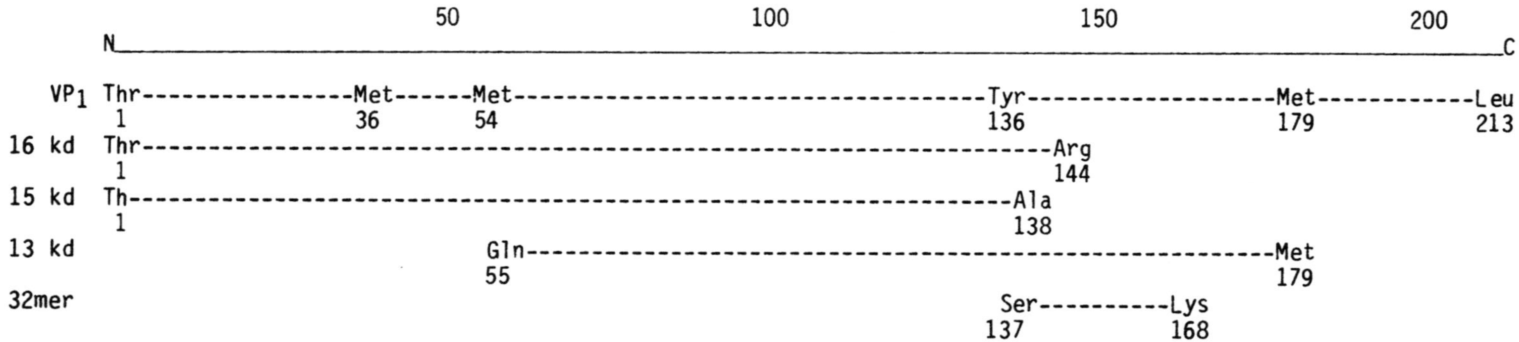

Figure 3. a. Linear diagram of type A_{12} VP_1-specific antigens. See text description of the 16 kd, 15 kd and 13 kd fragments and biosynthetic 32mer. Three methionyl residues and an iodinatable tyrosyl residue are shown on VP_1.

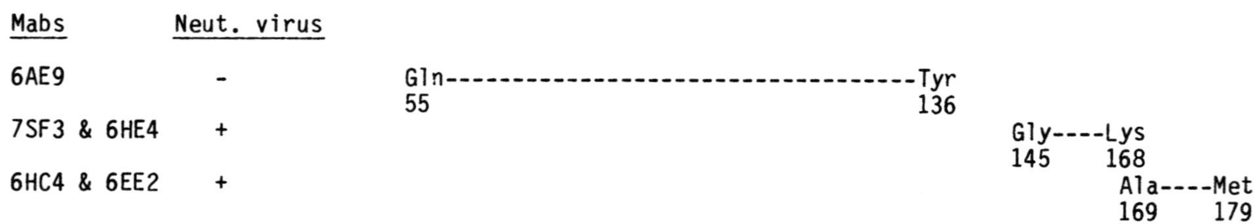

b. Antigenic regions defined by reactions (see text) of monoclonal antibodies (Mabs) with VP_1, the 16 kd and 13 kd VP_1 fragments and the 32mer shown in panel a. Mabs with 6 prefixes were generated against type A_{12} virus; 7SF3 was generated against the 13 kd fragment. All Mabs except 6AE9 neutralized virus. Adapted from Robertson, et al., in press, with permission.

Table 4. Specificity of Neutralization of Type A FMD Virus Field Isolates using Monoclonal Antibodies Elicited with Type A_{12} Virus[a]

Virus	Mab 1	Mab 3	Mab 4
A_{24} Colombia 8345	-	+	+
A_{24} Brazil '76	+	+	-
A_{24} Brazil '79	-	+	+
A_{27} Colombia 8046	-	+	+
A_{22}	-	+	+
A12	+	+	+

[a]Based on plaque reductions of 60% or more. From Morgan et al., 1984, with permission.

Epitopes on Type O_1 Virus and 12 S Subunits

Monoclonal antibodies from 137 hybridoma clones elicited with purified type O_1 Pacheco virus have been used to classify epitopes on 140 S viral particles and on 12 S^N and 12 S^A protein subunits (McCullough and Butcher, 1982). The 12 S^N subunits are isolated from infected cell lysates, whereas 12 S^A subunits are formed by the acid degradation of 140 S virions. The Mabs, all IgG, could be grouped into six reactivity classes: anti-140 S only, anti-12S^N only, anti-140 S/12 S^N, anti-140 S/12 S^A, anti-12 S^N/12 S^A and anti-140 S/12 S^N/12 S^A. Since there was no anti-12 S^A class, the processing of virions in animals to form 12 S^N subunits must be different from the acid degradation of purified virions. It was not determined whether any of the above epitope classes were present on isolated viral capsid proteins or fragments therefrom. Only Mabs having an anti-140 S component would be important in providing protective immunity, because 12 S subunits have been reported to be weak immunogens (Cartwright et al., 1982).

Sequence Variability in Epitopes

Amino acid residues 136-179 of type A_{12} USA VP_1 and comparative sequences of its A, B and C variants (Table 3) and serotypes O_1 Kaufbeuren, O_1 Campos and C_1 Obb are shown in Fig. 4 (Kleid et al., 1981; Rowland et al., 1983; Kurz et al., 1981; Robertson et al., 1983a; Beck et al., 1983). Several lines of evidence indicate that the 136-179 region of type A_{12} VP_1 contains prominent neutralization-specific epitopes: 1) the outer locations of VP_1 in the viral capsid, 2) the immunogenicity in animals and Mabs reactions of VP_1 fragments and biosynthetic VP_1-specific fusion proteins and 3) the variant and serotype-variability of amino acid sequences at specific loci within this region.

While not yet confirmed using Mabs, serotype-variable residues 136-144 would appear to have a neutralization-specific epitope. These residues appear to be exposed on the viral surface, because Tyr 136 is readily iodinated and Arg 144 is a site of cleavage by trypsin (Robertson et al., 1983a). Also all VP_1 fragments and biosynthetic peptides that span residues 136-144 are immunogenic: the 16 kd tryptic fragment (residues 1-144) and 13 kd CNBr fragment (residues 55-179) elicit high levels of neutralizing antibodies and protective immunity in animals Bachrach et al., 1979; 1982). The imunogenically-active biosynthetic 32mer (Fig. 3a) contains all of the residues 136-144 except Tyr 136. By comparison, an 8.5 kd fragment (residues 55-138 does not induce virus-neutralizing antibodies and does not compete against virions for VP_1-reactive antibodies (Robertson et al., 1983a). The putative epitope may be residues 137-143; its existence will have to be confirmed using appropriate synthetic peptides and neutralizing Mabs.

There is compelling evidence that the serotype-variable contiguous regions (145-168) and (169-179) possess neutralization-specific epitopes: 1) the cross-neutralization reactions of A_{12} variants USA, A, B and C (Table 3) and 2) the reactions of neutralizing Mabs, generated with A_{12} USA VP_1 or its 13 kd fragment, with VP_1-specific fragments and a biosynthetic 32mer (Fig. 3a & b). Moreover, amino acid residues 147, 152 and 170 of VP_1 of the A_{12} variants clearly reside in the neutralization-specific epitope regions circumscribed by the Mab reactions. The variants retain their A_{12} serotype even with radical substitutions at positions 152 (Pro is an alpha-helix breaker) and 170 (negatively charged Glu to an uncharged residue). The single residue substitutions at positions 147, 152 and 170 correlate with the production of type A_{12} variants, while multiple substitutions and insertion/deletions extending generally rightward from the same positions correlate with what has been considered to be changes in viral immunotype. The sequence variability patterns and lengths suggest that Mab-specified region 145-168 may have two 6 to 7 residue-long epitopes (ca. 146-151 and 152-157) and the 169-179 region only one (ca. 170-175). All three of these putative epitopes have one or two amino acid residues with charged side groups.

The idea that variant and serotype changes can both occur by amino acid substitutions in the same epitope is compatible with findings that cattle recovered from successive infections with two or more types of FMD virus begin to develop neutralizing antibodies and resistance to other viral types (Cottral and Gailiunas, 1971; Cottral, 1972). Complete resistance to infection was seen beginning with exposure to a fourth viral type.

CONCLUSION

Although type A and C capsid protein VP_1-specific fusion proteins have been demonstrated to be effective as vaccines against FMD in livestock, these moieties clearly do not have all of the neutralization-specific epitopes that are present on intact virions. Neutralization-specific epitopes unique to virions have been conclusively demonstrated using Mabs. Although other regions of VP_1 may have additional epitopes, those identified to date on VP_1 and its 13 kd fragment appear to be comprised of continuous amino acid sequences in the regions 55-136 (binding and 145-179 (neutralization). In addition, there is considerable evidence for a neutralization-specific epitope in the VP_1 136-144 sequence that appears to be exposed on the surface of FMD virus. Determination of the tertiary structures of VP_1 free in solution and in the viral surface may help in the identification and mapping of epitopes, especially of discontinuous epitopes comprised of distant residues brought together by molecular folding. However, predictions of epitopes for proteins of known tertiary structure such as lysozyme, myoglobin and hemoglobin have been largely unsuccessful (Atassi, personal commun.). If model building suggests structures for discontinuous epitopes, peptides could then be synthesized to simulate their surface features (Atassi and Sakata, 1980) and tested for immunogenicity and reactivity with mono- and poly-clonal antibodies.

```
                                                    144
               136                                   +
A12  USA       -Tyr-Ser-Ala-Ser-Gly-Ser-Gly-----Val-Arg-
        A                -Val-
        B                -Val-
        C                -Val-
                           +                       145
O1  K & C           -Asn-Arg-Asn-Ala-Val-Pro-Asn-Leu-
                                                   141
   C1 Obb           -Thr-        -Thr-----------------


               (145 _ 147                 152  +          + 157
A12  USA       -Gly-Asp-Phe-Gly-Ser-Leu-Ala-Pro-Arg-Val-Ala-Arg-Gln-
        A                -Ser-              -Leu-
        B                -Leu-              -Pro-
        C                -Ser-              -Ser-
               146      148 149      151 152       +          158
O1  K & C                -Leu-Gln-Val-      -Gly-Lys-         -Thr-
               142                                  +          154
   C1 Obb                -Leu-Ala-His-  -Thr-Ala-Thr-Arg-  -Gly-His-


               158                                         168)
A12  USA       -Leu-Pro-Ala-Ser-Phe-Asn-Tyr-Gly-Ala-Ile-Lys-
    A,B,C                                                   +
               159                                         169
O1  K & C                -Thr-
               155                                         165
   C1 Obb                -Thr-             -Phe-        -Val-


               (169 170           -                  + 179)
A12  USA       -Ala-Glu-Thr-Ile-His-Glu-Leu-Leu-Val-Arg-Met-
        A            -Val-
        B            -Val-
        C            -Val-
               170       +                                 180
O1  K & C            -Thr-Arg-Val-Thr-           -Tyr-
               166                                         176
   C1 Obb                       -Thr-
```

Figure 4. Amino acid residues 136-179 of type A_{12} VP_1 variants USA, A, B and C and comparative sequences for serotypes O_1 Kaufbeuren, O_1 Campos and C_1 Obb. Residues 136-144 of type A_{12} VP_1 are in a serotype-variable region and are on the viral surface; Tyr 136 is readily iodinated, and Arg 144 is the site of tryptic cleavage of in virion VP_1. Although not yet confirmed,

(Fig. 4. continued)

(continued)
these properties and the immunogenicities of VP_1 fragments indicate that the 136-144 region may have an epitope: a 16 kd fragment (residues 1-144) induces neutralizing antibodies and protective immunity, while an 8.5 kd fragment (residues 55-138) does not. The succeeding contiguous regions (145-168) and (169-179) of type A_{12} VP_1 appear to have epitopes reactive with neutralizing Mabs (Fig. 3b). Putative epitopes are singly underlined; their selection is more fully explained in text section Sequence Varability in Epitopes. The A_{12} variants differ by point mutations, while the serotypes generally differ by sequential mutations and insertion/deletions in the same regions. Only residues differing from the A_{12} USA variant are shown. ----- indicates deletions compared to O_1 Kaufbeuren and Campos. Doubly underlined type O_1 binding epitope (residues 146-152) on VP_1 and virus identified using overlapping hexapeptides (Geysen et al., 1984); see text CONCLUSION.

When this review was nearing completion, a report appeared on the mapping of epitopes of type O_1 virus, 12 S subunit particles and VP_1 molecules by binding a set of 208 overlapping hexapeptides covering the entire 213 amino acid long VP_1 singly to rabbit antiserum fractions specific for intact 140 S virus, 12 S subunits and VP_1 (Geysen et al., 1984). The dominant epitope mapped at residues 146-152 and was detected on both virus and VP_1 but not on 12 S subunits. This type O_1 epitope, -Gly-Asp-Leu-Gln-Val-Leu-Ala-, is co-sequential with a neutralization-specific of type A_{12} VP_1 (Fig. 4) determined by other methods.

Geysen and coworkers also determined the specificity of each of the seven positions within the type O_1 epitope. Viral-specific antiserum was reacted with a set of 120 synthetic hexapeptides having each of the first six amino acid residues of the epitope substituted in turn with the 20 possible amino acid residues. The specificity of the seventh position was determined similarly using 20 heptapeptides having the amino acids residues of the epitope in the first six positions and all 20 possible amino acid residues in turn in the seventh position. There was an absolute requirement for hydrophobic leucyl residues at positions 148 and 151 and a lesser requirement for glutaminyl residue 149 and alanyl residue 152. The synthetic approach to mapping epitopes at a resolution of single amino acids has implications for the improvements in the understanding of antibody specificity and the efficacy of vaccines.

REFERENCES

Atassi, M. Z., and Sakata, S., 1980, Antibodies to synthetic antibody-combining sites. Antibodies against a surface simulation peptide with antibody-combining activity toward lysozyme antigenic site 3 react with lysozyme antibodies, Biochem. Biophys. Acta., 624:573.

Bachrach, H. L., 1952, The determination of the sedimentation constant of a homogeneous component having the characteristics of the foot-and-mouth disease virus, Am. J. Vet. Res., 13:13.

Bachrach, H. L., 1968, Foot-and-mouth disease, in: "Annual Review of Microbiology," C. E. Clifton, S. Raffel and M. P. Starr, eds., v 22, p. 201, Annual Reviews Inc., Palo Alto.
Bachrach, H. L., 1977, Foot-and-mouth disease virus: Properties, molecular biology and immunogenicity, in: "Beltsville Symposium in Agricultural Research, I. Virology in Agriculture," J. A. Romberger, ed., pp. 3-32, Allenheld Osmun & Co., Montclair.
Bachrach, H. L., 1978, Foot-and-mouth disease: Worldwide impact and control measures, in: "Viruses and Environment," K. Maramorosch and E. Kurstak, eds., p. 229, Academic Press, NY.
Bachrach, H. L., Callis, J. J., Hess, W. R., and Patty, R. E., 1957, A plaque assay for foot-and-mouth disease virus and kinetics of virus reproduction, Virology, 4:224.
Bachrach, H. L., and Breese, S. S. Jr., 1958, Purification and electron microscopy of foot-and-mouth disease virus, Proc. Soc. Exp. Biol. Med., 97:659.
Bachrach, H. L., Hess, W. R., and Callis, J. J., 1955, Foot-and-mouth disease virus: Its growth and cytopathogenicity in tissue culture, Science 122:1269.
Bachrach, H. L., and McKercher, P. D., 1972, Immunology of FMD in swine: Experimental inactivated-virus vaccine, J. Am. Vet. Med. Assoc, 160:521.
Bachrach, H. L., Moore, D. M., McKercher, P. D., and Polatnick, J., 1975, Immune and antibody responses to an isolated capsid protein of foot-and-mouth disease virus, J. Immunol., 115:1636.
Bachrach, H. L., Moore, D. M., McKercher, P. D., and Polatnick, J., 1978, An experimental protein vaccine for foot-and-mouth disease, in: "Perspectives in Virology X," M. Pollard, ed., pp. 147, Raven Press, NY.
Bachrach, H. L., Morgan, D. O., McKercher, P. D., Moore, D. M., and Robertson, B. H., 1982, Immunogenicity and structure of fragments derived from foot-and-mouth disease virus capsid protein VP_3 and of virions having intact and cleaved VP3, Vet. Microbiol., 7:85.
Bachrach, H. L., Morgan, D. O., and Moore, D. M., 1979, Foot-and-mouth disease virus immunogenic capsid protein VP_T: N terminal sequences and immunogenic peptides obtained by CNBr and tryptic cleavages, Intervirology 12:65.
Bachrach, H. L., and Schwerdt, C. E., 1954, Purification studies on Lansing poliomyelitis virus II. Analytical electron microscopic identification of the infectious particle in preparations of high specific infectivity, J. Immunol., 72:30.
Bachrach, H. L., Swaney, J. B., and Vande Woude, G. F., 1973, Isolation of the structural polypeptides of foot-and-mouth disease virus and analysis of their C-terminal sequences, Virology 52:520.
Bachrach, H. L., Trautman, R., and Breese, S. S. Jr., 1964, Chemical and physical properties of virtually pure foot-and-mouth disease virus, Am. J. Vet. Res., 25:333.
Baxt, B., Morgan, D. O., Robertson, B. H., and Timpone, C. A., 1984, Epitopes on foot-and-mouth disease outer capsid protein VP1 involved in neutralization and cell attachment, J. Virol., 51:298.
Beck, E., Feil, G., and Strohmaier, K., 1983, The molecular basis of the antigenic variation of foot-and-mouth disease virus. EMBO, J., 2:555.
Bernard, S., Wantyghem, J., Grosclaude, J., and Laporte, J., 1974, Chromatographic separation of purified structural proteins from foot-and-mouth disease virus, Biochem. Biophys. Res. Commun., 58:624.

Bittle, J. L., Houghten, R. A., Alexander, H., Shinnick, T. M., Sutcliffe, J. G., Lerner, R. A., Rowlands, D. J., and Brown, F., 1982, Protection against foot-and-mouth disease by immunization with a chemically synthesized peptide predicted from the viral nucleotide sequence, Nature (London) 298:30.
Carrol, A. R., Rowlands, D. J., and Clarke, B. E., 1984, The complete nucleotide sequence of the RNA coding for the primary translation product of foot-and-mouth disease virus, Nucleic Acids Res., 12:2461.
Cartwright, B., Morrell, D. J., and Brown F., 1982, Nature of the antibody response to the foot-and-mouth disease virus particle, its 12S protein subunit and the isolated immunizing polypeptide VP1, J. Gen. Virol., 63:375.
Cottral, G. E., 1972, Foot-and-mouth disease neutralization test cross-reactions: A comparison of cattle and guinea pig serums, Proc. 76th Ann. Mtg. U. S. Animal Health Assoc., pp. 194-207.
Cottral, G. E., and Gailiunas, P., 1971, Experimental multiple infection of animals with foot-and-mouth disease viruses, Proc. 75th Ann. Mtg. U. S. Animal Health Assoc., pp. 441-465.
Forss, S., and Schaller, H., 1982, A tanden repeat gene in a picornavirus, Nucleic Acids Res., 10:6441.
Frascastorius, H., 1546, De allis differcattis contagionis, in: "De Sympathia et Antipathia Rerum Liber Unus, De contagione et contagiosis morbis et curatione, Bk 1, Chap. 12, Venecia.
Frenkel, H. S., 1951, Research on foot-and-mouth disease III. The cultivation of the virus on a practical scale in explantations of bovine tongue epithelium, Am. J. Vet Res., 12:187.
Geysen, H. M., Meloen, R. H., and Barteling, 1984, Use of peptide synthesis to probe viral antigens for epitopes to a resolution of a single amino acid, Proc. Natl. Acad. Sci. USA, 81:3998.
Kitamura, N., Semler, B. L., Rothberg, P. G., Larsen, G. R., Alder, C. J., Dorner, A. J., Emini, A., Homecak, R., Lee, J. J., Vander der Werf, S., Anderson, C. W., and Wimmer, E., 1981, Primary structure, gene organization and polypeptide expression of poliovirus RNA, Nature 291:547.
Kleid, D. G., 1983, Using genetically-engineered bacteria for vaccine production, Annals NY Acad. Sci., 413:23.
Kleid, D. G., Yansura, D., Small, B., Dowbenko, D., Moore, D. M., Grubman, M. J., McKercher, P. D., Morgan, D. O., Robertson B. H., and Bachrach, H. L., 1981, Cloned viral protein vaccine for foot-and-mouth disease: Responses in cattle and swine, Science, 214:1125.
Klump, W., Marquardt, O., and Hofschneider, P. H., 1984, Biologically-active protease of foot-and-mouth disease virus is expressed from cloned cDNA in Escherichia coli, Proc. Natl. Acd. Sci. USA, 81:3351.
Kurz, C., Forss, S., Kupper, H., Strohmaier, K., and Schaller, H., 1981, Nucleotide sequence and corresponding amino acid sequence of the gene for the major antigen of foot-and-mouth disease virus, Nucleic Acids Res., 9:1919.
Laporte, J., Grosclaude, J., Wantyghem, J., Bernard, S., and Rouge, P., 1973, Neutralization en culture cellulaire du pouvair infectieux du virus de la fievze aphteuse par des serums provenant de porcs immunises a l'aide d'une proteine virale purifiee, C. R. Acad. Sci., 276:3399.

Loeffler, F., and Frosch, P., 1897, Summarischer bericht uber ergebnisse der untersuchungen der kommission zur erforschung der maul-und klauenseuche bei den Institute fur Infektions-krankheiten in Berlin, Centr. Bakt. Parasitenk. Infekt. Abt., I., Orig., 22:257.

McCullough, K. C., and Butcher, R., 1982, Monoclonal antibodies against foot-and-mouth disease 146S and 12S particles, Arch. Virol., 74:1.

McKercher, P. D., Moore, D. M., Morgan, D. O., Robertson, B. H., Callis, J. J., Kleid, D. G., Shire, S. J., Yansura, D. G., Dowbenko, D., and Small, B., Dose response evaluation of a genetically-engineered foot-and-mouth disease virus polypeptide immunogen in cattle, Am. J. Vet. Res., in press.

Moore, D.M., 1983, Production of a vaccine for foot-and-mouth disease through gene cloning, in: "Genetic Engineering: Applications to Agriculture," Beltsville Symposium VII, L. D. Owens, ed., Rowman and Allanheld, Totowa.

Moore, D., Morgan, D., Robertson, B., McKercher, P., Patzer, E., Shire S., Kleid, D. G., 1983, A highly antigenic portion of FMDV O_1 VP_1 elicits bovine antibodies which protect mice but not cattle from FMDV infection, in: Abstracts "Modern Approaches to Vaccines," Cold Spring Harbor Laboratory, Cold Spring Harbor.

Morgan, D. O., McKercher, P. D., and Bachrach, H. L., 1970, Quantitation of the antigenicity and immunogenicity of purified foot-and-mouth disease virus vaccines for swine and steers, Appl. Microbiol., 20:770.

Morgan, D. O., Moore, D. M., Robertson, B. H., Shire, S. J., Bock L. A., and Kleid, D. G., 1983, Immunization of swine with defined segments on FMDV VP_1 produced by recombinant DNA procedures, Abs 211, 64th Cong. Res. Workers Animal Dis., Chicago.

Morgan, D. O., Robertson, B. H., Moore, D. M., Timpone, C. A., and McKercher, P. D., 1984, Aphthoviruses: Control of foot-and-mouth disease with genetic engineering vaccines, in, "Control of Virus Diseases," E. Kurstak, ed., Marcel Dekker, NY.

Mussgay, M., 1959, Uber den mechanismus der pH-inaktivierung des virus der maul-und-klauenseuche, Monatsh. Tierheilk., 11:185.

Palmenberg, A. C., Kirby, E.M., Janda, M. R., Drake, N. L., Duke, G. M., Potratz, K. F., and Collett, M. S., 1984, The nucleotide and deduced amino acid sequence of the encephalomycarditis viral polyprotein coding region, Nucleic Acids Res., 12:2969.

Pfaff, E., Mussgay, M., Bohm, H. O., Schulz, G. E., and Schaller, H., 1982, Antibodies against a preselected peptide recognize and neutralize foot-and-mouth disease virus, EMBO J., 1:869.

Robertson, B. H., Moore, D. M., Grubman, M. J., and Kleid, D. G., 1983a, Identification of an exposed region of the immunogenic capsid polypeptide VP_1 on foot-and-mouth disease virus, J. Virol., 46:311.

Robertson, B. H., Morgan, D. O., Moore, D. M., Grubman, M. J., Card, J., Fischer, T., Weddell, G. N., Dowbenko, D. J., and Yansura, D. G., 1983, Identification of amino acid and nucleotide sequence of the foot-and-mouth disease virus RNA polymerase, Virology 126:614.

Robertson, B. H., Morgan, D. O., and Moore, D. M., Location of neutralizing epitopes defined by monoclonal antibodies generated against the outer capsid polypeptide VP1 of foot-and-mouth disease virus A12, Virus Research, in press.

Rowlands, D. J., Clarke, B. E., Carrol, A. R., Brown, F., Nicholson, B. H., Bittle, J. L., Houghten, R. A., and Lerner, R. A., 1983, Chemical basis of antigenic variation in foot-and-mouth disease virus, Nature 306:694.
Rueckert, R. R., and Wimmer, E., 1984, Systematic nomenclature of picornavirus proteins, J. Virol., 50:957.
Strohmaier, K., Franze, R., and Adam, K. H., 1982, Location and characterization of the antigenic portion of the FMDV immunizing protein, J. Gen. Virol., 59:295.
Schmidt, S., 1936, Immunisierung des meerschuoinchons gegen drei verschiedene typen von maul-und-klauensechevirus vermittels eines trivalent aluminum-hydroxydadsorbates, Z. Immunitatforsch., 88:91.
Waldmann, O., and Kobe, K., 1938, Active immunization of cattle against FMD, Berlin. Tierarztl. Wochschr., 22:317.
Wild, T. F., Burroughs, J. N., and Brown F., 1969, Surface structure of foot-and-mouth disease virus, J. Gen. Virol., 4:313.
Wilson, T., 1984, Engineering tomorrow's vaccines, Biotechnology 2:29.

THE STRUCTURAL PROTEINS OF THE AUTONOMOUS PARVOVIRUS FELINE PANLEUKOPENIA VIRUS

Jonathan Carlson, Keith Rushlow, Alistair McNab and Scott Winston

Syngene Products and Research, Inc.
TechAmerica Research Center
225 Commerce Drive
Fort Collins, CO 80522

ABSTRACT

Approximately 80% of the genome of feline panleukopenia virus was cloned into the plasmid pBR322. The entire 3943 nucleotide sequence of the cloned portion of FPV was determined. This DNA includes the gene which codes for the structural proteins of the virus. Portions of this gene were expressed in *E. coli* as fusion proteins with bacterial proteins. Some of the fusion proteins were capable of raising neutralizing antibodies in guinea pigs. Through the use of deletion mapping, monoclonal antibodies, and synthetic peptides, attempts were made to localize the portion of the protein responsible for raising these antibodies.

INTRODUCTION

Feline panleukopenia virus (FPV) is an autonomous parvovirus. Parvoviruses have a small, linear single-stranded DNA genome and require proliferating cells for viral multiplication. Consequently, autonomous parvoviruses tend to cause diseases that affect tissues in which there is active cell division. For example, FPV attacks the lymph- and blood-forming organs of the cat, causing a decrease in the leukocytes. It also attacks the gastrointestinal mucosa, resulting in enteritis and diarrhea. In neonatal kittens FPV infects the developing cerebellum causing ataxia shortly thereafter. Other parvoviruses infect fetuses and neonates resulting in reproductive failure, birth defects, and developmental abnormalities (Siegl, 1976).

FPV is interesting because of the disease it causes in cats, and because of two closely related variants which have been described in the last forty years. In the late 1940s outbreaks of enteritis in mink raised on ranches in Canada were shown to be due to a parvovirus known as Mink Enteritis Virus (MEV). MEV is antigenically very closely related to FPV (Parrish and Carmichael, 1983). Restriction-endonuclease cleavage maps show a high degree of homology between FPV and MEV on a nucleic-acid level as well (Tratschin *et al*, 1982). Similarly, in the late 1970s, a parvovirus called Canine Parvovirus (CPV) was implicated in outbreaks of enteritis in dogs and myocarditis in puppies that were reported in several continents within a few months of each other. Again, CPV was closely related to FPV on both the protein and nucleic-acid levels shown by substantial antigenic cross-reactivity (Parrish and Carmichael, 1983) and similarity in restriction-endonuclease cleaveage maps (McMaster et al., 1981; Tratschin et al., 1982). Antibodies to CPV have not been reported in dog sera prior to 1976. Thus it appears that, although it is origin is unclear, CPV is a recently derived variant of FPV. (Carmichael and Binn, 1981; Tratschin et al. 1982).

Autonomous parvoviruses have a linear single-stranded DNA genome of approximately 5,000 bases. Most autonomous parvoviruses package only the negative strand of the genome into virions, although some package both strands into separate virions. The ends of the DNA molecules contain inverted repeat sequences and are folded into hairpin-like secondary structures. Shortly after infection the single-stranded genome is converted into a double-stranded replicative form by host cell DNA polymerases. This double-stranded replicative form of DNA functions as a template for DNA replication and transcription of virus mRNA in the infected cells.

Parvovirus virions contain three structural proteins. The largest protein (VP1) has a molecular weight of approximately 80,000 to 83,000 daltons and comprises 10-15% of the total viral protein. The rest of the viral protein is comprised of VP2' and VP2 which are approximately 64,000 to 67,000 and 61,000 to 64,000 daltons, respectively (Peterson et al., 1978). VP2 is derived from VP2' by a proteolytic cleavage after the virion is assembled. The fraction of VP2' which is converted to VP2 can vary greatly. The conversion of VP2' to VP2 involves the removal of approximately 30 amino acids from the amino terminus of VP2'. The entire amino acid sequence of VP2' is also contained within VP1 although there is no evidence that VP2' is derived from VP1 via proteolytic cleavage. The virion proteins comprise a nested set with the amino acid sequences of the smaller proteins contained within the largest protein (Tattersall et al., 1977). These proteins are assembled into an icosahedral virus particle containing no lipid.

We were interested in developing a vaccine against FPV and CPV which would not require large-scale tissue-culture growth of the virus. To this end, we cloned the structural protein coding region of the FPV genome by recombinant-DNA methods.

EXPERIMENTAL

ELISA

Antigen was dried down in the wells of polystyrene microtiter plates. The wells were blocked with 2% bovine serum albumin for 1 h at 37°C. The wells were washed three times with TEN buffer (0.05M TRIS, 0.001M EDTA, 0.15M NaCl [pH7.2]) and incubated with appropriate dilutions of antiserum or monoclonal antibody in TEN. After 1 h at 37°C, the wells were washed three times with TEN and peroxidase-conjugated secondary antibody was added. After incubation for 1 h at 37°C the wells were again washed and bound peroxidase was detected by the addition of 0.4mg/ml o-phenylenediamine and 0.01% H_2O_2 in citrate-phosphate buffer (pH5.0). After 15 min, 2.5N H_2SO_4 was added to stop the reaction. The absorbance at 490nm was measured on a Dynatech MR580 Microelisa plate reader.

Serum-Neutralization Assay for FPV

Crandell feline kidney cells (CRFK) were seeded in wells of a microtiter plate at a density of $5x10^3$ to 10^4 cells per well. Appropriate dilutions of antiserum in RPMI-10% FCS were incubated with a stock FPV suspension for 1 h at 37°C and then added to the wells. After 2 to 3 days at 37°C in a humidified incubator, 100ml of 2mM EDTA and 0.1% SDS were added per well to lyse the cells. Fifty ml of 0.6M NaOH was added to denature DNA and after at least five min, 50ul of 1.2M NaH_2PO_4 was added to neutralize the solution. One hundred ul of the solution was transferred to a nitrocellulose filter using a 96-well filtration manifold (Hybridot, BRL). This filter ("dot blot") was then dried at 80°C for 2 h. The blot containing the blotted DNA was incubated in a sealed plastic bag for 2 to 6 h at 42°C in 10ml of 50% formamide, 0.2% polyvinylpyrolidone (MW 360,000), 0.2% bovine serum albumin, 0.2% ficoll (MW 400,000), 0.5M TRIS-HCl (pH 7.5), 1M NaCl, 0.1% sodium pyrophosphate, 0.1% SDS, 10% sodium dextran sulfate (MW 500,000) and 100mg/ml denatured herring-sperm DNA. Nick-translated ^{32}P-labeled parvovirus DNA ($0.5-2x10^6$ cpm) was added to the bag in 2ml of 50% formamide, 0.2% polyvinylpyrolidone, 0.2% BSA, 0.2% ficoll, 0.5M TRIS-HCl (pH 7.5), 0.1% sodium pyrophosphate, 0.1% SDS, and 100ug/ml denatured herring-sperm DNA. Hybridization was performed at 42°C for at least 16 h. After hybridization, the blot was washed twice for 5 min each with 0.3M NaCl, 0.06M TRIS-HCl (pH7.5) and 0.002M EDTA at room temperature, twice for 30 min each with the same buffer containing 0.5% SDS at 60°C, and twice for 30 min each in 0.003M TRIS base at room temperature. The blot was dried on paper towels and hybridization was detected by autoradiography with Kodak X-Omat XAR-5 film. Neutralization of the virus is detected by a decrease in hybridization signal on the autoradiogram compared with control wells to which no serum was added.

Synthetic Peptides

Peptides 1 through 15 were synthesized by Richard Houghton, James Bittle, and Richard Lerner at the Research Institute of Scripps Clinic. Peptides 16 through 19 were synthesized by Peninsula Laboratories, Inc., San Carlos, California.

Peptides were coupled to either keyhole-limpet hemocyanin or thyroglobulin through an additional cysteine residue added to the peptide during synthesis. The coupling agent was m-maleimidobenzoyl-N-hydroxy-succinimide ester (MBS) (Lin et al., 1979). The procedure used for coupling was that of Green et al., 1982. Three injections of 100 to 250ug of protein-peptide conjugate were given to guinea pigs at two-week intervals. The first injection was given with Freund's complete adjuvant, the second with incomplete Freund's adjuvant and the third with $Al(OH)_3$. Serum samples were collected two weeks after the final injection.

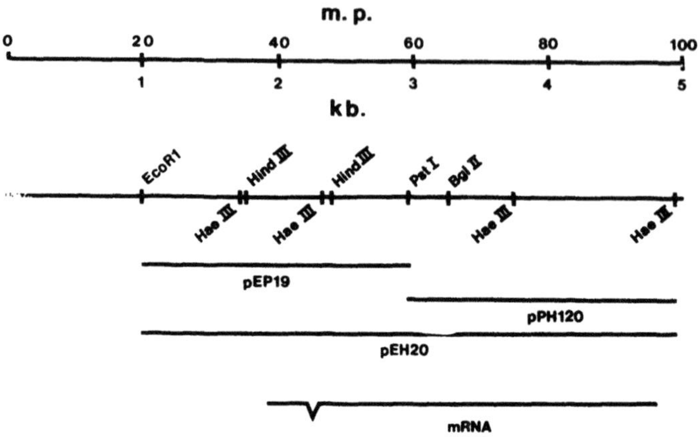

Fig. 1. Partial restriction map of FPV. The genome of FPV is represented in the top line with 0 as the 3' end and 100 as the 5' end. The length in kilobases is also indicated. A partial restriction map is shown in the next line. The portions of the genome continued in the plasmids pEP19, pPH120, and pEH20 are shown below the restriction map. The bottom line shows the portion of the genome contained in the major viral mRNA species. The polarity of the mRNA is 5' to 3' left to right.

RESULTS

The genome of FPV was cloned into the plasmid pBR322 in two steps (Carlson, et al., 1984). Initially, single-stranded DNA was isolated from purified virions and converted to double-stranded DNA using _E. coli_ DNA polymerase and deoxynucleotide triphosphates. This _in vitro_-synthesized DNA was then digested with the restriction endonucleases EcoR1 and PstI and the fragment between the EcoR1 site at map position 20 and the PstI site at map positive 39 was cloned into pBR322 to give the plasmid pEP19 (Fig. 1). The DNA in the right half of the genome was cloned from double-stranded replicative form DNA isolated from the nuclei of FPV infected cells. This DNA was digested with PstI and then partially cleaved with HaeIII. This DNA was ligated to pBR322 which had been cleaved with EcoR1, and was then reacted with DNA polymerase to fill in the ends, and cleaved with PstI. One clone from this ligation, pPH120, contained the DNA between the PstI site at map position 39 and the Hae III site at map position 96 (Fig. 1). A plasmid pEH20 containing all the FPV DNA from the EcoR1 site at map position 20 to the HaeIII site at map position 96 was constructed from the FPV inserts from pEP19 and pPH120.

The portion of the genome coding for the major mRNA in FPV-infected cells was determined using single-strand specific nuclease-protection experiments. Cytoplasmic RNA from FPV-infected CRFK cells was hybridized with single-stranded FPV DNA cloned into appropriate M13 phages. These hybrids were digested with mung-bean nuclease and the products were fractionated on alkaline agarose gels, blotted onto nitrocellulose filters, and hybridized with ^{32}P-labelled FPV probes. These experiments indicated that the major eytoplasmic FPV-specific RNA consists of a 5' exon of about 250-300bp spliced to a 3' exon of about 2,500bp. The 5' exon maps between the HindIII sites on the FPV genome and the 3' exon contains most of the right half of the FPV genome (Fig. 1). The major mRNAs for two other parvoviruses MVM and H1 map to the same region in their genome and are known to code for the viral structural proteins (Rhode and Paradiso, 1983; Pintel et al., 1983).

The entire sequence of pEH20 was determined and found to be 3943bp in length. As in MVM and H1 the sequence consists of two long, open, translational-reading frames (Rhode and Paradiso, 1983, Astell et al, 1983). The leftward open-reading frame contains the EcoR1 site which marks left end of pEH20 and continues to a termination codon about 1200 nucleotides downstream (Fig. 2). The nucleotide sequence of this open-reading frame is about 80% homologous with the analogous portions of the MVM and H1 genomes, and the amino-acid sequence predicted from the reading frame is about 80% homologous with those predicted from the H1 and MVM sequences. This portion of the genome in H1 and MVM codes for a nonstructural protein of unknown function which is antigenically conserved among several parvoviruses (Rhode and Paradiso, 1983; Cotmore et al., 1983).

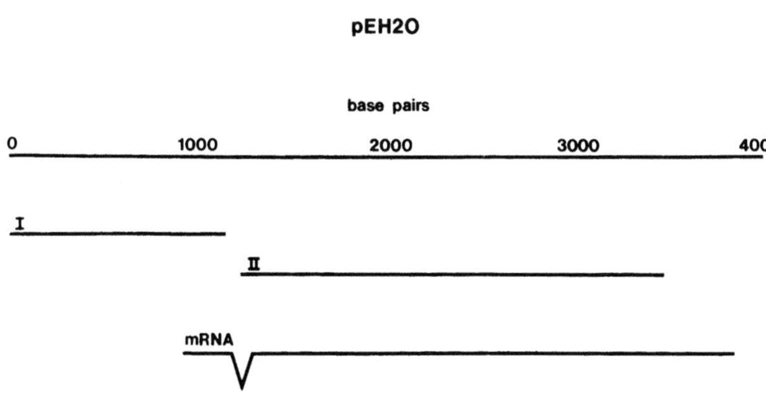

Fig. 2. Open translational reading frames of pEH20. The 3943bp nucleotide sequence of the portion of the FPV genome contained in pEH20 is represented by the top line. The positions of the leftward and rightward open translational reading frames are represented by the lines marked I and II. The position of the major mRNA is indicated by the bottom line.

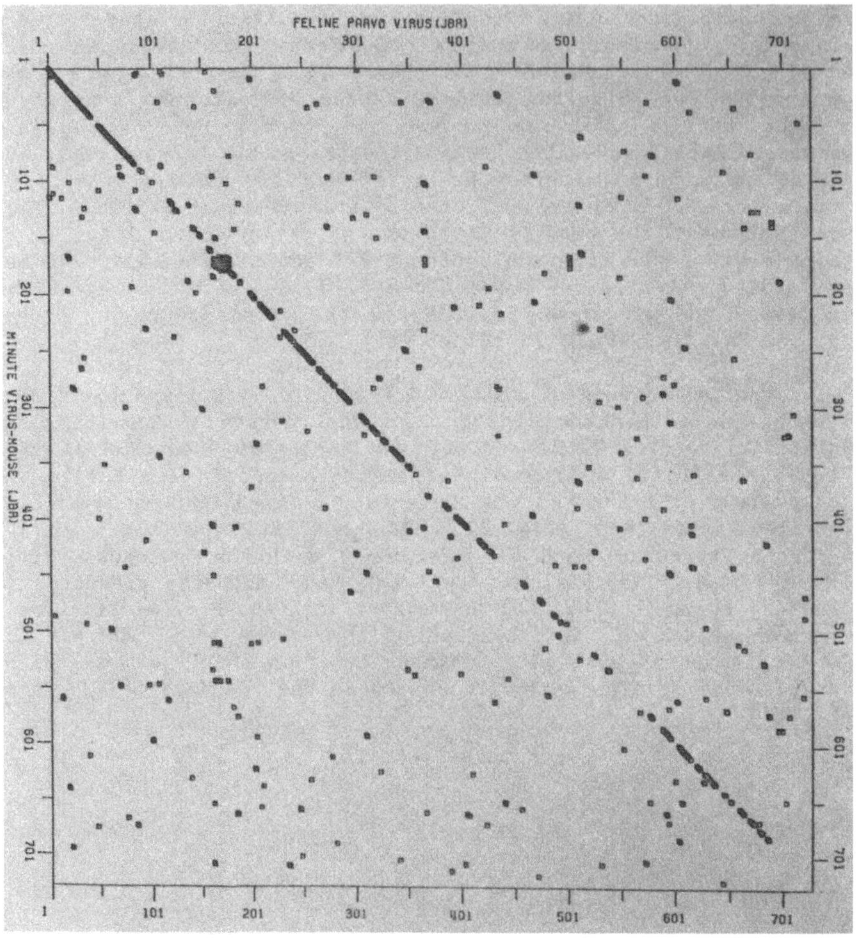

Fig. 3. Homology between the virion proteins of FPV and MVM. The predicted amino-acid sequence of VP1 from FPV was compared with the predicted sequence of VP1 from MVM using a matrix homology plot. The FPV amino-acid sequence on the horizontal axis was compared with the MVM amino-acid sequence four amino acids at a time. If three of the four matched, a box was drawn in that position.

The rightward open-reading frame of FPV begins approximately 100bp downstream from the terminator of the leftward reading frame and continues from about 2,200bp downstream to a termination codon about 900 bases to the left of the right-hand terminae of pEH20. This reading frame is only about 50% conserved on the nucleotide-sequence level between FPV and either H1 or MVM. Virtually all of this open-reading frame is included in the large 2.5kb exon in the major viral mRNA and codes for most of the viral structural proteins. Fig. 3 shows the comparison of the predicted sequences of VP1 from FPV and MVM using a matrix-homology plot. The prominent diagonal shows considerable homology between the proteins from the two viruses. The homology is more pronounced in the amino-terminal half of the protein than in the carboxyterminal half.

Examination of the FPV sequence and comparison with the published sequences for MVM and H1 allows us to postulate a mechanism based on RNA splicing for the generation of the two different viral structural proteins VP1 and VP2' (Fig. 4). The sequences of all three viruses contain promoters at about map position 38 (about nucleotide 900 in the pEH20 sequence) and ATG initiation codons about 300 and 800 base pairs downstream (nucleotides 1188 and 1689 in the pEH20 sequence). The first ATG is between two potential 5' splice junctions which are highly conserved among the three viruses, as is a potential 3' splice junction just downstream from the beginning of the long rightward open-reading frame. An mRNA in which the 5' splice junction at nucleotide 1182 on the pEH20 sequence is spliced to the 3' splice junction at nucleotide 1290 would eliminate the first ATG at position 1188 and translation would start at position 1689 to generate the smaller of the two proteins VP2' (Fig. 4). An mRNA in which the 5' splice junction at nucleotide 1218 is spliced to the 3' splice junction at 1290 would fuse the ATG at 1188 and nine codons to the long rightward open-reading frame to code for a protein the size of VP1 (Fig. 4). The relative positions of these splice junctions and initiation codons are highly conserved among the three viruses, and the molecular weights of the protein products predicted from these mRNA agree well with the molecular weights observed by polyacrylamide-gel electrophoresis.

We were interested in expressing the antigens of FPV in E. coli to produce a product which would raise neutralizing antibodies against FPV. We therefore constructed two kinds of expression plasmids (Carlson, et al., 1984) which expressed various portions of the long rightward open-reading frame as fusion proteins containing both FPV protein and bacterial protein. In one kind of plasmid the FPV open-reading frame was fused to the gene for E. coli β-galactosidase contained in the plasmid pLG400 (Guarente et al, 1980). The fusion was 17 codons upstream from the termination codon of the open-reading frame. To provide for efficient transcription and translation of this fusion gene,

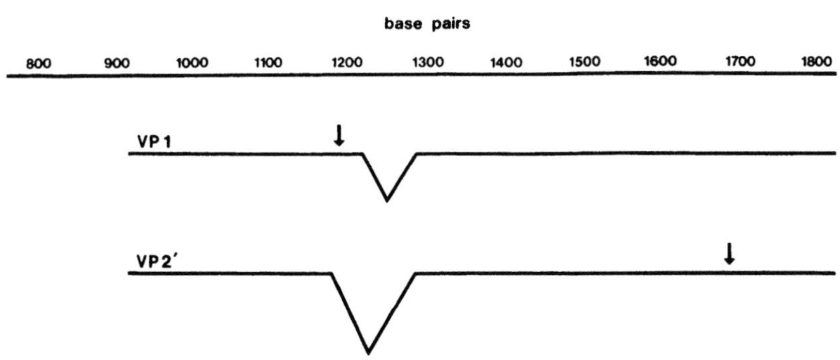

Fig. 4. Proposed mechanism for generation of mRNA species coding for VP1 and VP2' by alternate splicing pathways. The portion of the pEH20 sequence containing the putative promoter, splice junctions, and translational control sequences is represented by the top line. The numbering system is the same as in Fig. 2. ATG codons used for translational initiation of the two proteins are indicated by downward arrows.

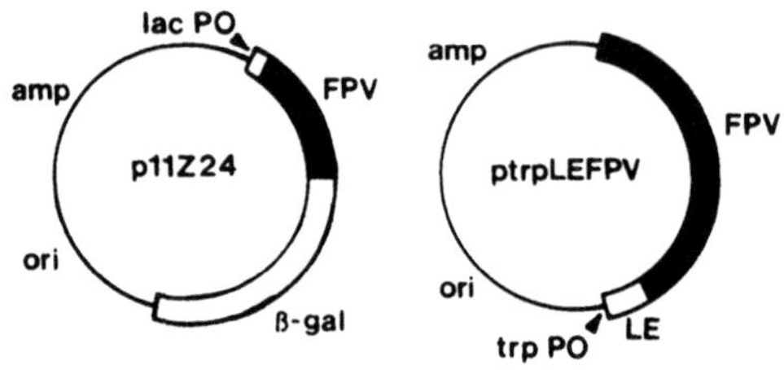

Fig. 5. Expression plasmids coding for fusion proteins containing FPV sequences. The plasmid 11Z24 contains 803bp of FPV DNA inserted into the amino-terminal end of ß-galactosidase as described in the test. Transcription and translation are under control of the lac operon control region. The plasmid ptrp LEFPV contains 2.3kb of FPV DNA inserted into the trp LE' gene as described in the text. Transcription and translation are under control of the trp operon control region. Genes coding for the fusion proteins are shown as thick lines. FPV DNA is represented by filled lines. Bacterial DNA is represented by open regions.

the lac promoter and first eight amino acids of E. coli ß-galactosidase were inserted at various positions in the FPV portion of the gene using BAL31 nuclease. These plasmids were introduced into a bacterial strain in which the ß-galactosidase gene was deleted and colonies which synthesized ß-galactosidase were selected on indicator plates. These bacteria synthesize tripartite fusion proteins containing the amino-terminal eight amino acids from ß-galactosidase followed by FPV DNA followed by the rest of ß-galactosidase (Fig. 5).

The second type of expression plasmid is based on the E. coli trp operon. FPV DNA from an HpaII site at map position to the right-hand terminus of pEH20 was ligated to an HpaII site in the trp LE' gene (Kleid et al., 1984). This resulted in the fusion of the fifty amino-terminal codons of the LE' gene to the carboxyterminal 596 codons of the rightward open-reading frame of FPV. Transcription and translation of this gene were mediated through the trp promoter and trp leader-peptide ribosome-binding site (Fig. 5).

Bacteria containing these plasmids synthesize fusion proteins which react with rabbit antiserum raised against SDS-disrupted FPV in both ELISA (Fig. 6) and immunoblot assay (Carlson et al., 1984). The FPV ß-galactasidase fusion protein synthesized by the plasmid 11Z24 comprises about 0.5% of the total cellular protein as determined by ELISA. Similarly, bacteria transformed with the plasmid p.trp LEFPV synthesize

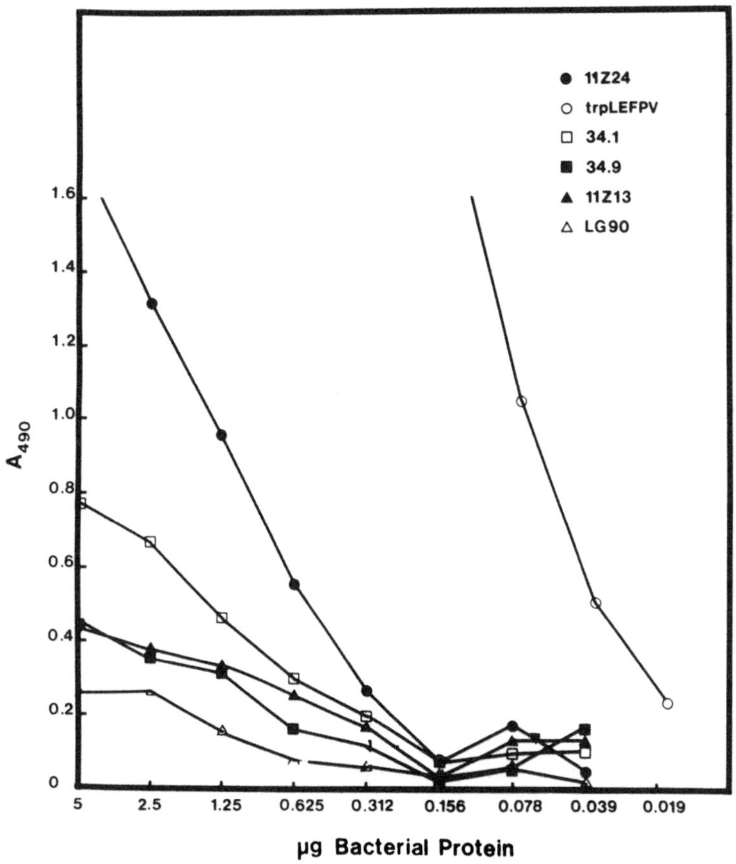

Fig. 6. Quantitation by ELISA of FPV antigen synthesized by bacteria containing various expression plasmids. Lysates from bacteria containing various expression plasmids were adsorbed to microtiter wells and reacted with a 1:100 dilution of rabbit antiserum raised against SDS-disrupted FPV as described in the experimental section. The absorbance at 490nm is plotted against the amount of total bacterial protein in the wells. Plasmids 11Z24, 11Z13, 34.9, and 34.1 are β-galactosidase fusion constructions containing approximately 800, 1100, 1400, and 1800bp of FPV DNA respectively. trp LEFPV is the LE' fusion construction. LG90 is the host bacterium containing no plasmid.

new protein species when the trp promoter is induced by tryptophan starvation (Carlson et al., 1984) products or prematurely terminated synthetic products. FPV antigenic protein comprises about 5% of the total cellular protein in these cells.

The fusion proteins from the plasmids 11Z24 and ptrp LEFPV are insoluble and sediment to the pellet fraction when lysate are centrifuged at 10,000xg. These pellets may be solubilized with either 0.1% SDS or 8M urea.

To determine whether these fusion proteins could raise antibodies to FPV, these solubilized crude cell pellets were used to immunize guinea pigs. Two injections of 0.5 to 1.5mg of total protein were administered two weeks apart. Complete Freund's adjuvant was used in the first injection. Incomplete Freund's adjuvant was used in the second injection. Serum collected from the animals prior to immunization and two weeks after the second injection was assayed for antibodies to FPV by two tests (ELISA and serum neutralization). Both 11Z24 and LEFPV fusion proteins were capable of raising antibodies which reacted with FPV in the ELISA assay at dilutions of greater than 1:1000. However, only the LEFPV fusion protein was able to reliably raise antibodies which were capable of neutralizing FPV in the serum-neutralization assay at dilutions of greater than 1:10.

Having established that the LEFPV fusion protein is capable of raising antibodies which neutralize FPV in a tissue culture-based serum-neutralization assay, we then tried to localize the epitopes responsible for eliciting the neutralizing antibodies on the viral protein. To do this, deletions of ptrp LEFPV that lacked portions of the LEFPV gene were constructed by digestion of the plasmid with various restriction enzymes. The deletion plasmids were constructed to remove increasing amounts of protein from the carboxyterminal end of the protein (Fig. 7). Bacteria transformed with these various deletion plasmids were grown and production of the various fusion proteins was induced by tryptophan starvation. Urea-solubilized cell pellets from these constructions were used to immunize guinea pigs as described above. All of the animals immunized with the deletion constructions developed antibodies against FPV as detected by both ELISA and SN assay. The shortest deletion NcoEcoΔ3 contained enough FPV DNA to synthesize 108 amino acids of FPV protein. Therefore, at least one epitope capable of raising neutralizing antibodies must be present in these 108 amino acids.

The fusion protein produced by ptrp LEFPV was tested for its ability to bind to a panel of eight monoclonal antibodies raised against canine parvovirus. These monoclonal antibodies all had a high reactivity to FPV with ELISA titers of greater than 1:100,000 (Teramoto et al., 1984). However, none of these monoclonal antibodies was able to bind to the LEFPV fusion protein at dilutions of >1:100. This suggests that either the conformation of the protein produced in bacteria is different from the protein in the native virus or the amino acids bound by these monoclonal antibodies are not present in the fusion proteins.

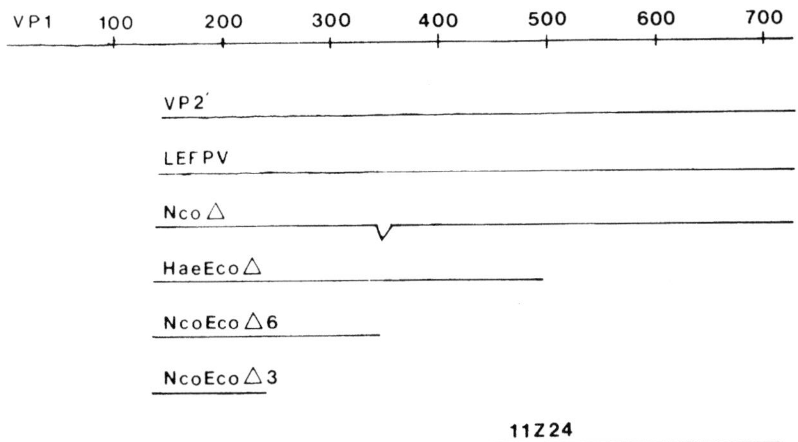

Fig. 7. Deletion analysis of FPV structural proteins. The 727 amino-acid sequence of VP1 is represented by the first line. The portion of that sequence contained in VP2' is represented in the second line. The portions of the VP1 amino-acid sequence in fusion proteins from various expression plasmids is shown in the next six lines.

Monoclonal antibodies were then isolated that would bind to both native and denatured virus. A panel of five such monoclonal antibodies resulted. Three were able to neutralize FPV. All five were tested for their ability to react with the fusion proteins depicted in Fig. 7. Two of the monoclonal antibodies reacted with the 11Z24 fusion protein as well as the LEFPV and NcoΔ fusion proteins. One of these reacted with the HaeEcoΔ construction as well. Therefore, these monoclonal antibodies map to the 11Z24 portion of the virion protein. Neither was capable of neutralizing FPV. The remaining three monoclonal antibodies reacted with all of the LEFPV-based deletion proteins but not with 11Z24. Thus, all three map to the 108 amino-acid portion of the FPV protein contained in the NcoEcoΔ3 construction. All three of these monoclonals are able to neutralize FPV although the titers are relatively low (1:100–1:200). Thus, once again we have localized at least one epitope capable of raising neutralizing antibodies to the 108 amino-acid sequence contained within the NcoEcoΔ3 clone.

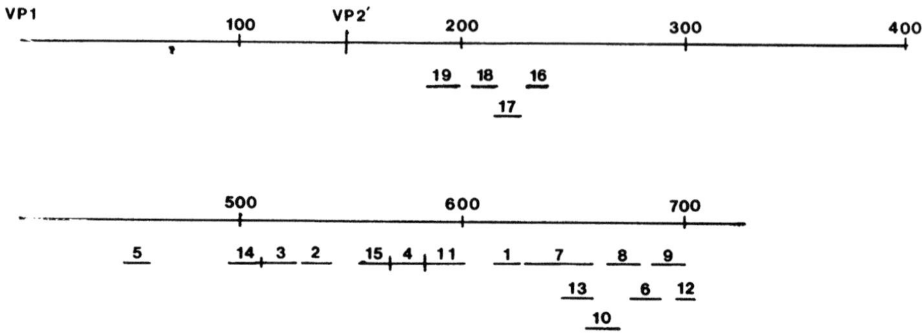

Fig. 8. Peptides synthesized from the predicted amino-acid sequence of FPV VP1. The 727 amino-acid sequence of VP1 is represented by the graduated line. The portions of the sequence which have been synthesized as peptides are indicated by short numbered lines below the graduated line.

We have also tried to localize epitopes capable of raising neutralizing antibodies through the use of synthetic peptides. Nineteen peptides have been synthesized from the predicted amino-acid sequence of the virion protein. These peptides range in length from 10 to 30 amino acids with most being approximately 15 amino acids in length. Peptides 1 through 15 fall within the sequence of the 11Z24 fusion protein. Peptides 16 through 19 come from the sequence contained in the NcoEco 3 fusion protein (Fig. 8). The peptides were conjugated to either keyhole-limpet hemocyanin or thyroglobulin through an additional cysteine residue at one end by MBS. These protein-peptide conjugates were used to immunize guinea pigs.

Antisera collected from these CsCl guinea pigs were tested for ability to react with nondisrupted CSCl-purified FPV or SDS-disrupted FPV by ELISA. In general, all of the peptides raised antisera which reacted with SDS-disrupted FPV at dilutions of greater than 1:80; however, only peptides 5 and 3 were able to raise antisera which reacted with nondisrupted FPV at dilutions of greater than 1:80. None of the antipeptide sera was capable of neutralizing FPV in a virus-neutralization assay at a dilution of 1:10. These results indicate that although the peptides are immunogenic, the amino-acid sequences in the intact virus particle are either not accessible or not in the appropriate conformation for binding to the antipeptide antisera.

DISCUSSION

The portion of the feline panleukopenia virus genome coding for the structural proteins was cloned and sequenced. Homology between the nucleotide sequences of FPV and two other autonomous parvoviruses, MVM and H1, suggested a mechanism for generating mRNA for the virion proteins VP1 and VP2' based on alternative splicing. The predicted amino-acid sequences of the proteins showed large stretches of homology on the protein level as well. This is interesting since that these viruses do not cross-react immunologically.

Portions of the structural protein gene were expressed as fusion proteins in E. coli. Immunization of guinea pigs with crude preparations of these proteins showed that some are capable of raising neutralizing antisera against FPV. However, serum-neutralization titers are low compared to what can be achieved with intact virus. The plasmid NcoEcoΔ3 contains only enough DNA to synthesize 108 amino acids of FPV protein. The fusion protein synthesized by this plasmid will raise neutralizing antisera against FPV. Three monoclonal antibodies raised against SDS-disrupted FPV bind to this fusion protein and are also capable of neutralizing FPV. Therefore, at least one epitope capable of raising neutralizing antibodies must be in this 108 amino-acid sequence.

Nineteen synthetic peptides from the FPV-protein sequence were tested for their ability to raise antisera against FPV. None was capable of raising neutralizing antibodies.

ACKNOWLEDGEMENTS

We express our gratitute to Richard Lerner, James Bittle, and Richard Houghton for synthesizing peptides 1 through 15. We also thank Jonathan Rothbard and Gary Schoolnik for the computer analysis of the parvoviral protein sequences. We are grateful to Leonard Guareate for plasmid pLG400, Jack Nunberg for ptrp LE', and Dave Hare and Jack Sadler for pDH68 containing the lac operon control region. This work was partially supported by USPHS grant number 1RH3AI20783 through the Small Business Innovation Research Program.

REFERENCES

Astell, C.R., Thomson, M., Merchlinsky, M., and Ward, D.C., 1983, The complete DNA sequence of minute virus of mice, an autonomous parvovirus, Nucleic Acids Research, 11:999-1018.

Carlson, J., Maxwell, I., Maxwell, F., McNab, A., Rushlow, K., Mildbrand, M., Teramoto, Y., and Winston, S., 1984, Expression of Feline Panleukopenia Virus Antigens in Escherichia coli, in: "Modern Approaches to Vaccines," R.M. Chanock and R.A. Lerner, eds., pp. 195-201. Cold Spring Harbor Laboratory, Cold Spring Harbor, New York.

Carmichael, L.E., and Binn, L.N., 1981, New Canine Enteric Viral Infections in: "Advances in Veterinary Science," pp. 1-37, Academic Press, New York.

Cotmore, S.F., Sturgenbecker, L.J., and Tattersall, P., 1983, The Autonomous Parvovirus MVM Encodes Two Nonstructural Proteins in Addition To Its Capsid Polypeptides, Virology, 129:333-343.

Green, N., Alexander, H., Olson, A., Alexander, S., Shinnick, T.M., Suteliffe, J.G., and Lerner, R.A., 1982, Immunogenic Structure of the Influenza Virus Hemagglutinea, Cell, 28:477-487.

Guarente, L., Lauer, G., Roberts, T.M., and Ptoshne, M., 1980, Improved Methods for Maximizing Expression of a Cloned Gene: a Bacterium That Synthesizes Rabbit β-Globin, Cell, 20:543-553.

Kleid, D.G., Yansura, D., Small, B., Dowbenko, D., Moore, D.M., Gaubman, M.J., McKercher, P.D., Morgan, D.O., Robertson, B.H. and Bachrach, H.L., 1981, Cloned Viral Protein Vaccine for Foot-and-Mouth Disease: Responses in Cattle and Swine, Science, 214:1125-1129.

Liu, F.T., Zinnecker, M., Hamaoka, T., and Katz, D.H., 1979, New Procedures for Preparation and Isolation of Conjugates of Proteins and a Synthetic Copolymer of D-Amino Acids and Immunochemical Characterization of Such Conjugates, Biochemistry, 18:690-697.

McMaster, G.K., Tratschin, J.D., and Siegl, G., 1981, Comparison of Canine Parvovirus with Mink Enteritis Virus by Restriction Site Mapping. J. Virol., 38:368-371.

Parrish, C.R. and Carmichael, L.E., 1983, Antigenic Structure and Variation of Canine Parvovirus Type-2, Feline Panleukopenia Virus and Mink Enteritis Virus, Virology, 192:401-414.

Peterson, J.L., Dale, R.M.K., Karess, R., Leonard, D., and Ward, D.C., 1978, Comparison of Parvovirus Structural Proteins: Evidence for Post-Translational Modification, in "Replication of Mammalian Parvoviruses," pp. 431-445. D.C. Ward and P. Tattersall, eds., Cold Spring Harbor Laboratory, Cold Spring Harbor, New York.

Pintel, D., Dadachanji, D., Astell, C.R., and Ward, D.C., 1983, The genome of minute virus of mice, an autonomous parvovirus, encodes two overlapping transcription units, Nucleic Acids Research, 11:999-1018.

Rhode, S.L. III, and Paradiso, P.R., 1983, Parvovirus Genome: Nucleotide Sequence of H-1 and Mapping of its Genes by Hybrid-Arrested Translation, J. Virol., 45:173-184.

Siegl, S., 1976, "The Parvoviruses," Springer-Verlag, Wein-New York.

Tattersall, P., Shatkin, A.J., and Ward, D.C., 1977, Sequence Homology Between the Structural Polypeptides of Minute Virus of Mice, J. Mol. Biol. 111:375-394.

Teramoto, Y.A., Mildbrand, M.M., Carlson, J., Collins, J.K. and Winston, S., 1984, Comparison of ELISA, DNA Hybridization, Hemog-glutination and Electron Microscopy for the Detection of Canine Parvovirus Infections, J. Clin. Microbiol. In press.

Tratschin, J.D., McMaster, G.K., Kronauer, G., and Siegle, G., 1982, Canine Parvovirus: Relationship to Wild-type and Vaccine Strains fo Feline Panleukopenia Virus and Mink Enteritis Virus, J. Gen. Virol. 61:33-41.

STUDIES OF TGEV SPIKE PROTEIN GP195 EXPRESSED IN *E. COLI* AND BY A TGE-VACCINIA VIRUS RECOMBINANT

Sylvia Hu, Joan Bruszewski, Ralph Smalling and Jeffrey K. Browne

Amgen, 1900 Oak Terrace Lane, Thousand Oaks California 91320-1789

ABSTRACT

The gene coding for the surface spike protein gp195 of TGEV has been cloned and expressed in *E. coli* in the form of fusion proteins. These proteins were isolated and used to immunize laboratory animals. All animals developed antibodies cross-reacting with the TGEV virion but failed to neutralize the virus. The entire gp195 gene was also inserted into vaccinia to generate a TGEV-vaccinia recombinant virus (vTGE) that expressed TGEV gp195. Animals vaccinated with vTGE produced neutralizing antibodies against both TGEV and vaccinia. These results suggest the potential use of the recombinant vTGE as a vaccine against TGEV infection.

INTRODUCTION

Transmissible gastroenteritis (TGE) is a worldwide enteric disease of swine. Pigs of all ages can be affected, but the disease is especially serious in newborn pigs, with a very high mortality rate. Current vaccines rely on the transmission of lactogenic immunity from vaccinated sows to nursing piglets and have not been very effective in practice (Bohl, 1981).

The disease is caused by a virus (TGEV) that belongs to the coronavirus family. Like other coronaviruses, the TGE virus particle contains three major structural proteins. The nucleocapsid protein, p50, (molecular weight 50,000 daltons) is associated with the genomic RNA, the matrix protein, gp31, is a glycoprotein that is mostly embedded in the viral membrane (Rottier, et al., 1984), and the spike protein, gp195, forms the characteristic knob-like structures which protrude from the surface of the viral envelope.

It has been shown that tunicamycin treatment of cells infected with murine or avian coronaviruses results in the production of non-infectious virions that lack the spike protein. (Holmes, et al., 1981; Niemann and Klenk, 1981; Rottier, et al., 1981; Stern and Sefton, 1982). Garwes et al. (1978/1979) reported that a TGEV subviral component consisting mainly of gp195 could elicit a neutralizing immune response when injected into pigs. These observations suggested that gp195 was important both for viral infectivity and for neutralization. We have reported (Hu, et al., 1984) the identification of six discrete TGE viral mRNAs with sizes ranging from 2 to 19 kb (kilobases). The gene coding for gp195 was mapped on the 5' half of the 9 kb mRNA. The entire gene of gp195 was cloned and expressed at high levels in E. coli. In this work, we describe the isolation of these proteins from E. coli and their antigenic properties, particularly whether they are capable of inducing a neutralizing response in animals.

Recently, the use of vaccinia virus as a vector for expressing foreign genes has led to the concept of a novel type of live vaccine. (Mackett et al., 1984). Recombinant vaccinia viruses containing the herpes simplex virus surface antigen (Paoletti et al., 1984) or the influenza hemagglutinin gene (Smith et al., 1984) have been shown to render the vaccinated animals resistant to the corresponding infectious virus challenge. The broad host range of vaccinia enables the application of such live vaccines in domestic animals such as pigs.

To test this possibility, we constructed a TGE-vaccinia hybrid virus by inserting the TGEV gp195 gene into vaccinia virus. The product of gp195 was detected in animal cells infected with the hybrid viruses by immunoprecipitation and by immunoperoxidase staining with TGEV specific antiserum. Mice inoculated with this TGE-vaccinia recombinant virus elicited neutralizing antibodies against both TGEV and vaccinia virus.

MATERIALS AND METHODS

Viruses and Cells

The virulent Miller strain TGEV (American Tissue Culture Collection VR-743) was propagated in swine testes (ST) cells as described by McClurkin (1965). Vaccinia virus used in this study was the WR strain (ATCC VR-119). African green monkey cells (CV-1) were used for the growth of vaccinia virus. Human TK^- (line 143) cells were used for selecting TK^- vaccinia recombinant virus.

Endoglycosidase H digestion

Endoglycosidase H (37IU/mg) was obtained from Miles. 50 µg of purified TGEV was digested with 5 mU of enzyme in 100 µl of 150 mM sodium citrate buffer (pH 5.5) at 37°C for 16 hrs. An aliquot of the product was tested on SDS-polyacrylamide gel for complete digestion.

Animal immunization

Rabbits, rats and mice have been used for testing the antigenicity of *E. coli* expressed TGEV proteins. Animals were given five subcutaneous immunizations at two week intervals. The first immunization was with complete, and subsequent four boosters were with incomplete Freund's adjuvant. Antiserum was collected after the third and last injection.

For the TGE-vaccinia recombinant virus studies, female Balb/C mice were inoculated by dabbing purified virus (at 10^9 pfu/ml) onto wounds made by scratching the base of the tail. Mice were bled at weeks 2, 4, 6 and 8 post-inoculation of vaccine and their sera were tested for antibodies against vaccinia and TGEV.

Antibody neutralization

Pre-immune and immune sera were heat-inactivated, mixed with an equal volume of virus, incubated at 37°C for 1 hour, and plated on cells. After one hour of adsorption, the monolayer was then washed twice with phosphate buffered saline (PBS) and overlaid with agar. Plaques were scored by staining with neutral red after 48 hr.

$[^{35}]$-methionine labeling of cells and immunoprecipitation.

About 8 to 10 hr after infection, cells were pre-incubated in methionine deficient medium for 30 min., then labeled with 50 µCi/ml $[^{35}S]$-methionine for 1 hr. The monolayer was washed with PBS and lysed with ice-cold disruption buffer containing 0.5% Nonidet P-40, 1% deoxy-cholate, 50mM NaCl, and 25 mM Tris (pH 8.0). Cell debris was removed by centrifugation at 6,000 g for 15 min. For immunoprecipitation, 250 to 500 µl of labeled cell lysate was mixed with 5 to 10 µl of antiserum and incubated at 37°C for 1 hr followed by overnight incubation at 4°C. 100 µl of formalin-treated staphylococcal cells (Pansorbin, Calbiochem-Behring) was added and the incubation was continued at 4°C for 45 min. The immunoprecipitate was washed three times in the solutions described by Oppermann et al. (1979) and analyzed by electrophoresis through a 6 to 18% gradient polyacrylamide gel using the buffer system described by Laemmli (1970). Fluorography of the gel was performed using EN^3hance (New England Nuclear) following manufacturer's procedure.

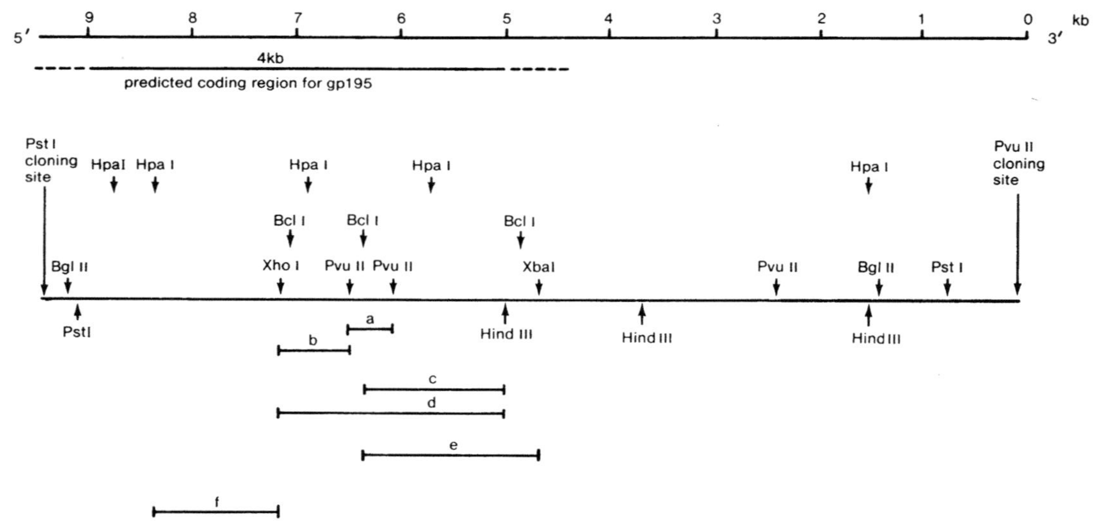

Figure 1. Restriction endonuclease cleavage map of a 9.4 kb TGEV cDNA clone. Nucleic acid sequence analysis has revealed one continuous, unique, translation open-reading frame coding for gp195, which starts at 8 bases downstream from the second 5' HpaI site and ends at 80 bases upstream from the XbaI site. Fragments a through f have been subcloned for expression in *E. coli*.

Immunoperoxidase staining

Infected cells were fixed with 2% para-formaldehyde followed by cold acetic acid-ethanol (5/95%) and processed for immunoperoxidase staining using (i) a porcine hyperimmune serum to TGEV (M-20, from Dr. L. Saif, Ohio State University) diluted 1:200, and (ii) a rabbit HRP (horse radish peroxidase) conjugate against porcine immunoglobulin, diluted 1:200 (Dako). When the test serum is of mouse origin, a goat HRP conjugate against mouse IgG (Boehringer-Mannheim) was used.

RESULTS

E. coli expression of TGEV gene gp195

We have previously (Hu et al., 1984) reported the cloning and expression of the TGEV gp195 gene in *E. coli*. Briefly, each restriction enzyme cleavage fragment (depicted as a-->f in Figure 1) of gene gp195 was ligated to another fragment with the same translation reading frame which codes for the amino terminus of a second protein. This resulted in the formation of fusion proteins. Such fusion proteins were produced under the regulation of the copy number runaway plasmid pCFM414 in *E. coli* at high levels. In all cases, these proteins were sequestered in an insoluble matrix that became visible under light microscope at 2 to 3 hours after induction by temperature shift. Upon breakage of the *E. coli* cells with a French press apparatus, these proteins can be separated from most of the soluble cellular proteins by a short, low speed centrifugation. The pellet was harvested and subjected to various purification procedures described below.

Isolation of TGEV fusion proteins from E. coli

The inclusion bodies with which these TGEV fusion proteins were associated contain many other less abundant proteins, lipids and nucleic acids. A number of dissociation conditions were tested to release the proteins of interest from the inclusion bodies. We noticed that different fusion proteins could have quite variable biochemical properties thus requiring different isolation schemes.

Methods that have been used include:

(i) Alkali solubilization: The optimal pH required to solubilize each protein from the inclusion bodies without irreversibly denaturing it was determined by analyzing the insoluble pellet and soluble supernatant fractions after each alkali treatment. Usually at ~pH 12, these fusion proteins can be solubilized whereas the other abundant bacterial membrane proteins remain in the pellet. An example is shown in Figure 2.

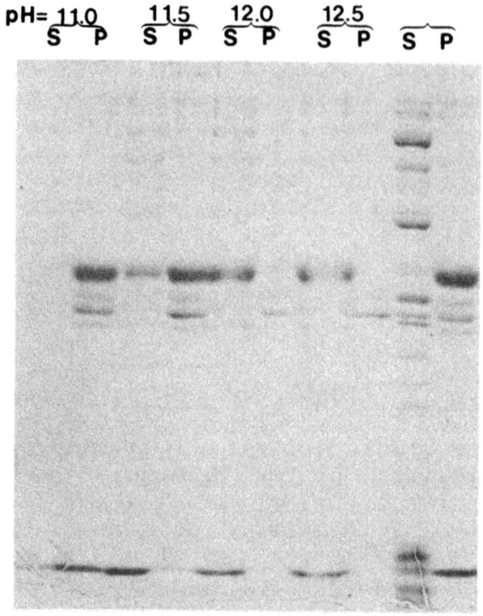

Figure 2. SDS-polyacrylamide gel depicting the solubilization process of E. coli expressed, TGE gp195 (fragment b) by alkali pH. The two lanes to the far right represent the protein content in the pellet (P) and the supernatant (S) fractions after a low speed centrifugation of the total E. coli cell extract. This pellet fraction was then adjusted to alkali pH as indicated and centrifuged to separate the soluble (S) from the insoluble (P) components.

(ii) Acid solubilization: We found that some of the proteins could tolerate and be solubilized in 50% acetic acid. After removing the insoluble material by centrifugation, the supernatant was applied on a gel filtration column such as Sephadex, and the proteins were eluted with 10% acetic acid. The fractions containing the proteins of interest were pooled and recovered by dialysis extensively against buffer.

(iii) A denaturing reagent such as urea could be used in conjunction with a reducing agent for solubilizing the proteins. Further purification could be achieved by the use of ion-exchange columns such as DEAE-and/or CM-Trisacryl (Reactifs IBF). Urea was then removed by dialysis against buffer.

(iv) Some of the proteins could be preferentially solubilized from the inclusion bodies with 1% Sarkosyl under non-reducing conditions (Figure 3). Solubilized samples were fractionated through a gel filtration column using 1% Sarkosyl as eluent. High molecular weight aggregates came through in the excluded volume. Sarkosyl associated with the proteins from pooled fractions was removed by passing through an ion-retarding column, such as AG11-A8 (Bio-Rad).

We have frequently experienced aggregation problems while isolating these proteins. This is not unexpected because these proteins contain many hydrophobic regions, as predicted from the nucleic acid sequences, which is typical for membrane associated proteins. Furthermore, these proteins, when expressed in E. coli, lack the carbohydrate moieties that are fairly abundant in the native viral protein. Our previous experiments involving tunicamycin inhibition or endoglycosidase H digestion of gp195 (Hu et al., 1984) showed that about 50,000 daltons of this

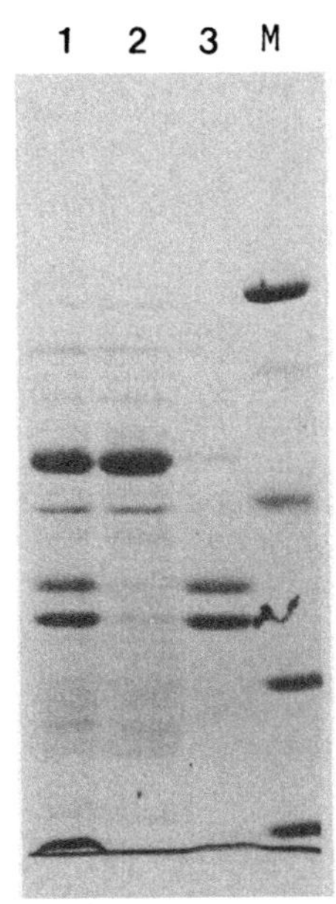

Figure 3. SDS-polyacrylamide gel pattern of (1) total proteins of inclusion bodies from E. coli expressing TGEV gp195 (fragment f). (2) Soluble and (3) insoluble components of (1) after the treatment with 1% sarkosyl in the absence of reducing reagent.

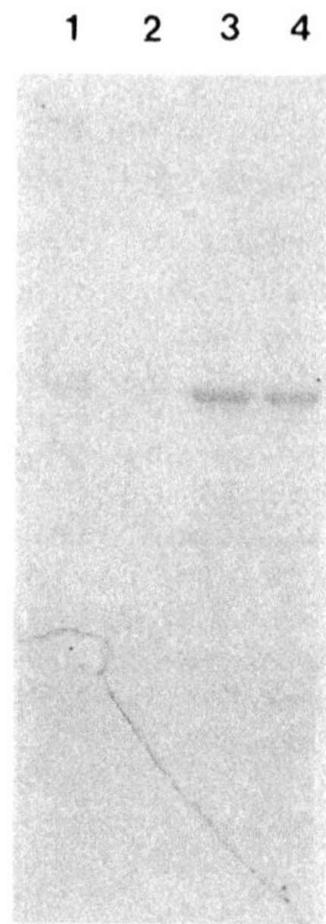

Figure 4. Nonreducing SDS-polyacrylamide gel of TGEV gp195 (fragment e) when the sample contains protein aggregates (lane 1,2) or monomers (lane 3,4).

protein was composed of carbohydrates. These carbohydrates could be necessary for maintaining gp195 or its fragments in solution. To determine the extent of aggregation occurring during purification of these proteins, we constantly compared their electrophoresis patterns on reducing gels versus non-reducing gels. The appearance of high molecular weight bands migrating as multimers, or material which failed to enter the non-reducing gel, was taken as indications of aggregates. One example of such comparison is depicted in Figure 4. The procedure that resulted in the lowest level of aggregates was chosen to isolate each protein.

(v) Preparative SDS-polyacrylamide gel electrophoresis: This procedure was only used to prepare a small amount of pure protein. Coomassie blue stained bands containing the proteins of interest were excised from the gel and subjected to electroelution. The recovered proteins were then dialyzed extensively to reduce the amount of associated SDS.

Neutralizing response induced by deglycosylated TGEV

We have previously demonstrated (Hu et al., 1984) that endoglycosidase H treatment converts gp195 and gp31 into proteins of apparent molecular weight 145,000 and 29,000 daltons respectively. In order to determine whether the neutralization site of gp195 resides in the carbohydrate side chains or in the polypeptides, we immunized mice with TGEV that had been digested to completion with endoglysodidase H. All immunized animals developed neutralizing antibodies against TGEV. This result is consistent with the hypothesis that the polypeptide structure represents the neutralization target. The carbohydrates could still be important for the proper folding of gp195 when the protein is initially synthesized. However, our data indicate that once the gp195 is assembled into the virion, removal of the carbohydrate does not disturb the neutralization site(s) of the molecule.

Antigenicity of E. coli expressed gp 195

To analyze the antigenic properties of the TGEV gp195 gene product expressed in various systems, we used two antisera directed specifically against TGEV. The polyvalent neutralizing serum, M-20, was from a hyperimmunized pig recovered from TGEV infection. In addition, we have isolated a mouse hybridoma clone, 5H2-7, from which monoclonal antibodies with neutralizing reactivity against TGEV were obtained. M-20 serum reacts with all of the three major structural proteins of TGEV, namely, p50, gp31 and gp195, whereas the monoclonal antibodies only react with gp195. In a standard ELISA test, all of the gp195 fusion proteins synthesized in *E. coli* could be recognized by M-20 antiserum, typical level of reactions are shown in Figure 5a. Two of the products (coded by fragment *b* and *d*) also reacted well with the monoclonal antiserums (Figure 5b). These two fragments have in common the region between the XhoI site and the nearest PvuII site. We believe that this region contains the neutralizing epitope recognized by this monoclonal antibody.

Analysis of immune sera

Sera from animals immunized with various *E. coli* expressed gp195 fusion proteins were collected. All of them contain antibodies that react with the immunogen as well as with native TGEV virion in an ELISA test (Figures 6a and 6b). Furthermore, these sera were all capable of immunoprecipitating specifically gp195 from [^{35}S]-methionine labeled, TGEV infected cell lysates, an example is shown in Figure 7. However, none of the immune sera had significant levels of neutralizating antibody. We suspected that the attachment of carbohydrates is crucial for folding gp195 into its native structure and that without these carbohydrates, these E. coli expressed fusion proteins were not in the correct conformation around the neutralization sites to induce the appropriate immune responses.

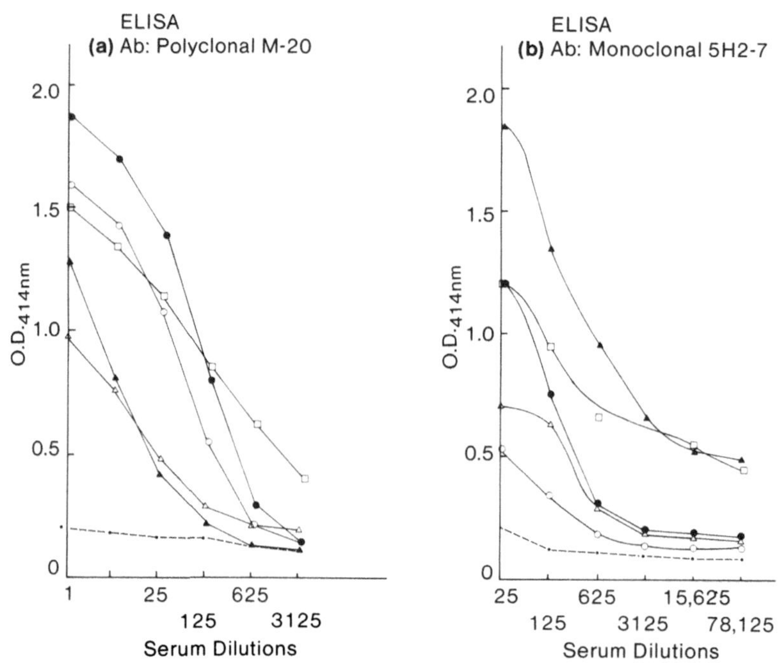

Figure 5. Binding characteristics of different E. coli expressed TGEV gp195 fragment to a polyvalent antiserum M-20 (a) or a monoclonal antiserum 5H2-7(b). Serial five-fold dilutions of antibody were incubated in the wells of microtiter plates containing adsorbed proteins from TGEV virion (□), gp195 fragment b (▲), d (●), e (o), f (△), or control (-----) E. coli lysates. Bound antibodies were detected with HRP-labeled antiserum as described in the text.

Expression of gp195 in TGE-vaccinia recombinant virus (vTGE)

To construct a TGE-vaccinia recombinant virus, the DNA fragment carrying the entire TGEV gp195 gene (starting at the second 5' HpaI site and ending at the unique XbaI site as shown in Figure 1) and including the putative signal sequences was inserted into the BamHI cloning site of a

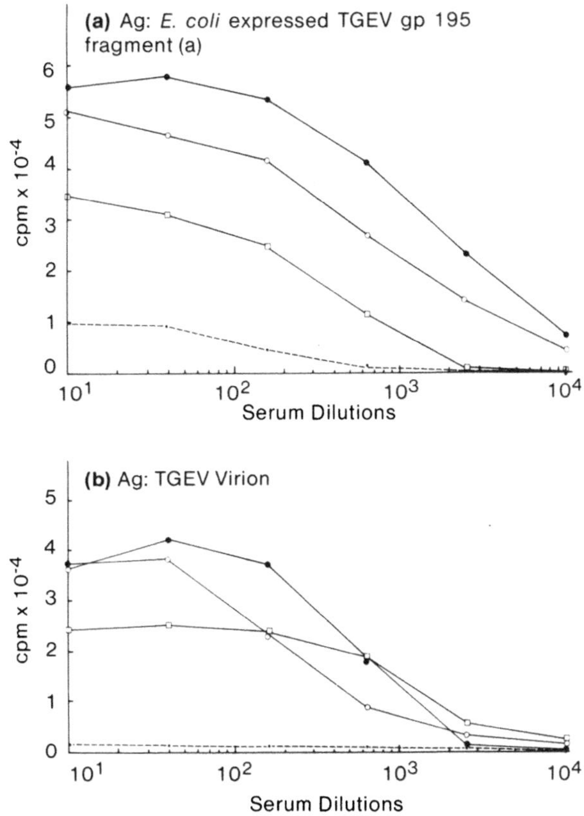

Figure 6. Specific binding of antibodies from two rabbits immunized with E. coli expressed TGEV gp195 fragment a (●, o). The microtiter plates were preabsorbed with either the immunogen (a) or intact TGEV virion (b). The polyvalent antiserum M-20 was used as positive control (□) and the preimmune rabbit serum was the negative control (-----). Bound antibodies were quantitized with ^{125}I-labeled staphalococcal protein A.

vaccinia expression vector, pGS20 (Mackett et al., 1984) after BamHI linker was added to the insert DNA. The insert was oriented to place its first ATG immediately downstream from the vaccinia 7.5 K gene promoter. This chimeric gene is in turn flanked by the vaccinia TK (thymidine kinase) gene (Figure 8). Recombinant viruses were selected on the basis of their TK$^-$phenotype after transfecting the constructed plasmid DNA into TK$^-$cells previously infected with the wild type vaccinia virus.

Cultures individually infected with viruses picked from separate TK$^-$plaques were further examined for the expression of gp195 by immunoperoxidase staining using

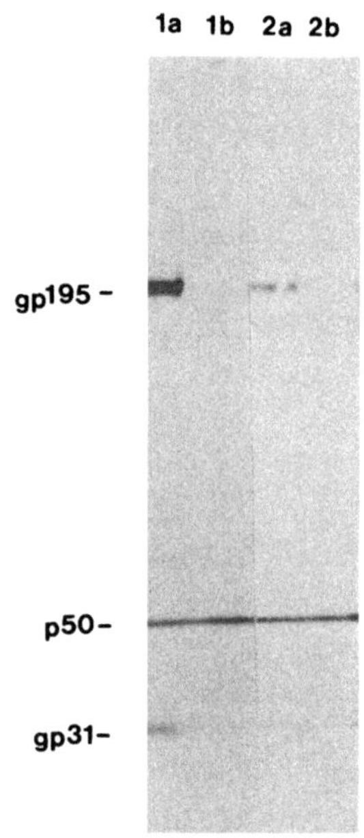

Figure 7. SDS-polyacrylamide gel analysis of TGEV proteins immunoprecipitated from infected cells with neutralizing pig serum M-20 (1a) and normal pig serum (1b), or with serum from rabbit immunized with E. coli expressed TGEV gp195 (2a) and preimmune rabbit serum (2b). Notice that M-20 recognizes all of the three major TGEV proteins whereas the rabbit immune serum only recognizes gp195.

TGEV-specific antiserum. Positive staining appeared at 12 hours after infection in the cytoplasm of 1/12 of the cultures indicating the infection by TGE-vaccinia recombinant virus (vTGE). Vaccinia viruses which obtained the TK^-phenotype due to spontaneous mutation in the viral genome do not lead to positive staining of infected cells and can be easily recognized and discarded. The genuine vTGE virus was then subjected to several rounds of plaque purification. The stability of the recombinant virus and its ability to express TGEV gp195 was not altered after multiple passages in culture. The infectivity of vTGE is comparable to that of the wild type vaccinia virus in a number of cell lines that have been tested.

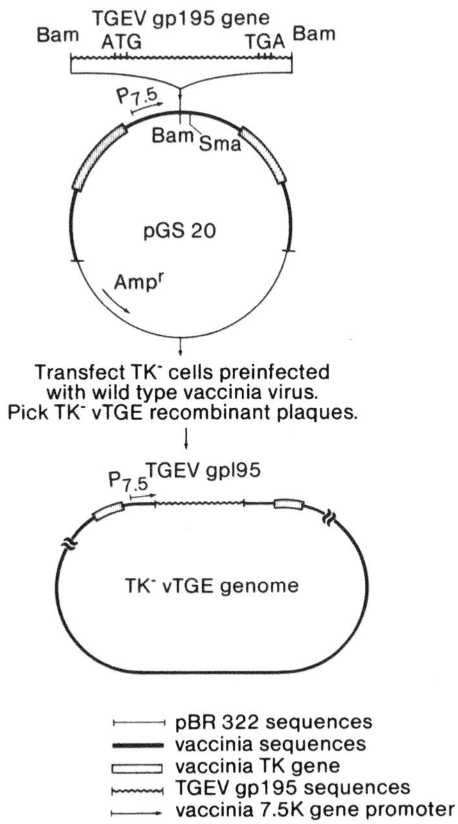

Figure 8. Construction of a TGE-vaccinia recombinant virus. The complete TGEV gpl95 gene is inserted into vaccinia virus under the control of the 7.5K gene promoter. Recombinant viruses were generated as the result of homologous recombination between the transfer vector and the genomic DNA of vaccinia and was selected on the basis of their TK^-phenotype.

The product of the inserted gp195 gene in vTGE-infected cells was analyzed by immunoprecipitation. As shown in Figure 9, a unique protein of apparent molecular weight 185K daltons was specifically precipitated by TGEV antiserum. The discrepancy in molecular weight with the native gp195 could reflect the difference in glycosylation between the two virus systems.

The 185K dalton protein represents a minor component when compared with the vaccinia virus-derived proteins. This is not surprising since the 7.5K gene promoter is not a strong promoter. Nevertheless, the amount of the 185K dalton protein is still detectable even in the whole cell lysate as shown in Figure 9.

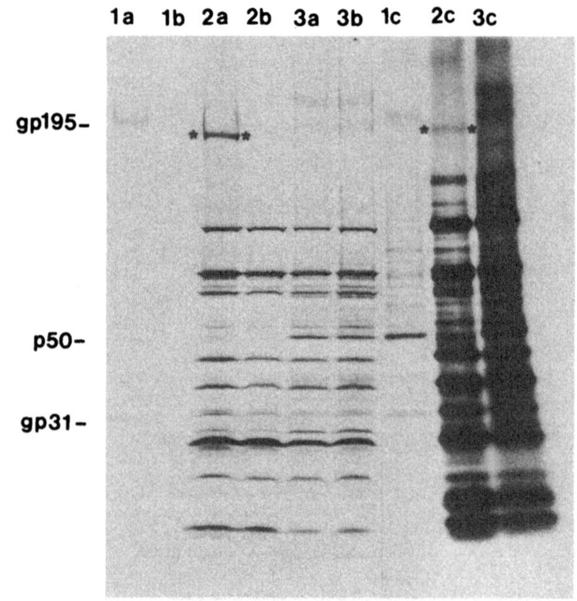

Figure 9. SDS-polyacrylamide gel analysis of immune-precipitated proteins from (1) ST cells infected with TGEV (2) CV-1 cells infected with recombinant vTGE and (3) CV-1 cells infected with vaccinia virus. Lane a's were precipitated with antiserum M-20, lane b's were precipitated with normal control serum, lane c's were total cytosol extract from corresponding infected cells.

TABLE 1

	# Weeks Post inoculation*	2	4	6	2	4	6
Mice # (Jackson Balb/C)	Neutralizing titer**	@ vaccinia			@ TGEV		
1898	vaccinia	40	100	400	<20	<20	<20
1899	vaccinia	60	100	400	<20	<20	<40
1989	vaccinia	70	150	380	<20	<20	<20
1990	TGE/vaccinia recombinant	60	130	345	<20	40	110
1991	TGE/vaccinia recombinant	70	70	75	<20	60	50
1992	TGE/vaccinia recombinant	50	60	350	<20	65	120
1993	TGE/vaccinia recombinant	40	70	370	<20	50	60
1994	TGE/vaccinia recombinant	65	60	250	<20	50	60
1995	TGE/vaccinia recombinat	55	175	400	<20	50	160
1996	TGE/vaccinia recombinant	60	135	350	<20	50	120

* Inoculation was done by dabbing a tail scratch with purified viruses with titer of 10^9 pfu/ml.

** Neutralizing titer was determined by plaque reduction assay, shown as the reciprocal of serum dilution that reduces 50% of the plaques.

Preimmune serum from each mouse was included in each assay as control.

We also used the technique of immunoperoxidase staining to analyze the distribution of gp195 in the cells. Both the polyvalent M-20 and the monoclonal 5H2-7 sera stained positively fixed cells that have been infected with TGEV or the recombinant vTGE. However, gp195 was found only on the surface of TGEV infected cells but not on vTGE infected cells, suggesting a possible difference in the transport pathway of gp195 in these two systems. We are currently investigating this hypothesis.

Immune response of mice vaccinated with vTGE

To determine whether vTGE could elicit a neutralizing antibody response in animals, mice were inoculated with vTGE or wild type vaccinia viruses by scarification. All mice developed local skin lesions typical of vaccinia infection, which healed quickly with no other clinical symptoms.

Figure 10 shows the relative binding titer of the representative mouse sera to TGEV and vaccinia virus at six weeks after inoculation using a standard ELISA test.

The neutralizing titer of these sera was determined by plaque reduction assay. As summarized in Table 1, neutralizing titer against vaccinia virus became detectable in all animals at the second week and continued to increase for at least six weeks. Mice inoculated with vTGE all showed concurrent increasing neutralizing titer against TGEV. The neutralizing titer against TGEV has been low, in the range of 50 to 150 but is consistently higher than the control group. Subsequent booster scratches did not raise the titer significantly.

DISCUSSION

To develop a subunit vaccine for any viral disease, it is crucial to first identify the viral antigen(s) that is essential for viral infectivity, particularly in the step of adsorption to the receptors of the target cells. Furthermore, such antigen itself must be able to induce neutralizing antibody production in animals and to confer protection against infection by the virus.

Subviral component of TGEV derived from purified virus that demonstrated these properties has been reported (Garwes et al. 1978/1979; Gough et al., 1983) but the immunogens in these studies have not been well characterized and the quantity required for successful protection was high.

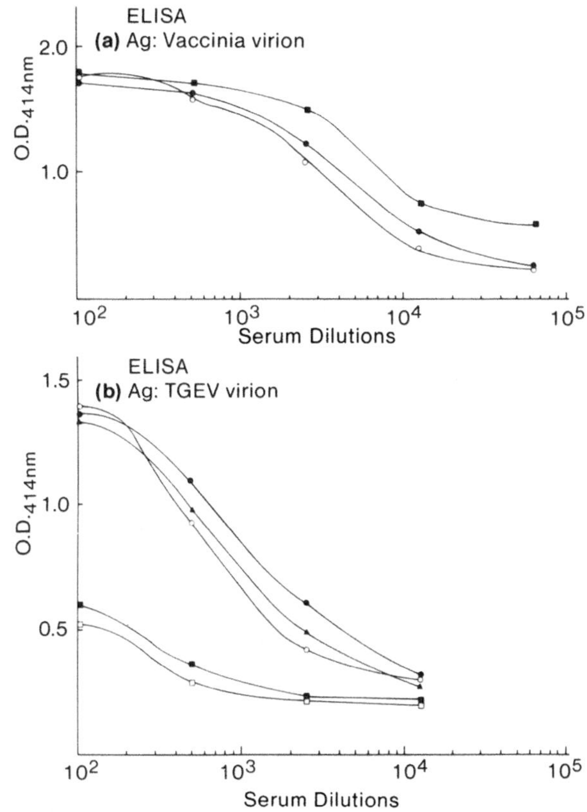

Figure 10. Antibody titer of mice inoculated with TGE-vaccinia recombinant virus (vTGE) (● , ○ , ▲) or a wild type vaccinia virus (■ , □) at six weeks using a standard ELISA test. Only the mice inoculated with vTGE elicit binding activity to TGEV in their sera, whereas the binding titers to vaccinia are comparable among all the mice.

The spike protein gp195 seems to be the most likely candidate for such a subunit vaccine because of its most accessible location on the virion. Our success with the TGE-vaccinia recombinant virus (vTGE) confirmed that gp195 alone will give rise to a TGEV-neutralizing immune response in animals.

We have prepared hybridoma cell lines derived from mice inoculated with TGEV and screened them with an in vitro neutralization assay. We found that all of the hybridoma clones that were secreting neutralizing antibodies reacted specifically with gp195. Our observation seems parallel to the studies reported by

Collins et al. (1982) on mouse hepatitis virus, namely, neutralization was mediated only by monoclonal antibodies which precipitated the spike protein (GP-1). Clearly, the spike protein plays an important role in virus attachment to the target cells in both cases.

It is disappointing that although the *E. coli* produced TGEV-fusion proteins elicited antibodies that reacted with the immunogen and intact viruses, they do not neutralize the infectivity of the virus. It appears that the structure of the *E. coli* produced protein is different and results in altered, "inactivated" neutralization epitopes. One possible explanation is that the addition of carbohydrates is essential for membrane glycoproteins such as gp195 to maintain their conformation. We are currently expressing gp195 in yeast and mammalian systems in an attempt to obtain material in its native conformation for testing.

It has been noticed that natural infection of sows with virulent TGE virus stimulates production of effective secretory antibodies but intramuscular inoculation of the pathogen generally does not (Bohl et al, 1972; Bohl and Saif, 1975). Since suckling pigs are protected only as a result of the frequent ingestion of colostrum or milk that contains TGEV-neutralizing antibodies, antigenic stimulation of the intestinal tract of the sows is a prerequisite for conferring lactogenic immunity to the piglets.

Recently, Smith et al. (1983) reported that hamsters intradermally vaccinated with a recombinant vaccinia virus carrying the influenza hemagglutinin (HA) gene produced sufficient local immunity to render the animals resistant to intranasal challenge of infectious influenza virus. Such protection is also believed to be greatly dependent on the secretory IgA antibodies. Their data encourages an approach in the control of TGEV infections by the use of the TGE-vaccinia recombinant virus. We are currently carrying out protection studies in pigs to examine the efficacy of the use of vTGE as a vaccine.

ACKNOWLEDGEMENT

We thank Dr. Margery Nicolson for advising the hybridoma work, Dr. Bernard Moss for providing pGS20 plasmid and Dr. Linda Saif for M-20 pig serum. We also thank Ms. Joan Bennett for preparation of the manuscript.

REFERENCES

Bohl, E.H., Gupta, P.K.P., Olquin, M.V.F., and Saif, L., 1972. Antibody responses in serum, colostrum and milk of swine after infection or vaccination with transmissible gastroenteritis virus. Infect. Immun. **6**:289-301.

Bohl, E.H., and Saif, L.J., 1975. Passive immunity in transmissible gastroenteritis of swine: Immunoglobulin characteristics of antibodies in milk after inoculating virus by different routes. Infect. Immun. **11**:23-32.

Bohl, E.H., 1981. Transmissible gastroenteritis, in: "Disease of Swine", A.D. Leman et al. (eds.), Iowa State Press, Ames, Iowa. pp. 195-208.

Collins, A.R., Knobler, R.L., Powell, H., and Buchmeier, M.J., 1982. Monoclonal antibodies to murine hepatitis virus-4 (strain JHM) define the viral glycoprotein responsible for attachment and cell-cell fusion. Virology **119**:358-371.

Garwes, D.J., Lucas, M.H., Higgins, D.A., Pike, B.V.,and Cartwright, S.F., 1978/1979. Antigenicity of structural components from porcine transmissible gastroenteritis virus. Vet. Microbiol. **3**:179-190.

Gough, P.M., Ellis, C.H., Frank, C.J., and Johnson, C.J., 1983. A viral subunit immunogen for porcine transmissible gastroenteritis. Antiviral Res. **3**:211-221.

Holmes, K.V., Doller, E.W., and Sturman, L.S., 1981. Tunicamycin resistant glycosylation of a coronavirus glycoprotein: Demonstration of a novel type of viral glycoprotein. Virology **115**:334-344.

Hu, S., Bruszewski, J., Boone, T., and Souza, L., 1984. Cloning and expression of the surface glycoprotein gp195 of porcine transmissible gastroenteritis virus. In: "Modern Approaches to Vaccines", R.M. Chanock and R.A. Lerner (eds.). Cold Spring Harbor Laboratory, New York., pp. 219-223.

Laemmli, U.K., 1970. Cleavage of structural proteins during the assembly of the head of bacteriophage T4. Nature **227**:680-685.

Mackett, M., Smith, G.L., and Moss, B., 1984. General method for production and selection of infectious vaccinia virus recombinants expressing foreign genes. J. Virol. **49**:857-864.

McClurkin, A.W., 1965. Studies on transmissible gastroenteritis of swine I. The isolation and identification of a cytopathogenic virus of transmissible gastroenteritis in primary swine kidney cell cultures. Can. J. Comp. Med. **29**:46-53.

Niemann, H., and Klenk, H.-D., 1981. Coronavirus glycoprotein El, a new type of viral glycoprotein. J. Mol. Biol. **153**:993-1010.

Oppermann, H., Levinson, A.D., Varmus, H.E., Levintow, L., and Bishop, J.M., 1979. Uninfected vertebrate cells contain a protein that is closely related to the product of the avian sarcoma virus transforming gene (src). Proc. Natl. Acad. Sci. (U.S.A.). **76**:1804-1808.

Paoletti, E., Lipinskas, B.R., Samsonoff, C., Mercer, S., and Panicali, D., 1984. Construction of live vaccines using genetically engineered poxviruses: Biological activity of vaccinia virus recombinants expressing the hepatitis B virus surface antigen and the herpes simplex virus glycoprotein D. Proc. Natl. Acad. Sci. (U.S.A.) **81**:193-197.

Rottier, P.J.M., Horzinek, M.C., and van der Zeijst, B.A.M., 1981. Viral protein synthesis in mouse hepatitis virus strain A59-infected cells: Effect of tunicamycin. J. Virol. **40**:350-357.

Rottier, P., Brandenburg, D., Armstrong, J., van der Zeijst, B., and Warren, G., 1984. Assembly in vitro of a spanning membrane protein of the endoplasmic reticulum: The E1 glycoprotein of coronavirus mouse hepatitis virus A59. Proc. Natl. Acad. Sci. (U.S.A.). **81**:1421-1425.

Smith, G.L., Murphy, B.R., and Moss, B., 1983. Construction and characterization of an influenza vaccinia virus recombinant that expresses the influenza hemagglutinin gene and induces resistance to influenza virus infection in hamsters. Proc. Natl. Acad. (U.S.A.). **80**:7155-7159.

PILI OF ENTEROTOXIGENIC *ESCHERICHIA COLI* FROM PIGS AND CALVES

Richard E. Isaacson

Pfizer Central Resarch
Eastern Pont Road
Groton, Connecticut 06340

INTRODUCTION

The ability to colonize mucosal surfaces is an important attribute that is required for the virulence of most bacterial pathogens. Both invasive and noninvasive bacteria must proliferate and ultimately achieve sufficient populations on mucosal surfaces to cause disease. Various physiologic functions of mammals tend to reduce the size of bacterial populations in vivo. When bacteria are potentially pathogenic, these functions provide a defensive mechanism against disease. The flow of fluids and particulate matter in small intestines due to peristalsis and villus pumping helps to keep bacterial populations low. In the small intestines of man where bacterial populations are normally small, any overgrowth that results from a cesation of flow has the potential to be coupled to a disease process. In the large intestine, where the rate of flow becomes quite slow, bacterial populations are naturally higher and ecologically complex.

The bacterial attributes that contribute to colonization of mucosal surfaces are probably varied amongst bacterial species, but must include a mechanism for associating with the tissue surface. Enterotoxigenic *Escherichia coli* (ETEC), the etiologic agent of enterotoxic colibacillosis (acute diarrheal disease) of neonatal animals and man, colonizes small intestines by direct adherence to mucosal surfaces (Moon, et al., 1979). Figure 1 is a scanning electron micrograph that vividly demonstrates what is meant by colonization. The micrograph is of a section of small intestine taken from a neonatal pig that had been colonized by an ETEC strain. The mucosal surface is densely covered with a mat of rod shaped structures which are the challenge organism. Occasional patches of microvillus membrane may be seen where gaps in the bacterial layer occur. Specific bacterial organelles, called somatic pili, act as the adhesins on ETEC allowing them to adhere to mucosal surfaces. By adhering to the microvillus membrane of small intestines, the ETEC become resistant to host clearance. These organisms may then proliferate at that site and subsequently cause disease.

Pili are defined as filamentous, hairlike, non-flagellar structures on the outer surface of bacteria (Briton, 1959). Figure 2 is a transmission electron micrograph of two piliated *E. coli* cells that were negatively stained with phosphotungstate. The pili are the thin filaments radiating outward from the cell surface. Pili have been divided into two general groups: sex pili and somatic pili. Sex pili have a specific function being required for conjugal exchange of genetic information between gram-negative bacteria. Although the precise functions of some somatic pili are known, in most cases they are simply recognized as morphologically or serologically distinct structures. The term fimbriae has been widely used as a synonym for somatic pili (Duguid, et al., 1955).

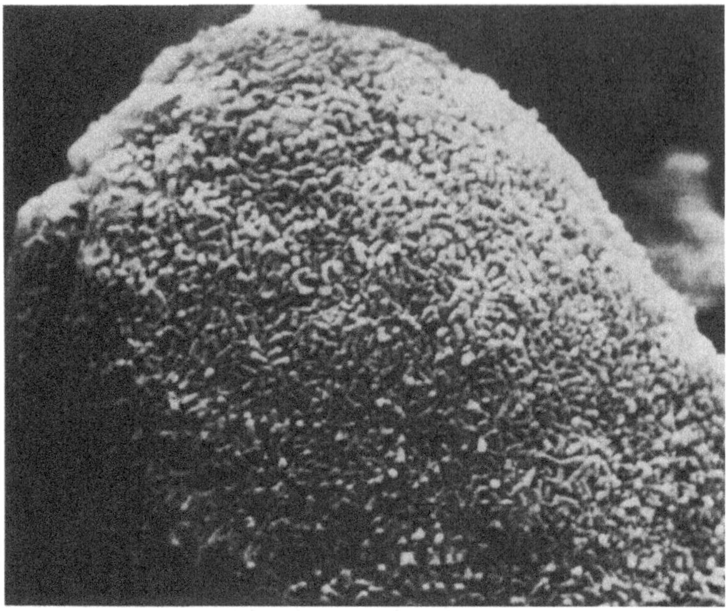

Figure 1. Scanning electron micrograph of a small intestinal villus covered with ETEC.

Pili known to act as adhesins (also called pilus-adhesins) on ETEC that cause disease in pigs are K88, K99, 987P and F41, in lambs and calves are K99 and F41 and in man are CFA/I and CFA/II (for a recent review see Jones and Isaacson, 1983). Each is antigenically unique and, with the exception of 987P, has unique hemaglutination properties. Although most ETEC express a single pilus-adhesin, some strains have been identified that express two or more (Jayappa, et al., 1983). K88 (Bak, et al., 1972), and K99 (Smith and Lingood, 1971) are both plasmid encoded while the genetic loci of 987P and F41 are unknown although assumed to be chromosomal. Therefore, a strain producing K88, K99, and 987P or F41 could occur in nature. In fact, most if not all, freshly isolated F41 positive ETEC also produce K99 (Isaacson and Richter, 1983).

EVIDENCE FOR PILI AS ADHESINS

K88 was the first pilus-adhesin to be identified. It was initially identified as an antigen that was epidemiologically associated with ETEC strains that caused disease in pigs (Orskov, et al., 1961). The evidence that K88 was required for virulence by promoting colonization was based, in part, on data from genetic manipulations (Smith and Linggood, 1971a). K88 negative ETEC strains were selected after treating a virulent, wild-type strain with the plasmid curing agent acridine orange. Loss on the K88 plasmid and thus K88 resulted in the concomitant reduction of virulence even though the cured strain retained its ability to produce enterotoxins. The reduction in virulence was shown to be related to an inability of the strain to colonize porcine small intestines. Pigs were challenged orally with either the wild-type strain or the cured strain and 18 hours later the concentration of each strain in small intestines was determined. Pigs challenged with the K88 positive wild-type strain contained 10^9 K88 positive organisms per gram of tissue while pigs challenged with the K88 negative strain contained 10^5-10^6 of the K88 negative organism per gram of tissue. Reintroduction of the K88 plasmid into the cured strain completely restored colonization capacity and virulence.

Analogous experiments subsequently were performed with K99 (Smith and Linggood, 1971) and 987P (Isaacson, et al., 1977) producing ETEC. For the 987P strains spontaneous mutants were selected rather than curing specific plasmids. The requirement of pili for colonization

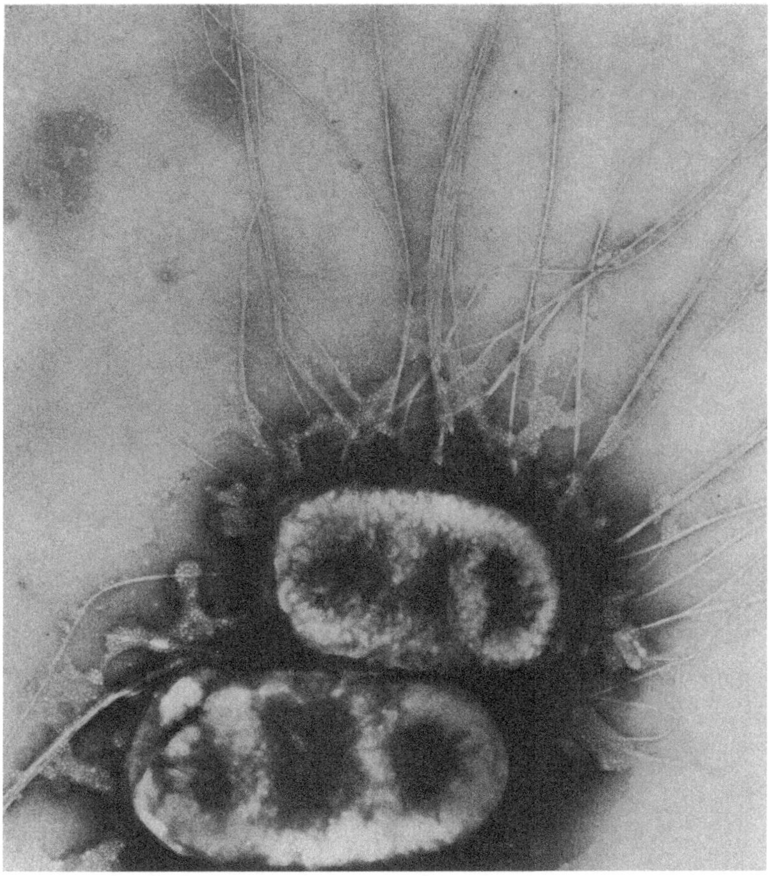

Figure 2. An electron micrograph of two negatively strained *E. coli* cells. The filamentous structures radiating outward from the cell surface are pili.

can be visualized in Figure 3. Panel A is a photomicrograph of a section of small intestine removed from a neonatal pig 18 hours post challenge with a virulent ETEC strain. The section was stained with a fluorescent antibody specific for the challenge organisms. A confluent layer of fluorescence from villus tip to crypt is seen which is characteristic of colonized tissue. Panel B, on the other hand, is tissue from a pig challenged with a nonpiliated mutant. The lack of bacterial layers by the mutant is in marked contrast to its wild-type parent. A few free ETEC cells can be seen in the lumen.

Although the results from *in vivo* virulence studies clearly demonstrate a role for pili in intestinal colonization, they may not be used to support a specific role for these structures. It is assumed that pili function *in vivo* as adhesins. This hypothesis is supported, in part, by the transmission electron micrograph shown in Figure 4. Shown is an ETEC cell adherent to porcine small intestines. Since the tissue was processed by standard methods for electron microscopy, which differ from the methods used for negative staining, it can only be assumed that the filamentous structures that radiate outward from the bacterial cell and contact the brush border membrane of the gut tissue are pili.

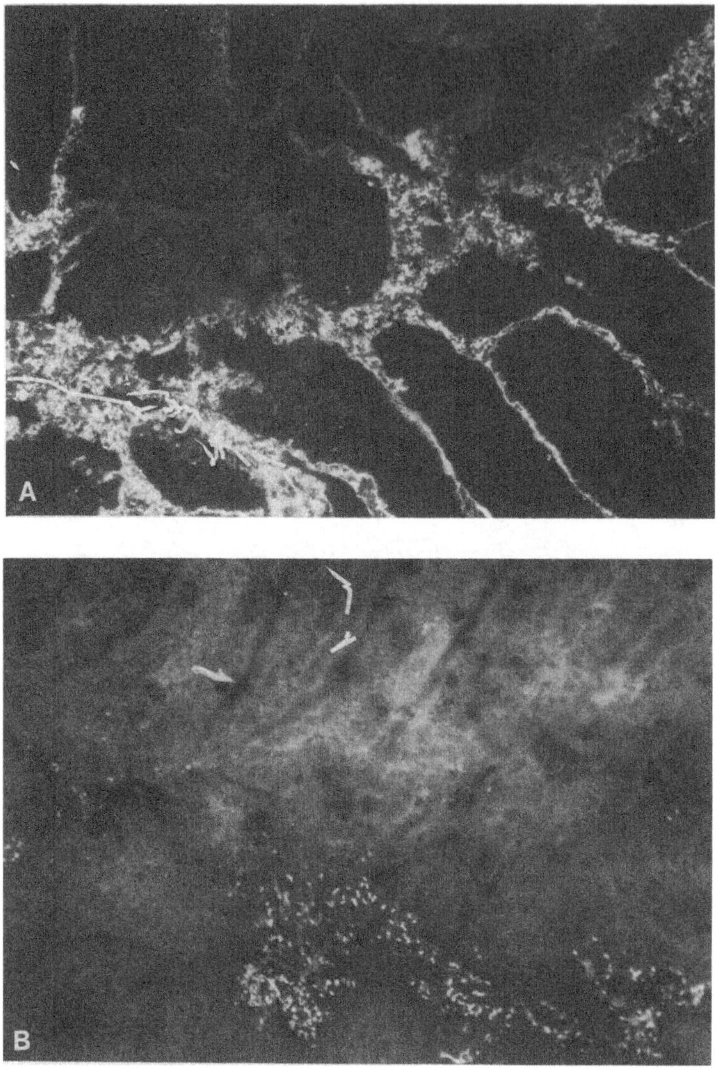

Figure 3. Photomicrographs of segments of ileum from pigs challenged with a virulent ETEC strain (A) or a non-piliated, non-virulent but otherwise isogenic ETEC strain (B). Sections were stained with fluorescein labeled antibody. Only the piliated organism appears to be adherent to the tissue while the non-piliated mutants cells are found only in the lumen.

The best evidence supporting the hypothesis that some pili are adhesins was derived from *in vitro* adhesion experiments. Piliated *E. coli* producing K88, K99, or 987P pili adhere to intact, isolated enterocytes obtained from neonatal pigs while isogenic but nonpiliated strains do not (Wilson and Hohmann, 1978, Nagy, et al., 1977, and Isaacson, et al., 1978). Dose-dependent adherence of a K99 positive *E. coli* strain and its nonpiliated derivative, for example, is shown in Figure 5. The addition of increasing amounts of piliated but not nonpiliated *E. coli* results in increasing numbers of bacteria adhering to enterocytes until a saturation point is achieved. Nonagglutinating Fab fragments prepared against purified pili inhibit the adherence of *E. coli* producing homologous pili to enterocytes (Table 1) (Isaacson, et al., 1978). Inhibition of adhesion is not, however, prevented by Fab fragments against heterologous pili nor by Fab fragments against surface antigens other than pili (eg. anti 1475 shown in Table 1). Thus it can be concluded that only piliated *E. coli* are adherent and that pilus-specific antibodies neutralize the adhesiveness of the pili.

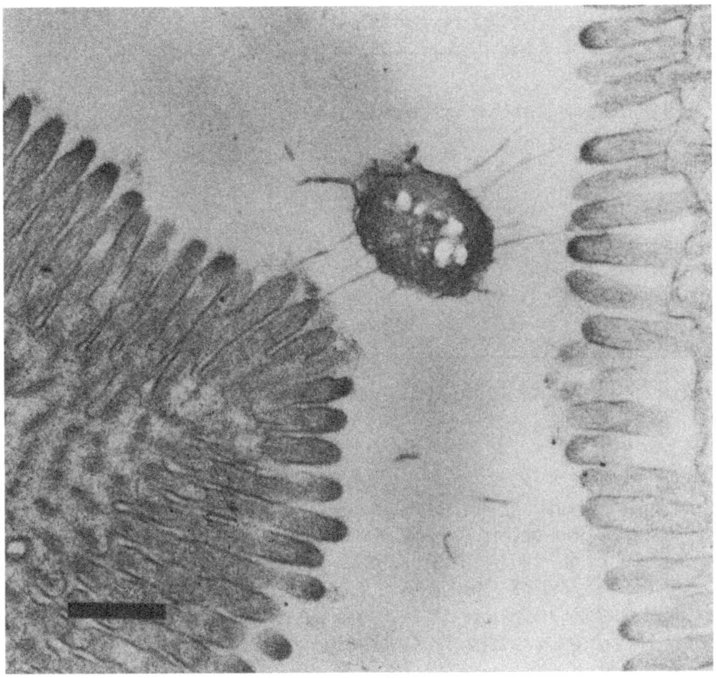

Figure 4. Electron micrograph of a section of porcine small intestine colonized by an ETEC strain. Filaments on the surface, that may be pili, come into direct contact with the brush border membranes of the tissue.

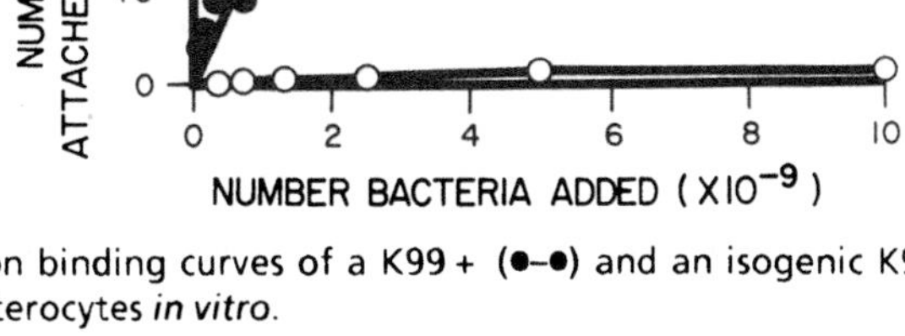

Figure 5. Saturation binding curves of a K99 + (●–●) and an isogenic K99- (o–o) *E. coli* to isolated porcine enterocytes *in vitro*.

TABLE 1. Effect of Fab fragments on pilus-mediated in vitro adhesion of *E. coli* strains to porcine enterocytes

Pilus *E. coli* strain produces	Specificity of Fab fragment (2 mg added per assay)	Number of bacterial cells attached per enterocyte[a]
	None	39.5
	1475 (non-pilus surface antigens)	39.2
K99	K99	4.3
	987P	40.3
	Type I	40.6
	None	40.5
987P	K99	39.3
	987P	4.1
	Type I	36.6

a. Each number is the arithmetic mean of three determinations. From Isaacson et al., 1978.

Purified pili compete with *E. coli* cells producing the same pili for binding sites on enterocytes. Competition occurs only when the competitor pilus is the same as on the bacterium (Isaacson, et al., 1978). That is, 987P pili prevent attachment of 987P positive *E. coli* to enterocytes but not of K99 positive *E. coli*. These results are consistent with the existence of pilus-specific receptors on the enterocyte cell surface. They also demonstrate the uniqueness of each pilus receptor. K99 recognizes a receptor that is different from the 987P or K88 receptors.

Although the remaining discussion will specifically deal with pili, it is worth noting at this point that the mammalian host plays an active rather than a passive role in the colonization process. The degree of intestinal motility initially may predispose the host to infection. However, the host must also provide specific receptors that are recognized by the pili. For example, age dependent resistance and genetic resistance to infection by K88 positive ETEC has been well documented (Rutter, et al., 1975, Sellwood, et al., 1975, and Runnels, et al., 1980). Thus host factors are specifically required for pathogenesis.

CHARACTERIZATION OF PILI

Morphology

Somatic pili are readily observed by electron microscopy and in general can be divided into two morphologically distinct groups: rigid and flexible. Pili found on animal pathogens that are exemplary of the rigid group are *E. coli* type I pili (also called common pili) and 987P pili (Figure 6). These pili have apparent axial holes and uniform cross-sectional diameters of 7 nm (Brinton, 1965, and Isaacson and Richter, 1981). The length of these pili is variable and may be as long as 4 um. The rigid pili have a strong tendency to aggregate into parallel bundles and angle layered crystals which is enhanced by salt (Brinton, 1965).

The *E. coli* K88 and K99 pili are exemplary of a group of pili that are morphologically flexible (Figure 6). K88 and K99 are difficult to visualize on bacterial cells. K99 positive cells often appear to have a fuzzy coat especially when fixed and observed in ultra-thin sections (Moon, et al., 1977 and Isaacson, 1981). Like pili in the rigid pilus group, K99 also forms aggregates (Isaacson, 1977). Aggregated, purified K99 can readily be seen by electron microscopy (Isaacson, 1977, de Graaf, et al., 1981, and Wadstrom, et al., 1978). The measured diameters for K99 vary between 2 and 8.4 nm. The larger structures probably are aggregates; at 8.4

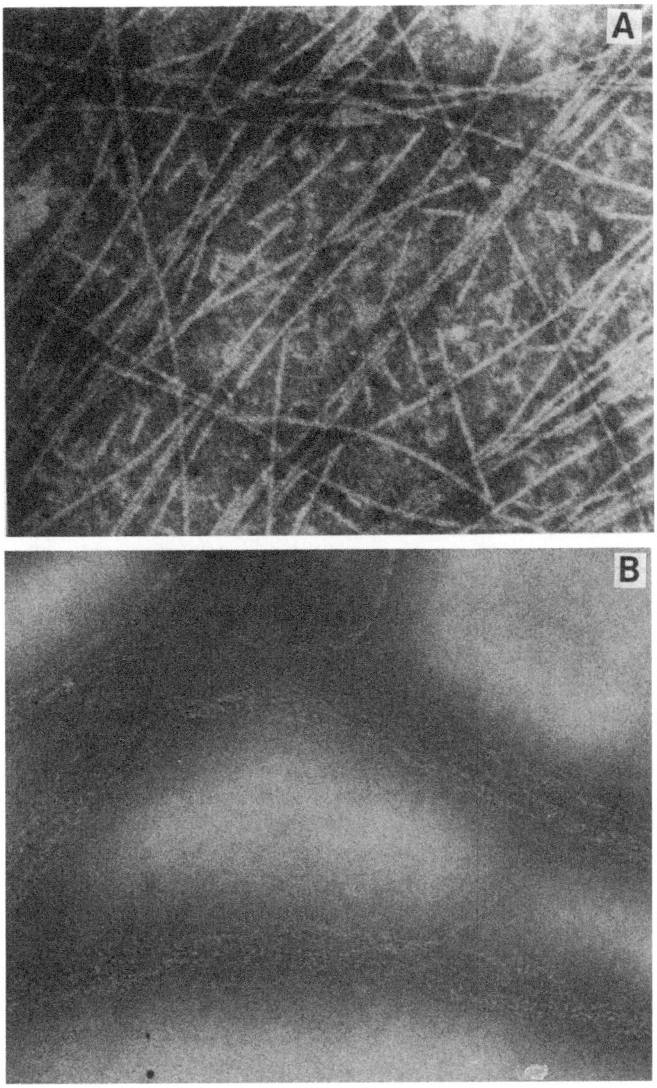

Figure 6. Electron micrographs of purified 987P (A) and K99 (B). Preparations were negatively strained with 1% sodium phosphotungstate.

nm the aggregate is composed of four strands. The aggregates of K99 have several different characteristics compared to aggregates of rigid pili. Most obvious of these differences is that the K99 aggregates do not have an ordered parallel alignment, but rather appear to wrap around each other to form loose, helical structures.

Pili are polymers composed of identical subunits called pilin (Jones and Isaacson, 1983). The number of subunits defines the molecular weight and length of an intact pilus. The forces holding pilin subunits together are unknown but do not include covalent bonds. Pili readily dissociate into free pilin in the presence of sodium dodecyl sulfate, guanidine, or dilute acid. It is reasonable to believe that hydrogen and hydrophobic bonds are important in pilus formation and maintenance.

TABLE 2. Molecular weights, isoelectric points and non-amino acid substitutions on various *E. coli* pilin

Pilus	Subunit Molecular weight (SDS-Page)	pI	Non-amino acid substitutions	
K88ab	26,200	4.2	None	Mooi and de Graff, 1979
K99	18,200	10.1	Phosphatidyl ethanolamine	Isaacson, 1977,1981
F41	29,500	–	None	deGraaf and Roorda, 1982
987P	20,000	3.9	Hexosamine	Isaacson and Richter, 1981
Type I	17,100	3.7	None	Brinton, 1965

The molecular weights of pilin molecules from animal ETEC strains are listed in Table 2. Pilin molecules are primarily composed of amino acids. Several nonamino acid substitutions have, however, been detected on certain pili. 987P pili have one or two residues of a hexosamine per pilin molecule (Isaacson and Richter, 1981) while K99 appears to contain phosphatidyl ethanolamine (Isaacson, unpublished). Whether these substituents are integral parts of the pili or simply remnants of cellular attachment retains to be determined. However, periodate cleavage of the hexosamine on 987P pili abolishes the capacity of the pili to subsequently bind to enterocytes.

The amino acid compositions of some pilin proteins are shown in Table 3. Each contains about 50% nonpolar amino acids. Frequently, such figures have been cited in assigning hydrophobic properties to pili (which has been suggested as a property necessary for adhesive function). The determination of hydrophobicity based on nonpolar amino acid content leads to rather erroneous results since the amount of hydrophilic amino acids and the degree of hydrophobicity assigned to each amino acid is not taken into account. The relative hydrophobicity of the pilin molecules were calculated using the values of amino acid hydrophobicity of Nozoaki and Tanford (1971) and are listed in Table 3. Values range from -459 to -609 cal/mole/amino acid. For comparison, human serum albumin has a calculated hydrophobicity of -769 cal/mole/amino acid and is composed of 55% nonpolar amino acids. Human serum albumin is therefore more hydrophobic than the pili.

The calculation of protein hydrophobicity based upon amino acid composition results in a value that assumes a uniform distribution of hydrophobic amino acids which also is unrealistic (see Figure 8 below for confirmation). Other methods have been developed and used to measure the relative hydrophobicity of piliated and nonpiliated bacteria. Type I pili, for example, reduce the electrophoretic mobility of *E. coli* cells and alter the phase partitioning of cells in polyethylene glycol-dextran (Brinton, 1965). Pili also promote the adsorption of cells to hydrophobic gels such as phenyl or octyl sepharose (Smyth, et al., 1978 and Wadstrom, et al., 1978).

Hydrophobic domains at the receptor recognition site may promote the interaction with the receptor. These domains may, however, represent minor regions of the pilus with respect to overall surface area and hydrophobic content and thus, although critically important for the biological activity of the pilus, may not contribute sigificantly to the overall hydrophobicity of the protein. Consequently, the results obtained by measuring pilus hydrophobicity may not be useful in defining the specific interactions between pilus and receptor or for that matter the interactions between pilin molecules. An analysis of hydrophobic regions based on the primary structure of K99 is described below (and in Figure 8).

Charge interactions between bacteria and target cell are particularly significant in the adhesive process (Jones and Isaacson, 1983). The charge of a pilus may contribute to the overall attractive and repulsive forces between bacterium and target cell. Pili that reduce or

TABLE 3. Amino acid composition and hydrophobicity of various *E. coli* pili

Amino Acid	K88ab	K99	F41	987P	Type I
Lysine	12	9	12	10	3
Arginine	7	4	4	2	3
Histidine	0	0	6	0	2
Aspartic	28	25	27	34	20
Threonine	27	20	18	28	20
Serine	16	12	33	23	10
Glutamic	16	7	24	16	13
Proline	6	3	11	8	2
Glycine	35	15	39	26	17
Alanine	27	46	16	26	34
Cysteine	0	2	ND	2	2
Valine	18	14	17	15	13
Methionine	2	3	3	1	0
Isoleucine	10	13	10	12	4
Leucine	18	7	14	17	10
Tyrosine	8	0	13	1	0
Phenylananine	10	0	9	3	8
Tryptophan	ND	3	ND	1	0
Hydroxylysine	ND	3	10	ND	ND
Percent nonpolar amino acids	52.5	54.7		48.4	54.6
Hydrophobicity (-cal/mole/amino acid)	609	577		459	511

ND. Not determined
References are the same as in Table 2.

reverse the overall surface charge (a negative charge) on a bacterial cell contribute to the adhesive process, in part, by reducing the repulsive forces between negatively charged bacterial cells and target cells. Most pili, however, are negatively charged structures with isoelectric points in the range of pH 3.7-4.2 (see Table 2). The *E. coli* K99 pilus is an exception with an isoelectric point of pH 10.1 (Isaacson, 1977). This property may make K99 a particularly useful adhesin. K99 positive *E. coli* may adhere to gut epithelial cells totally on the basis of charge interaction. Although, as discussed above, it is presumed that K99 specific receptors exist on the epithelial cell surface, the charge interaction between K99 and the target cell may well increase the stability of the bacterial layer thereby increasing the probability of permanent, irreversible attachment.

Although the pilus structure does not contain interpilin covalent bonds, it is likely that in some pili, intrapilin covalent bonds are essential. For example, K99 (Isaacson, et al., 1981) and 987P (Isaacson and Richter, 1981) molecules contain disulfide bonds that allow these molecules to take on three different conformations depending upon the integrity of the disulfide bond. Native molecules undergo a conformational change upon cleavage of the disulfide bond. Molecules treated with the reducing agent 2-mercaptoethanol have a slower mobility in sodium dodecyl sulfate polyacrylamide gels than non-treated pilin. If the disulfide bonds are permitted to re-form, pilin molecules may regain their original conformation or assume a third conformation. These differences again can be demonstrated by the differences in electrophoretic mobility. Pilin in the third conformation migrates slower than native molecules and faster than molecules with disrupted disulfide bonds.

Antigenicity

The proteinaceous nature of pili makes them good antigens. The detection of pili on bacteria has made use of their antigenic properties. Detection has been performed by a variety of serologic tests including serum agglutination and enzyme linked immunosorbent assays (Jones and Isaacson, 1983).

Interestingly, antigenic cross reactions among pili are limited and antigenic diversity is the rule. K88, K99, 987P, type I pili, and the human ETEC adhesins CFA/I and CFA/II are serologically unrelated as shown in Table 4. This heterogeneity even appears to be true in the case of closely related pili. The morphology, genetic regulation, and N-terminal amino acid sequences of 987P and type I pili, for example, are identical yet they are antigenically unrelated. Within the pili in the so-called type I family (including type I pili), antigenically unrelated variants exist (Klemm, et al., 1982 and Orskov, et al., 1980). The pili of *Neisserria gonorrhoeae* also show a similar antigenic diversity. The P9 strain of *N. gonorrhoeae* also show a similar antigenic diversity. The P9 strain of *N. gonorrhoeae* has now been shown to produce individually four different pili (α, β, γ, δ) (Lambden, et al., 1981 and Heckels, 1982). The α and β pili have subunits with identical molecular weights and nearly identical amino acid compositions and tryptic peptides, yet they do not cross react in serum agglutination tests. The shift from one antigenic variant to another is quite rapid and can be demonstrated after serial passage *in vitro*.

Analysis of cyanogen bromide fragments of the P9 pili by immunologic and biochemical techniques have provided some insight into the mechanism of antigenic variation (Schoolnik, 1982). At the amino terminus is a highly conserved sequence that may be involved in pilin assembly. A middle conserved sequence specifies receptor binding activity. At the carboxy terminus is a highly variable, type specific antigen, changes in which do not cause functional or structural alterations of the pilus. However, the carboxy terminal region is immunodominant. Thus, polyclonal sera prepared against *N. gonorrhoeae* pili are almost exclusively specific for this region. Therefore, closely related pili, based on sequence homololgy, may appear to be antigenically unrelated. Since the pili of *N. gonorrhoeae* are essential for its survival in nature, this means of antigenic variation provides a mechanism to avoid host immune reponses yet retain a functionally active adhesin. Whether a similar mechanism is responsible for the variation observed in the *E. coli* type I family remains to be determined.

Within the K88 group, three antigenic variants have been recognized and analyzed. Each variant possesses a common a antigen and a b, c or d antigen (Guinee and Jansen, 1975 and Guinee, et al., 1980).

TABLE 4. Antigenic specificity of different *E. coli* pili

Strain (pilus)	Antiserum endpoint[a]					
	K99	K88ac	987P	CFA/I	CFA/II	Type I
1474 (K99)	256	--[b]	--	--	--	--
1476 (K88ac)	--	256	--	--	--	--
987 (987P)	--	--	45,665	--	--	--
I95 (CFA/I)	--	--	--	2048	--	--
I127 (CFA/II)	--	--	--	--	256	--
Bam (Type I)	--	--	--	--	--	1024

a. Reciprocal of the highest dilution that results in the agglutination of the piliated cells. Assay is by agglutination on slides.
b. No detectable agglutination with the antiserum.

On the other hand, using absorbed anti K99 antisera, it was shown that K99 from geographically and temporally diverse sources are identical (Isaacson, et al., 1978). This result has also been confirmed by gene-specific DNA hybridization studies (Isaacson and Richter, 1983). Thus, antigenic variation within pilus class is not necessarily the rule.

Analysis of pilin with regard to function and antigenicity has been facilitated by peptide analysis and by determining amino acid sequences. For example, the amino acid sequence of K88ac has been determined by Edman degradation (Klemm, 1981) and confirmed by translation of its DNA sequence (Gaastra, et al., 1981). The primary sequence was analyzed and used to compare K88ac to K88ad (Gaastra, et al., 1983). The analysis was used to predict the regions that were likely associated with the a, c, and d antigens. Since it was believed that the variable regions of K88 (ie. b, c, and d antigens) were involved in receptor recognition, this analysis was particularly useful in predicting specific functions along the polypeptide chain.

The primary sequence of K99 also has been deduced from its DNA sequence and is shown in Figure 7. Mature K99 contains 175 amino acids and is synthesized with a 22 amino acid leader sequence. The amino terminal amino acid is asparagine. The cysteine residues located at residues number 15 and 52 form a disulfide bridge. The amino acid sequence was subsequently analyzed for hydrophobic and hydrophilic regions using the algorithm of Hopp and Woods (1981) (Figure 8). The peaks labeled I and II represent sequences predicted to be epitopes since these represent hdyrophilicity maxima. Using the algorithm of Chou and Fasman (1978) to predict secondary structure it was determined that the two regions predicted to be epitopes also had the potential for turns which may increase their

```
AAAAAATAATATCGAACAATGAAGAATCTAG ATG AAA AAA ACA CTG CTC GCT ATT ATC TTA GGT GGT
                                Met Lys Lys Thr Leu Leu Ala Ile Ile Leu Gly Gly

ATG GCT TTT GCG ACT ACC AAT GCT TAT GCG AAT ACA GGT ACT GTC AAT GGT AAT GGC AAA
Met Ala Phe Ala Thr Thr Asn Ala Ser Ala Asn Thr Gly Thr Asn Thr Phe Asn Gly Lys
                                         1                                   10

ATA ACG AGT GCT ACT TGT ACA ATT GAG AAT GAG GTC AAT GGT AAT CGT ACA GAT CTG GCT
Ile Thr Ser Ala Thr Cys Thr Ile Glu Pro Glu Val Asn Gly Asn Arg Thr Asp Leu Ala
                                     20                                      30

GCC ATT AGC GGT CAT GGC ACT GTA GTG GAT TTT AAA CTA AAA CCA GCG CCC GGC AGT AAT
Ala Ile Ser Gly His Gly Thr Val Val Asp Phe Lys Leu Lys Pro Ala Pro Gly Ser Asn
                                     40                                      50

GAC TGC CTA GCG AAA ACA AAT GCT CGT ATT GAC TGG TCT GGT TCT ATG AAC AGT TTA GGT
Asp Cys Leu Ala Lys Thr Asn Ala Arg Ile Asp Trp Ser Gly Ser Met Asn Ser Leu Gly
                                     60                                      70

TTT AAT AAT ACA GCT TCA GGA AAT ACT GCT GCT AAG GGA TAC CAT ATG ACT TTG CGC GCA
Phe Asn Asn Thr Ala Ser Gly Asn Thr Ala Ala Lys Gly Tyr His Met Thr Leu Arg Ala
                                     80                                      90

ACA ACG TTG GAA TGG TTC GGT GGT GTC AAT ATT ATG TTG GCC CGC AAT ATT ATC GAT GGA
Thr Thr Leu Glu Trp Phe Gly Gly Val Asn Ile Met Leu Ala Arg Asn Ile Ile Asp Gly
                                     100                                     110

ATA TTA TAT GAA TTG AAA AAC ATC GAT TTT ATA ATT TTA TTG TTG TGG CGA ATA GCA GTA
Ile Leu Tyr Glu Leu Lys Asn Ile Asp Phe Ile Ile Leu Leu Leu Trp Arg Ile Ala Val
                                     120                                     130

AAT CTT CCG CTT CGA AAA GAA TCG AAC CTG CAA TTT ATA CGA CAT AAA TTT GGC CAA AAC
Asn Leu Pro Leu Arg Lys Glu Ser Asn Leu Gln Phe Ile Arg His Lys Phe Gly Gln Asn
                                     140                                     150

ACA GAA GAT ATT GAG TTA TTT TTC AAT AGC TCA ACA GTG TTG CCT AGT AAT TAT ACT GTA
Thr Glu Asp Ile Glu Leu Phe Phe Asn Ser Ser Thr Val Leu Pro Ser Asn Tyr Thr Val
                                     160                                     170

GAT GTA AAG TTA AAA TGATGAAGTTATTGGTCGTGCAAAGCTAGAAGTGAGCCAAGATGAT
Asp Val Lys Leu Lys
```

Figure 7. Nucleotide sequence of the K99 structural gene and its deduced amino acid sequence. The first amino acid in mature protein is asparagine (at position number 1).

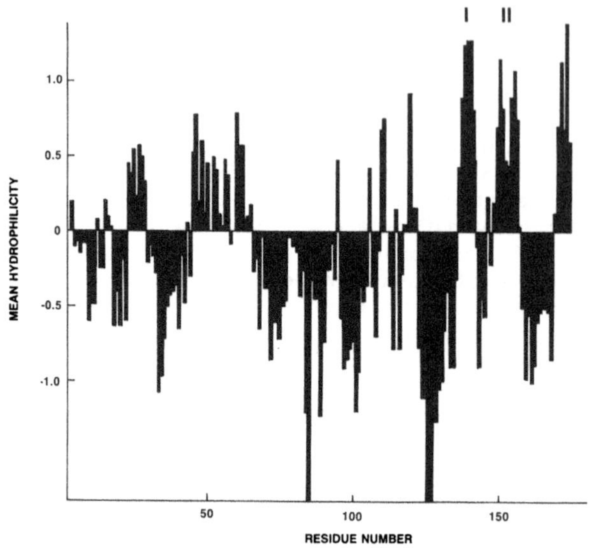

Figure 8. Hydrophilicity plot of K99. Peaks I and II represent hydrophilicity maxima and therefore may be major epitopes on this protein.

potential antigenicity. Interestingly, the disulfide bridge should bring the hydrophilic regions bounded by residues 21 to 28 and form 44-64 into close proximity and thus could generate a third rather large hydrophilic domain. The regions of hydrophobicity (negative numbers in Figure 8), particularly the regions from residue 79 to 110 and from 122 to 136 may represent receptor recognition sites or pilin recognition sites (ie. involved in pilin polymerization to pili).

Vaccination

The role of pili and thus adherence in the pathogenesis of ETEC induced diarrheal disease has been discussed. Although the pilus genes have been demonstrated as a requirement for virulence and that pili may act as adhesins *in vitro*, the only direct evidence of the adhesive function of pili *in vivo* has been based on the results of vaccination studies.

Earlier work had demonstrated that pilus-adhesins are expressed *in vivo* (Isaacson, et al., 1978 and Moon, et al., 978). Sections removed from the small intestines that were colonized with ETEC can be stained by pilus-specific fluorescent antibodies. Thus, since pili are expressed *in vivo* and if they truly serve as adhesins then specific antibody delivered to the mucosal surface should prevent colonization and disease.

This hypothesis was tested by vaccination of pregnant pigs with purified K99 or 987P (Morgan, et al., 1978 and Nagy, et al., 1978). Within hours of birth, baby pigs were challenged with a virulent ETEC strain and the degree of protection confered by consumption of pilus-specific antibodies in colostrum was determined. Protection was measured based on a reduction in mortality when compared to pigs from dams that were not vaccinated (control). In one experiment 12 out of 44 pigs challenged with a 987P positive ETEC strain in the control group died while 0 out of 38 pigs from dams that had been vaccinated with 987P died. Protection against the 987P positive ETEC challenge was not confered by vaccination with K99 (7 out of 41 pigs died). Several points are demonstrated in this experiment. Firstly, vaccination with 987P pili prevents severe fatal ETEC diarrrhea by 987P producing strains. Secondly, protection does not extend to ETEC strains producing other pili such as K99. Doing the converse experiment, it was shown that

K99 prevents disease by K99 producing ETEC but not by 987P producing ETEC. Thus protection was dependent upon homology of pili in the vaccine and on the challenge ETEC strain.

Protection against ETEC strains of several serotypes was solely dependent upon the type of pili produced. Thus, equal protection was observed when pigs from dams vaccinated with 987P pili were challenged with an 09:K103:987P strain or a 020:K101:987P strain (Morgan, et al., 1978).

The route of vaccination did not appear to be important. Protection was confered regardless of whether the pilus antigen was presented parenterally or orally (Moon and Isaacson, 1978).

A multivalent pilus vaccine (EcoBac) was ultimately prepared by mixing equal amounts of K88ac, K99, and 987P. The vaccine was administered parenterally to dams four weeks and two weeks prior to the anticipated dates of farrowing. The vaccine confered good protection against fatal diarrheal disease caused by K88, K99, or 987P producing ETEC (Table 5). Since all naturally occurring F41 positive ETEC also product K99, protection against these organisms should also be confered. Although 54% of pigs in the control group died regardless of the ETEC challenge strain only 5% in the vaccinated group died.

Colostrum and milk samples were collected from dams and analyzed for pilus-specific antibodies (Table 6). Elevated titers were detected in the colostrum of vaccinates collected on the day of farrowing. However, the titers had dropped to marginally elevated levels in milk collected on day five. By day ten, antibody titers in the milk of vaccinates and controls were the same.

CONCLUDING REMARKS

The evidence presented overwhelmingly supports the hypothesis that some pili act *in vivo* as adhesins. The importance of pilus-adhesins in pathogenesis of ETEC mediated diarrheal disease is clear. For ETEC to colonize and cause disease in small intestines they must adhere to the tissue surface. Although *E. coli* cells that cannot ahdere in the small intestines do not, as a result, die they must, however, compete against much larger and complex bacterial populations in the large intestine, their normal habitat. Thus, the ecologic importance of adherence is implicit.

The highly antigenic nature of ETEC pili has greatly facilitated intervention in the disease process. Antibodies against ETEC pili prevent the attachment of these organisms to small intestinal mucosa. Antibody class and source seem to unimportant; only the antibody specificity. Consequently, monoclonal antibodies against K99 prepared from mice are equally effective in preventing colonization and disease in calves as are the antibodies present in colostrum.

TABLE 5. Protection confered by vaccination with EcoBac

Vaccine	Challenge Strain-Pilus Type		
	K88	K99	987P
Saline Control	54/89 (60.6)[a]	54/102 (52.9)	46/93 (49.5)
EcoBac Three pilus vaccine	5/68 (7.3)	8/104 (7.7)	1/115 (0.8)

a. Per cent mortality.

TABLE 6. Pilus-specific antibody titers in colostrum and milk in response to vaccination with EcoBac

Antibody Class		K88		K99		987P	
Day Post Farrow		Control	Vaccinate	Control	Vaccinate	Control	Vaccinate
	IgM	88	250	86	228	8	142
0	IgG	22	105	9	174	7	373
	IgA	80	113	6	24	36	224
	IgM	4	37	2	9	1	37
5	IgG	1	3	1	7	<1	8
	IgA	44	16	4	10	11	26
	IgM	2	5	3	24	8	9
10	IgG	1	2	1	2	<1	2
	IgA	30	13	5	7	43	10

Whether the exploitation of adhesive properties of other bacteria will be successful in the control of other infectious diseases remains to be seen. It is clear, however, that this approach to infectious disease control is feasible and in many regards is quite desirable when compared to conventional antibotic therapies.

REFERENCES

Bak, A. L., Christiansen, G., Christiansen, C., Stenderup, A., Orskov, I., and Orskov, F., 1972, Circular DNA molecules controlling synthesis and transfer of the surface antigen (K88) in *Escherichia coli*, J. Gen. Microbiol. 73:373.

Brinton, C. C., 1959, Non-flagellar appendages of bacteria, Nature 183:782.

Brinton, C. C., 1965, The structure, function, synthesis and genetic control of bacterial pili and a molecular model for DNA and RNA transport in gram negative bacteria, Trans. N.Y. Acad. Sci. 27:1003.

Chou, P. and Fasman, G. D., 1978, Prediction of tne secondary structure of proteins from their amino acid sequence, Adv. Enzymol. 47:45.

de Graaf, F. K., Klemm, P., and Gaastra, W., 1981, Purification, characterization, and partial covalent structure of *Escherichia coli* adhesive antigen K99, Infect. Immun. 33:877.

de Graaf, F. K. and Roorda, I., 1982, Production, purification, and characterization of the fimbral adhesive antigen F41 isolated from calf enteropathogenic *Escherichia coli* strain B41M, Infect. Immun. 36:751.

Duguid, J. P., Smith, I. W., Dempster, G., and Edmunds, P. N., 1955, Non-flagellar filamentous appendages ("fimbriae") and haemagglutinating activity in *Bacterium coli*, J. Pathol. Bacteriol. 70:335.

Gaastra, W., Mooi, F. R., Stuitje, A. R., and de Graaf, F. K., 1981, The nucleotide sequence of the gene encoding the K88ab protein subunit of porcine enterotoxigenic *Escherichia coli*, FEMS Microbiol. Lett. 12:41.

Gaastra, W., Klemm, P., and de Graaf, F. K., 1983, The nucleotide sequence of the K88ad protein subunit of porcine enterotoxigenic *Escherichia coli*, FEMS Microbiol. Lett. 18:177.

Guinee, P. A. M. and Jansen, W. H., 1979, Behavior of *Escherichia coli* antigens K88ab, K88ac, and K88ad in immunoelectrophoresis, double diffusion and hemagglutination, Infect. Immun. 23:700.

Guinee, P. A. M., Mooi, F. R., and Jansen, W. H., 1980, Preparation of specific *Escherichia coli* K88 antisera by means of purified K88ab and K88ad antigens, Zbl. Bakt., Hyg., I. Abt. Org. A. 242:182.

Heckels, J. E., 1982, Role of surface proteins in pathogenic adhesion, in "Microbiology 1982" D. Schlessinger, ed., American Society for Microbiology, Washington.

Hopp, T. P. and Woods, K. R., 1981, Prediction of protein antigenic determinants from amino acid sequences, Proc. Natl. Acad. Sci. 78:3824.

Isaacson, R. E., 1977, K99 surface antigen of *Escherichia coli*: purification and partial characterization, Infect. Immun. 15:272.

Isaacson, R. E., 1981, Pili of enterotoxigenic *Escherichia coli,* in "Third International Symposium on Neonatal Diarrhea", S. D. Acres, ed., Veterinary Infectious Disease Organization, Saskatoon.

Isaacson, R. E., Colmenero, J., and Richter, P., 1981, *Escherichia coli* K99 pili are composed of one subunit species, FEMS Microbiol. Lett. 12:229.

Isaacson, R. E., Fusco, P. C., Brinton, C. C., and Moon, H. W., 1978, *In vitro* adhesion of *Escherichia coli* to porcine small intestinal epithelial cells: pili as adhesive factors, Infect. Imun. 21:392.

Isaacson, R. E., Moon, H. W., and Schneider, R. A., 1978, Distribution and virulence of *Escherichia coli* in the small intestines of calves with and without diarrhea, Am. J. Vet. Res. 39:1750.

Isaacson, R. E., Nagy, B., and Moon, H. W., 1977, Colonization of porcine small intestines by *Escherichia coli*: colonization and adhesion factors of pig enteropathogens that lack K88, J. Infect. Dis. 135:531.

Isaacson, R. E. and Richter, P., 1981, *Escherichia coli* 987P pilus: purification and partial characterization, J. Bacteriol. 146:784.

Isaacson, R. E. and Richter, P., 1983, Molecular and genetic analysis of K99 plasmids, in "Fourth International Symposium on Neonatal Diarrhea", S. D. Acres, ed., Veterinary Infectious Disease Organization, Saskatoon.

Jayappa, H. G., Strayer, J. G., Goodnow, R. A., Fusco, P. C., Wood, S. W., Haneline, P., Cho, H.-J., and Brinton, C. C., 1983, Experimental infection and field trial evaluations of a multiple-pilus phase-cloned bacterin for the simultaneous control of neonatal colibacillosis and mastitis in swine, in "Fourth International Symposium on Neonatal Diarrhea", S. D. Acres, ed., Veterinary Infectious Disease Organization, Saskatoon.

Jones, G. W. and Isaacson, R. E., 1983. Proteinaceous bacterial adhesins and their receptors, CRC Crit. Rev. Microbiol. 10:229.

Klemm, P. 1981, The complete amino acid sequence of the K88 antigen – a fimbrial protein from *Escherichia coli*, Europ. J. Biochem. 117:617.

Klemm, P., Orskov, I., and Orskov, F., 1982, F7 and type I-like fimbriae from uropathogenic *Escherichia coli* strains – protein chemical and immunological aspects, FEMS Microbiol. Lett.

Lambden, P. R., Robertson, J. N., and Watt, P. J., 1981, The preparation and properties of α and β pili from variants of *Neisseria gonorrhoeae* P9, J. Gen. Microbiol. 124:109.

Mooi, F. R. and de Graaf, F. K., 1979, Isolation and characterization of K88 antigens, FEMS Microbiol. Lett. 5:17.

Moon, H. W. and Isaacson, R. E., 1978, Immunization against enterotoxigenic *Escherichia coli*: response of piglets given live piliated or non-piliated *Escherichia coli* orally, in "Proceedings of the 14th Joint Conference on Cholera", National Institutes of Health, Bethesda.

Moon, H. W., Isaacson, R. E., and Pohlenz, J., 1979, Mechanisms of association of enteropathogenic *Escherichia coli* with intestinal epithelium, Am. J. Clin. Nutr. 32:119.

Moon, H. W., McClurkin, A. W., Isaacson, R. E., Pohlenz, J., Skartvedt, S. M., Gillette, K. G., and Baetz, A. L., 1978, Etiologic diagnosis of diarrheal disease in calves: mixed infections, J. Am. Vet. Med. Assoc, 173:577.

Moon, H. W., Nagy, B., and Isaacson, R. E., 1977, Intestinal colonization and adhesion by enterotoxigenic *Escherichia coli*: some ultrastructural observations on adherence to ileal epithelium of the pig, J. Infect. Dis. 136:S124.

Morgan, R. L., Isaacson, R. E., Moon, H. W., Brinton, C. C., and To, C. C., 1978, Immunization of suckling pigs against enterotoxigenic *Escherichia coli*-induced diarrheal disease by vaccinating dams with purified 987 or K99 pili: protection corrleates with pilus homology of vaccine and challenge, Infect. Immun. 22:771.

Nagy, B., Moon, H. W., and Isaacson, R. E., 1977, Colonization of porcine intestine by enterotoxigenic *Escherichia coli*: selection of piliated forms *in vivo*, adhesion of piliated forms to epithelial cells *in vitro*, and incidence of a pilus antigen among porcine enteropathogenic *E. coli*, Infect. Immun. 16:344.

Nagy, B., Moon, H. W., Isaacson, R. E., To, C. C., and Brinton, C. C., Immunization of suckling pigs against enterotoxigenic *Escherichia coli* infection by vaccinating dams with purified pili, Infect. Immun. 21:269.

Nozaki, L. and Tanford, C., 1971, The solubility of amino acid and two glycine peptides in aqueous ethanol and dioxane solutions, J. Biol. Chem. 246:2211.

Orskov, I., Orskov, F., and Birch-Andersen, A., 1980, Comparison of *Escherichia coli* fimbrial antigen F7 with type I fimbriae, Infect. Immun. 27:657.

Orskov, I., Orskov, F., Sojka, W. J., and Leach, J. M., 1961, Simultaneous orrurrence of *E. coli* B and I antigens in strains from diseased swine, Acta Pathol. Microbiol. Scand. 53:404.

Runnels, P. L., Moon, H. W., and Schneider, R. A., 1980, Development of resistance with host age of adhesion of K99 + *Escherichia coli* to isolated intestinal epithelial cells, Infect. Immun. 28:298.

Rutter, J. M., Burrows, M. R., Sellwood, R., and Gibbons, R. A., 1975, A genetic basis for resistance to enteric disease caused by *E. coli*, Nature 257:135.

Schoolnik, G. K., 1982, Mechanism of binding of gonococcal pili, in "Microbiology 1982", D. Schlessinger, ed., American Society for Microbiology, Washington,

Sellwood, R., Gibbons, R. A., Jones, G. W., and Rutter, J. M., 1975, Adhesion of enteropathogenic *Escherichia coli* to pig intestinal brush-borders; the existence of two pig phenotypes, J. Med. Microbiol. 8:405.

Smith, H. W. and Linggood, M. A., 1971, Further observations on *Escherichia coli* enterotoxins with particular regard to those produced by atypical piglet strains and by calf and lamb strains: the transmissible nature of these enterotoxins and of a K antigen possessed by calf and lamb strains, J. Med. Microbiol. 5:243.

Smith, H. W. and Linggood, 1971a, Observations on the pathogenic properties on the K88, HLY, and ENT plasmids of *Escherichia coli* with particular reference to porcine diarrhea, J. Med. Microbiol. 4:467.

Smyth, C., Jonsson, P., Olsson, E., Soderlund, O., Rosengren, J., Hjesten, S., and Wadstrom, T., 1978, Differences in hydrophobic surface characteristics of porcine enteropathogenic *Escherichia coli* with and without K88 antigen as revealed by hydropohobic interaction chromatography, Infect. immun. 22:462.

Wadstrom, T., Smyth, C. J., Faris, A., Jonsson, P., and Freer, J. H., 1978, Hydropobic adsorptive and hemagglutinating properties of enterotoxigenic *Escherichia coli* with different colonizing factors: K88, K99, and colonization factor antigens and adherence factor, in "Second International Symposium on Neonatal Diarrhea", S. D. Acres, ed., Veterinary Infectious Disease Organization, Saskatoon.

Wilson, M. R. and Hohmann, A. W., 1974, Immunity to *Escherichia coli* in pigs: adhesion of enteropathogenic *Escherichia coli* to isolated intestinal epithelial cells, Infect. Immun. 10:776.

IMMUNE RESPONSES OF CATTLE AND MICE TO THE G GLYCOPROTEIN OF VESICULAR STOMATITIS VIRUS

Tilahun Yilma, Roger G. Breeze, Sandra Ristow,
John R. Gorham and Steve R. Leib

Department of Veterinary Microbiology and Pathology
Washington State University, Pullman, WA 99164-7040

ABSTRACT

A subunit vaccine for vesicular stomatitis was developed from a purified vesicular stomatitis virus preparation by selectively removing the immunogenic G glycoprotein of the virus with the dialyzable, non-ionic detergent, ß-D-octylglucoside. Cattle immunized intramuscularly with a single dose of 112 μg of G glycoprotein preparation in complete Freund's adjuvant did not develop vesicular disease following challenge by intralingual inoculation of 400 times the infectious dose of the virus. Similarly, mice vaccinated subcutaneously with a single dose of 10 μg of G glycoprotein preparation, with or without complete Freund's adjuvant, were protected from lethal encephalitis caused by vesicular stomatitis virus. A subunit vaccine for vesicular stomatitis of cattle, horses, and swine avoids the hazards associated with attenuated and inactivated vaccines, such as vaccine breaks, reversion to virulence, or introduction of virus into potential wild reservoirs or arthropod hosts. Further, it is possible to distinguish serologically animals vaccinated with the subunit preparation from those that have had the clinical disease or that have been vaccinated with whole virus. This is an essential consideration both for epidemiological studies and for disease control or establishment of quarantine programs.

INTRODUCTION

Vesicular stomatitis is a contagious viral disease primarily affecting cattle, horses, and swine. It is characterized by vesicular lesions on the tongue and other areas of the oral cavity, and the teats of the bovine udder. In pigs, vesicles often appear on the snout, and in both cattle and swine, lesions may develop along the coronary band and in the interdigital space of the foot. The disease is also of public health concern, since humans can be infected with the virus and exhibit influenza-like illness (Hanson, 1952; Patterson et al., 1958). Outbreaks of vesicular stomatitis have a major economic impact on the livestock industry and also cause serious problems for the Animal and Plant Health Inspection Service (APHIS) of the United States Department of

Agriculture (USDA) in recognizing and controlling outbreaks of more devastating vesicular diseases, such as foot-and-mouth disease.

Experimental vesicular stomatitis in mice, unlike the vesicular disease of other species, is characterized by a rabies-like encephalitis, usually followed by severe flaccid paralysis of the hind limbs and death 6-12 days after intravenous challenge with the virus. The mouse has provided a very useful system for studying the parameters of the immune response to the disease (Olitsky et al., 1936; Wagner, 1974; Zinkernagel, 1978; LeFrancois and Lyle, 1983a,b).

The natural spread and transmission of vesicular stomatitis have been an enigma. This may be related to the fact that vesicular stomatitis virus belongs to the family Rhabdoviridae, a group notorious for its virtually unlimited host range, encompassing homeothermic and poikilothermic vertebrates, invertebrates, and plants (Bishop, 1979). Usually the disease appears in the summer months and disappears after the first killing frost, suggesting transmission by an arthropod vector. However, an outbreak of vesicular stomatitis in cattle was responsible for significant economic loss to farmers in the 13 western states of the U.S. in the winter of 1982-83: unlike previous epizootics, the disease persisted when transmission by an arthropod vector appeared unlikely (King, 1983). Nevertheless, several investigators have shown experimentally that vesicular stomatitis virus (VSV) can be transmitted by horseflies, mosquitoes, and sandflies (Ferris et al., 1955; Tesh et al., 1972). Furthermore, VSV has been isolated from mites, biting gnats, and mosquitoes (Jonkers et al., 1964; Jonkers et al., 1965; Shelokov and Peralta, 1967; Sudia et al., 1967). Experimental replication of the virus in animal- and plant-feeding insects has also been demonstrated (Liu and Zee, 1976; Lastra and Esperza, 1976). Transovarial transmission of VSV in phlebotomine sandflies at rates of 20 to 30% has been shown experimentally. However, this rate of transmission is not adequate to sustain the virus in the insect population under natural conditions (Tesh et al., 1972). Further, no vertebrate species has been found to develop sufficient viremia to infect a biting arthropod. There are clearly gaps in understanding of transmission of VSV, but once introduced into a herd the disease appears to be transmitted by contact through abrasions of the oral mucosa, coronary bands, and teats, or by mechanical means such as milking machines (Hanson, 1952; Mason, 1978).

Experimental attenuated and inactivated VSV vaccines have been developed, but vaccination for the disease is usually not permitted in the U.S. In countries where whole-virus vaccines have been tried, their value in providing long-term immunity has been disappointing. As a result of field trials in Panama, yearly vaccination of cattle with attenuated vaccine was recommended to provide adequate immunity (Lauerman and Hanson, 1963). In another study in Guatemala, cattle vaccinated with a VSV-New Jersey (NJ) strain of low antigenicity were not protected and developed serious disease (Correa, 1964).

There are several reasons why a G glycoprotein subunit vaccine is preferable to attenuated or inactivated whole-virus VSV vaccine. First, antibody to the "G" glycoprotein has been shown to be virus neutralizing (Kelley et al., 1972) and protective in mice (Dietzschold, et al., 1974), unlike antibody to the other known proteins of VSV (L, N, NS, M). Second, dangers associated with vaccine breaks, reversion to the virulent form of the virus, and accidental introduction into potential wild reservoirs or arthropod hosts are avoided by use of a subunit vaccine. Finally, such a vaccine permits utilization of a specific serological test to distinguish animals given this vaccine from animals exposed to whole virus, either naturally or by vaccination. This is an essential consideration for epidemiological studies and for disease control or quarantine programs during outbreaks of the disease.

The development of an effective subunit vaccine that has been shown to protect cattle from clinical disease and mice from lethal infection is described below. The potential of such a vaccine in disease prevention and the development of a rapid diagnostic kit for distinguishing vaccinates from those naturally infected is discussed.

MATERIALS AND METHODS

Cells and Virus

Bovine turbinate and baby hamster kidney (BHK-21) cells were propagated in a modified Dulbecco's minimum essential medium (DMEM) supplemented with 10% fetal calf serum (FCS) and antibiotics (growth medium). Virus was produced in either BHK-21 cells propagated in roller bottles with growth medium or suspension cultures of BHK-21 cells (gift of John Rose of the Salk Institute) propagated in Joklik's minimum essential medium (JMEM) supplemented with 2% FCS and antibiotics.

The VSV-NJ was isolated (WADDL #82-11475E) from cattle during a vesicular stomatitis outbreak in the state of Idaho. Stock virus was cloned by plaque purification, amplified by a passage in bovine turbinate cells, and then grown in the same cells using a multiplicity of infection of 0.1 before propagation in large quantities in BHK-21 cells. Initial passage of VSV in bovine turbinate cells before propagation in BHK-21 cells increased the yield approximately 100-fold, presumably due to minimal generation of defective interfering (DI) particles (Wu, et al., 1984).

Propagation and Purification of VSV

Propagation and purification of VSV and the extraction of G glycoprotein have been described in detail previously (Volk et al., 1982). Briefly, plaque-purified VSV was propagated to high titer in BHK-21 suspension culture or roller bottles at a multiplicity of 0.01 plaque forming units (PFU)/cell in order to obtain a virtually pure preparation of infectious (B) particles and to avoid the generation of DI or (T) particles (Huang and Wagner, 1966). Virus was harvested after a general cytopathic effect was evident, which was usually 18-20 hours postinfection. Virus was titered by PFU assay in BHK-21 cells. For purification, supernatant containing virus was clarified by centrifugation, first at 1,000 x g for 10 minutes and then by a second centrifugation at 10,000 x g for 10 minutes. Virus was then pelleted through a 50% glycerol cushion at 85,000 x g for 90 minutes. The pellet was resuspended in TNE buffer (0.01M Tris, pH 7.4, 0.1M NaCl, 0.001M EDTA), sonicated at 40 watts for 9 seconds to eliminate clumping, layered on a 10-40% (W/W) sucrose gradient, and centrifuged at 50,000 x g for 90 minutes. The light-scattering viral band was collected and resuspended in 5 times the volume of TNE buffer. Virus was pelleted at 150,000 x g for one hour. The virus pellet was resuspended in TNE buffer prior to dialysis overnight against 10 mM Tris, pH 7.5, at 4°C. All virus preparations were stored at -70°C until used.

Isolation and Purification of VSV G Glycoproteins

The G glycoprotein of VSV has been prepared and selectively solubilized with the nonionic detergent, Triton X-100 (Kelley et al., 1972). However, the low critical micellar concentration of the detergent makes its removal from the isolated G glycoprotein very difficult, and residual detergent interferes with consequent reconstitution of the glycoprotein. Therefore, ß-D-octylglucoside, a nonionic detergent with high critical micellar concentration, was used. In other studies, ß-D-octylglucoside has been used successfully for the isolation of the G glycoprotein of VSV with 97% purity (Volk et al., 1982; Petri and Wagner, 1979) and to solubilize other membrane proteins (Stubbs and Litman, 1978).

The G glycoprotein was extracted with 0.06M ß-D-octylglucoside from purified virions at a protein concentration of 1 mg/ml in 10mM Tris (pH 7.4) by incubating for one hour at room temperature. Other components of the virus were removed by centrifugation at 150,000 x g, and the pellet was stored at -70°C for later use. The supernatant fluid containing the G glycoprotein was dialyzed for 24 hours at 4°C against 4 liters of 10mM Tris and dried by lyophilization. The relative purity of the isolated G glycoprotein preparation was examined by polyacrylamide gel electrophoresis (PAGE) followed by staining with Coomassie brilliant blue (Laemmli, 1970). The G glycoprotein was compared with total viral proteins, as well as with total viral proteins from which G glycoprotein had been removed (Figure 1). The concentration of the G glycoprotein preparation was determined by BioRad™ protein assay (BioRad, Richmond, CA).

Serum Neutralization Test

Humoral responses of cattle and mice to VSV were determined by serum neutralization tests. Briefly, doubling dilutions of serum in 96-well microtiter plates were incubated with an equal volume of 100 times the 50% tissue culture infective doses ($TCID_{50}$) of virus at 37°C for one hour. BHK-21 cells were added and the plates incubated for 48 hours in a humidified CO_2 incubator. Serum neutralization titers were expressed as the reciprocal of the dilution of sample that gave complete protection against the cytopathic effect of 100 $TCID_{50}$ of virus.

Immunization of Cattle

All cows used in this study were demonstrated to be seronegative to VSV-NJ by a serum neutralization test. Seven cows were vaccinated intramuscularly with 112 µg of purified G glycoprotein of VSV-NJ in complete Freund's adjuvant. Three cows in group 1 (13Y-15Y), were vaccinated with a G glycoprotein preparation simply admixed with complete Freund's adjuvant, while the mixture used to vaccinate the four cows in group 2 (19Y-22Y) was completely emulsified. Four unvaccinated cows served as controls. After vaccination, blood was regularly collected from all cows for serological testing as indicated in Tables 1 and 2.

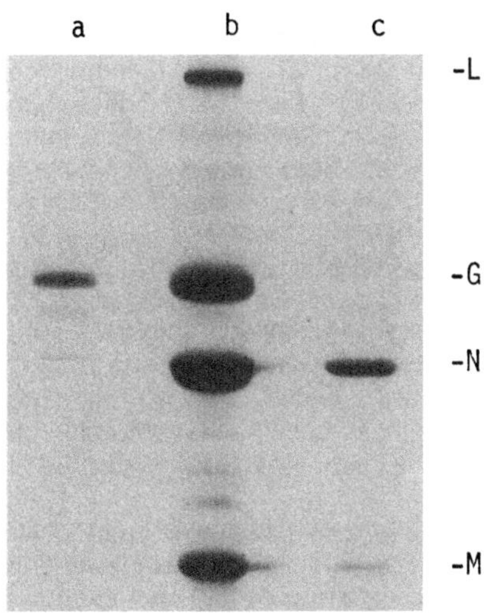

Fig. 1. A Coomasie brilliant blue-stained SDS-polyacrylamide gel electrophoresis of proteins of VSV before and after treatment of purified virus with ß-D-octylglucoside. Octylglucoside-extracted G glycoprotein of VSV (Lane a). Purified virus before treatment with octylglucoside (Lane b). Octylglucoside-treated virus after removal of G glycoprotein (Lane c).

Immunization of Mice

One group of Balb/c mice was vaccinated with a single, subcutaneous injection of 10 μg of purified G glycoprotein alone and a second group with 10 μg of G glycoprotein emulsified in a mixture of equal volumes of complete and incomplete Freund's adjuvant. Half of the mice in each group received a booster vaccination with an identical and equal amount of antigen preparation 28 days after primary vaccination. Serum samples were collected at 6, 14, 28, 42 and 84 days postvaccination. Mice were challenged intravenously with 10^8 PFU of VSV in 0.05 ml of DMEM 44 days after primary vaccination.

Virus Challenge of Cattle

To reproduce the clinical disease of vesicular stomatitis in cattle, the virus must be administered on the tongue by intradermal inoculation or scarification (Hanson, 1952). A prelminary study using three cows showed that the minimum dose required to produce disease by this method was 10^2 PFU virus in 0.1 ml of DMEM. (In another cow, administration of 10^8 PFU virus intravenously failed to produce the disease.) Since VSV has been shown to undergo attenuation in tissue culture (Palma and Huang, 1974), vesicular fluid obtained from the cows in the preliminary study was used in test challenges to ensure full virulence of the virus, The fluid contained 10^{11} PFU/ml of VSV. It was diluted ten-fold in DMEM and frozen in 0.1-ml aliquots at -70°C until used. For the challenge experiments, group 1 cattle were inoculated in each of the four quadrants of the tongue with 10^4 PFU of VSV-NJ in 0.1 ml of DMEM (Table 1). Group 2 cattle were inoculated with 10^2 PFU (two upper quadrants) and 10^3 PFU (two lower quadrants) per inoculation site. Cattle were examined daily for several days after challenge for the appearance of vesicles and other clinical signs of vesicular stomatitis.

Virus challenge of cattle and mice was conducted in P3 quarantine facilities under license from the USDA-APHIS and the Washington State Veterinary Service. All cattle remained in quarantine until 30 days after the last oral lesion (if any) had healed.

Virus Challenge of Mice

Mice have been shown to succumb to lethal encephalitis with hind leg paralysis after exposure to virus by intracerebral, intranasal, and intravenous routes (Olitsky et al., 1936; Wagner, 1974; Huang, 1977). In a preliminary experiment, administration of 10^8 PFU of VSV-NJ in the tail vein gave greater than 50% mortality between 6 and 12 days postexposure. In this study, both control and vaccinated mice were challenged with 10^8 PFU of VSV-NJ in 0.1 ml of DMEM by injection in the tail vein. Mice were checked daily for twelve days after challenge for death or signs of encephalitis (characterized by convulsions and flaccid paralysis of the hind legs). All mice that died less than six days after challenge were discarded from the group.

Table 1. Serum Neutralization Titers of Vaccinated Cattle before and after Challenge with Vesicular Stomatitis Virus-New Jersey serotype (VSV-NJ).

		Serum neutralization titers[a] after vaccination[b]							Serum neutralization titers after challenge[c]			
Cow number	Vaccine	Day 0	Day 4	Day 7	Day 14	Day 21	Day 28	Day 34	Day 41	Day 48	Day 80	Lingual vesicular lesions
5Y	Control	0	0	0	0	0	not done	0	960	1,920	640	+
6Y	Control	0	0	0	0	0	not done	0	240	960	1,280	+
13Y	G	0	0	20	240	240	320	480	20,480	51,200	3,840	-[d]
14Y	G	0	0	10	320	320	480	960	640	1,920	640	-
15Y	G	0	0	5	320	480	640	960	1,280	1,920	640	-

[a]Serum neutralization titers are expressed as the reciprocal of the dilution of sample that gave complete protection against the cytopathic effect of 100 $TCID_{50}$ of VSV-NJ.
[b]Cows were vaccinated intramuscularly with 112 µg G glycoprotein VSV-NJ in complete Freund's adjuvant.
[c]Cows were challenged on Day 34 by intradermal inoculation of 10^4 PFU of VSV-NJ per site on the two upper and two lower quadrants of the tongue.
[d]Cow 13Y had a small vesicle on the lower right quadrant of the tongue but was completely protected on all of the other 3 sites of injection.

RESULTS

G Glycoprotein Preparation

As shown in Figure 1, the G glycoprotein preparation contained a single, major band with an apparent molecular weight of 66,000, in close agreement with the molecular weight range of 65,000-67,000 reported previously (Knipe et al, 1975; Little and Huang, 1978).

Immune Response and Results of Viral Challenge in Cattle

Group 1. The immune response of cattle vaccinated with G glycoprotein was assessed by analyzing the titer of serum neutralizing antibodies at regular intervals postvaccination and after challenge with infectious virus. The serum neutralization titers of group 1 cattle and controls are summarized in Table 1 for days 0, 4, 7, and at weekly intervals thereafter. No detectable serum neutralizing antibody was found on days 0 and 4. There were, however, low level serum neutralizing titers of 20, 10, and 5 for G-vaccinated cows 13Y, 14Y, and 15Y respectively, with an average titer of 12, by seven days postvaccination. A gradual rise in serum neutralization titers was observed until the day of challenge (day 34). Serum neutralization titers for group 1 cows 13Y, 14Y, and 15Y were 480, 960, and 960 respectively, with an average titer of 800, on the day of challenge. As shown in Table 1, none of the control animals had detectable levels of serum neutralizing titers against VSV before challenge.

Within 48 hours after intralingual challenge with infectious virus, both control cows had vesicular lesions at all four sites of virus inoculation. Of those vaccinated with G glycoprotein, two cows (14Y and 15Y) were completely protected at all four inoculation sites. Cow 13Y had a small vesicle at one injection site (10^4 PFU, right lower quadrant) but no vesicular lesions were present at the other three sites.

Group 2. As shown in Table 2, serum neutralization titers were much higher for cows in group 2 than for those in group 1 at seven days postvaccination, except for cow 19Y, which had a serum neutralization titer of 15. Cows 20Y, 21Y and 22Y had serum neutralization titers of 120, 80 and 320 respectively at this time. The mean titer for all four test cows of group 2 was 134, which can be compared to a mean titer of 12 for those in group 1. At day 35 postvaccination, group 2 cows had a mean serum neutralization titer of 3280 compared to 800 for group 1: the increased mean titer was the result of very much higher titers in cows 21Y and 22Y. The greater response of cows in group 2 was presumably due to emulsification of the antigen-adjuvant mixture used for vaccination. On the day of challenge (day 44), the average serum neutralization titer for the four G-vaccinated animals was 5280 and the range was from 1280 to 15,360.

Cows in group 2 were challenged by intradermal inoculation of 10^2 or 10^3 PFU of VSV on the two upper and two lower quadrants of the tongue, respectively. Both control cows, 23Y and 26Y, developed vesicular lesions at all inoculation sites by 24 hours after challenge. None of the four cows (19Y, 20Y, 21Y, and 22Y) vaccinated with G glycoprotein developed any vesicles at any of the four inoculation sites. Cow 21Y developed a small erythematous spot, which failed to develop further, at one of the 10^3 PFU injection sites.

Table 2. Serum neutralization Titers of Vaccinated Cattle before and after Challenge with Vesicular Stomatitis Virus-New Jersey Serotype (VSV-NJ)

		Serum neutralization titers[a] after vaccination[b]							Serum neutralization titers after challenge[c]			Lingual vesicular lesions	
Cow number	Vaccine	Day 0	Day 7	Day 14	Day 21	Day 28	Day 35	Day 44	Day 51	Day 58	Day 85	10^2 PFU	10^3 PFU
23Y	control	0	0	0	0	0	0	0	960	1920	600	+[d]	+
26Y	control	0	0	0	0	5	5	30	640	960	1600	+	+
19Y	G	0	15	640	640	480	640	1280	640	1280	600	-[d]	-
20Y	G	0	120	1280	2560	2560	960	1560	3840	1280	400	-	-
21Y	G	0	80	640	1920	5120	7680	1920	6400	15360	3200	-	-[e]
22Y	G	0	320	1280	3840	5120	3840	15360	3840	15360	2400	-	-

[a]Serum neutralization titers are expressed as the reciprocal of the dilution of serum that gave complete protection against cytopathic effects of 100 $TCID_{50}$ of VSV-NJ.

[b]Cows were vaccinated intramuscularly with a single dose of 112 μg G glycoprotein of VSV-NJ in complete Freund's adjuvant.

[c]Cattle were challenged 44 days after the primary vaccination by intradermal inoculation of 10^2 or 10^3 PFU of VSV-NJ on the two upper and two lower quadrants of the tongue, respectively.

[d]+ = Presence of vesicular lesion. - = Absence of vesicular lesion

[e]Cow 21Y developed a small erythema at one of the inoculation sites (10^3) but the lesion failed to develop further.

Serum neutralization titers in control and test animals were analyzed after challenge with infectious VSV. In group 1, control cows (5Y and 6Y) had serum neutralization titers of 960 and 240 one week after challenge, and 640 and 1280 about seven weeks after challenge. In contrast, G-vaccinated cows (13Y, 14Y, and 15Y) had serum neutralization titers of 20480, 640, and 1280 one week after challenge, and 3840, 640, and 640 seven weeks after challenge. The greater anamnestic response observed in cow 13Y was probably due to partial replication of the virus in the injection site at which the small vesicle was observed. The serum neutralization titers of G-vaccinated cows in group 2, except for cow 21Y, did not rise significantly from the titer at challenge on day 44 (Table 2). This is presumably due to a superior level of immunity and lack of replication of the infectious virus used for challenge. Control cows 23Y and 26Y developed peak titers of 1920 and 1600 respectively. Cow 21Y, which developed an area of erythema at one inoculation site, demonstrated an increase in titer from 1920 to 15,350 between days 44 and 58. Although there was no clinical disease or gross vesicular lesion in this animal, it appears likely that there was limited viral replication in the one injection site.

Immune Responses and Results of Viral Challenge in Mice

The immune response of mice to the G glycoprotein of VSV and the results of protection studies after challenge with infectious virus are summarized in Table 3. Serum neutralization titers of the pooled samples of 27 mice given G glycoprotein without adjuvant were 10, 80, 590, and 1040 for days 6, 14, 28 and 42 respectively. Fifteen mice from this group received a booster injection on day 28. By day 42, their pooled serum neutralization titer had risen to 4750, which was about 4 times that of mice given a single vaccination.

The serum neutralization titers of the pooled sera of 28 mice vaccinated with 10 μg of G glycoprotein in Freund's adjuvant were 20, 1400, 5670, and 26770 for days 6, 14, 28, and 42 respectively. Thirteen mice from this treatment group received a booster injection of 10 μg G glycoprotein in Freund's adjuvant on day 28. Pooled sera from these mice had a neutralization titer of 56,220 on day 42, a two-fold increase over those receiving a single vaccination.

With one exception, all mice given G glycoprotein, as a single or double dose and with or without Freund's adjuvant, were completely protected when challenged on day 44 with 10^8 PFU of VSV intravenously. (One mouse of 12 that received a single vaccination of 10 μg of G glycoprotein without adjuvant died upon challenge.) In the unvaccinated control group, 12 out of 21 mice developed severe flaccid hind limb paralysis and died from encephalitis.

DISCUSSION

Although VSV infection has a fundamentally different pathogenesis in cattle and mice, the results of this study show that both species can be protected against VSV by vaccination with a G glycoprotein subunit preparation of the virus. In mice, VSV infection is essentially a central nervous system disease, and the vaccine must protect against encephalitis rather than vesiculation on squamous epithelial surfaces. A further significant difference in the challenge system is that the virus is introduced directly into the bloodstream in mice while it is injected intradermally in the tongue in cattle. Therefore, it is anticipated that the results of the protection studies in cattle and mice would differ. In mice, introduction of challenge virus into the bloodstream results in the immediate interaction of the virus with specific antibody and T-cells before it reaches the target organ, the brain. Furthermore, since the incubation period in mice is 6-12 days, the virus has ample time to induce a good anamnestic response. In cattle, the relatively avascular target organ, the epidermis, presumably does not allow much interaction with specific antibody and T-cells. In addition, the short incubation period (<48 hours) does not permit development of a good anamnestic response before vesiculation. Consequently, the minimum protective dose required for mice should be lower than that for cattle.

Table 3. Average Serum Neutralization Titers of Vaccinated Mice before and after Challenge with Vesicular Stomatitis Virus-New Jersey Serotype (VSV-NJ)

Mouse group	Serum neutralization titer (day) 6	14	28	42	84	Number of vaccinated mice challenged with VSV-NJ	Number of mice dead from VSV infection
Control	0	0	0	0	7760	21	12
G only(1X)	10	80	590	1040	8000	12	1
G only(2X)				4750	1030	15	0
G+CF(1X)	20	1400	5670	26770	13310	15	0
G+CF(2X)				56220	14290	13	0

Mice were vaccinated with 10 μg purified G glycoprotein VSV-NJ alone or in Freund's adjuvant (CF). Booster vaccinations (2X) were given 28 days after the primary vaccination (1X). Serum neutralization titers are expressed as the reciprocal of the dilution of serum that gave complete protection against the cytopathic effects of 100 $TCID_{50}$ of VSV-NJ. Control and vaccinated mice were challenged 44 days after the primary vaccination with 10^8 PFU VSV-NJ by intravenous administration in the tail vein.

In mice, the test system was verified in that 12 of 21 unvaccinated controls died of the paralytic VSV syndrome described previously by others (Olitsky et al., 1936; Wagner, 1974; Huang, 1977) after intravenous challenge with virulent virus. In contrast, with the exception of one mouse, vaccinated mice were protected from the disease. Protection against encephalitis did not appear to differ with the vaccine used: G glycoprotein given alone was as effective as G glycoprotein with complete Freund's adjuvant and one dose of either vaccine protected just as well as two doses. Equivalent protection of mice vaccinated with G glycoprotein alone or mixed with complete Freund's adjuvant was an unexpected finding. Further experiments are in progress to examine the duration of protection and the minimum dosage of G glycoprotein necessary to induce an adequate immune response under both experimental vaccination schedules. The 10 μg G glycoprotein arbitrarily chosen as a vaccine dose for mice may well turn out to be very much greater than the essential immunogenic dose. It will be particularly interesting to determine, by reducing the dose, the levels at which G glycoprotein alone is unable to protect when given in a single or double treatment schedule.

It has been known for some time that only the G glycoprotein of VSV is capable of inducing neutralizing antibody (Kelley, et al., 1972) and protective response in mice (Dietzschold, et al., 1974). In this study we were not able to correlate the level of serum neutralizing antibody to protection because mice in all vaccine treatment groups were essentially fully protected. However, there were obvious differences between vaccines in serological response. G glycoprotein given alone, as a single or double dose, induced a lower level of antibody than G glycoprotein administered as a single or double treatment in complete Freund's adjuvant. A single vaccination with either vaccine preparation resulted in lower levels of neutralizing antibody than two doses.

Cattle were protected after a single dose of 112 μg G glycoprotein in complete Freund's adjuvant against 400 times the amount of VSV shown to cause disease. That the demonstration of protection was so clear-cut was remarkable considering the test system used. Challenge virus was delivered in high titer into the relatively avascular dermal layer of the tongue. Based on previous report (Henderson and Galloway, 1953) and personal observations of other vesicular diseases, including foot-and-mouth disease, we expected to find a small localized lesion at the inoculation site. Hence, levels of protection would have to be assessed by the vaccine's ability to prevent secondary lesions (in the feet, for example) or, less precisely, by comparison of the size and nature of oral lesions. In foot-and-mouth disease vaccine testing, challenge virus delivered by any route, including intralingual, intranasal, intravenous, or aerosol, results in both oral and secondary lesions (Sellers and Parker, 1969; Sutmoller and McVicar, 1976; Yilma, 1980). However, we were unable to produce any lesions by intravenous administration of high-titered VSV (10^8 PFU), or any secondary lesions by intradermal inoculation of the tongue. Fortunately, vaccinated cows, except for cow 13Y, were protected against oral lesions, even at the site of challenge inoculation. This made a subjective evaluation unnecessary, since the response was so obviously different from that of control cows, in which large vesicles developed and progressively denuded the entire epithelium of the tongue. Cow 13Y had a small lesion in one of the four challenge inoculation sites. It is reasonable to consider this cow to be protected against VSV in any field situation if one takes into account the nature of the experimental challenge system and the enormous dose of virulent virus administered. In fact, in challenge studies done on vesicular diseases such as foot-and-mouth disease, protected cattle typically exhibited small lesions at the inoculation site (Henderson and Galloway, 1953). The significant observation in this study is the total absence of such lesions among the other six vaccinates, showing an extremely high level of protection.

Protection in cattle appeared to be related to the level of serum neutralizing antibody. The one vaccinated cow that developed a lesion (13Y) had the lowest neutralizing antibody titer (480) on the day of challenge. This cow also had over a hundred-fold rise in neutralizing antibody titer after challenge, presumably due to increased antigenic mass as the result of viral replication at one of the four sites of inoculation. The corollary of this is that five of the other six vaccinates (excluding 21Y) had no appreciable change in titer after challenge. Their level of protection must have been such that viral replication was prevented, as might be deduced also from the absence of gross vesicular lesions.

We have not tested a lower dose of G glycoprotein in complete Freund's adjuvant than that reported here. Complete Freund's adjuvant would not be acceptable for a commercial field vaccine in cattle because it would interfere with testing for tuberculosis. However, since it was not known what the immune response of cattle to the G glycoprotein would be, the best adjuvant available was used for this initial study. We are currently evaluating in cattle the efficacy of various dosages of G glycoprotein alone and in combination with different adjuvants used in commercial bovine vaccines.

Development of an effective G glycoprotein subunit vaccine for cattle has important implications for control of the disease in endemic and epidemic situations. Two major objections, lack of safety and diagnostic capacity, to the use of conventional whole-virus VSV vaccines can be overcome by the use of a subunit vaccine. A G glycoprotein vaccine has none of the hazards associated with current attenuated or inactivated whole-virus vaccines. These include: reversion of attenuated strains to virulence in natural hosts, vectors or reservoirs in the field; accidental introduction of virus into wild reservoirs and/or arthropod vectors; and incomplete inactivation of whole-virus vaccine. Animals given G glycoprotein vaccine express antibody only to this single protein, whereas animals that have recovered from a natural infection or that have been vaccinated with whole-virus preparations have antibody to other viral proteins. This has allowed the development of a "rapid diagnostic kit" that uses monoclonal antibodies to differentiate these serological responses and to identify G glycoprotein-vaccinated cattle (manuscript in preparation). Unlike whole-virus vaccines, therefore, the G glycoprotein vaccine permits the use of serological tests to distinguish vaccinates from whole-virus exposed animals. This is essential in epidemiological studies and in regulation of import and export of livestock.

ACKNOWLEDGMENTS

We express our thanks to the staff of USDA-APHIS, Olympia Washington and to the Washington State Veterinarian for speedy assistance in certification and licensing. We are grateful to Dr. John Rose, Salk Institute, for the gift of BHK-21 cells; Dr. Jay Weidner, for assistance in various aspects of laboratory and challenge studies; and Dr. Sally Owens and Kjerstie Guenther for critical review of the manuscript.

This work was supported by USDA-SEA contract #83-CRSR-2-2189 and funds obtained under PL-1433 for Animal Health and Disease Research.

REFERENCES

Bishop, D. H. L., 1980, "Rhabdoviruses, Volume II," David H. L. Bishop, ed., CRC Press, Inc., Boca Raton, Florida.

Correa, W. H., 1964, Prophylaxis of vesicular stomatitis: A field trial in Guatemalan dairy cattle, Am. J. Vet. Res., 1300-1302.

Dietzschold, B., Schneider, L. G., and Cox, J. H., 1974, Serological characterizations of the three major proteins of vesicular stomatitis virus, J. Virol., 14:1-7.

Ferris, D., Hanson, R. P., Dicke, R. J., and Robers, R. H., 1955, Experimental transmission of vesicular stomatitis virus by diptera, J. Infect. Dis., 96:184-192.

Hanson, R. P., 1952, The natural history of vesicular stomatitis, Bacteriol. Rev., 16:179-204.

Henderson, W. M., and Galloway, I. A., 1953, The use of culture virus in the preparation of foot-and-mouth disease vaccine, J. Hyg. (Camb.), 51:546-558.

Huang, A. S., 1977, Viral pathogenesis and molecular biology, Bacteriol. Rev., 41:811-821.

Huang, A. S., and Wagner, R.R., 1966, Defective T particles of vesicular stomatitis virus. II. Biologic role in homologous interference, Virology, 30:173-181.

Jonkers, A. H., Shope, R. E., Aitken, T. H. G., and Spence, L., 1964, Cocal virus, a new agent in Trinidad related to vesicular stomatitis virus, Am. J. Vet. Res., 25:236-241.

Jonkers, A. H., Spence, L., and Aitken, T. H. G., 1965, Cocal virus epizootiology in Bush Bush Forest and the Nariua Swamp, Trinidad, W.I.; further studies, Am. J. Vet. Res., 26:758-763.

Kelley, J. M., Emerson, S. U., and Wagner, R. P., 1972, The glycoprotein of vesicular stomatitis virus is the antigen that gives rise to and reacts with neutralizing antibody, J. Virol., 10:1231-1235.

King, L. J., 1983, "Summary of vesicular stomatitis meeting. Jan. 11-12, 1983, Hyattsville, Maryland," Emergency Programs, APHIS, USDA, NVSL, Ames, Iowa.

Knipe, D., Rose, J. K., and Lodish, H. F., 1975, Translation of individual species of vesicular stomatitis virus mRNAs, J. Virol., 15:1004-1011.

Laemmli, V.K., 1970, Cleavage of structural proteins during the assembly of the head of the bacteriophage T4, Nature 227:680-685.

Lastra, J. R., and Esperza, J., 1976, Multiplication of VS virus in the leafhopper Peregrinus maidis (Ashm), a vector of a plant rhabdovirus, J. Gen. Virol., 32(1):139-142.

Lauerman, L. H., and Hanson, R. P., 1963, Field trial of live virus vaccination procedure for prevention of vesicular stomatitis in dairy cattle: Evaluation of emergency vaccination in Georgia, Proc. U.S. Livestock Sanitary Assoc., 67:483-490.

LeFrancois, L., and Lyles, D. S., 1983, Antigenic determinants of vesicular stomatitis virus: Analysis with antigenic variants, J. Immunol., 130:394-398.

LeFrancois, L., and Lyles, D. S., 1983, Cytotoxic T lymphocytes reactive with vesicular stomatitis virus: Analysis of specificity with monoclonal antibodies directed to the viral glycoprotein, J. Immunol., 130:1408-1412.

Little, S. P., and Huang, A. S., 1978, Shedding of the glycoprotein from vesicular stomatitis virus infected cells, J. Virol., 27:330-339.

Liu, I. K. M., and Zee, Y. C., 1976, The pathogenesis of vesicular stomatitis virus, serotype Indiana, in Aedes aegypti mosquitoes, Am. J. Trop. Med., 25:177-185.
Mason, J. H., 1978, The epidemiology of vesicular stomatitis, Boll. Centro Panmericano Fiebre Aftosa, 29(30):35-53.
Olitsky, P. K., Sabin, A. B., and Cox, H. R., 1936, An acquired resistance of growing animals to certain neurotropic viruses in the absence of humoral antibodies or previous exposure to infection, J. Exp. Med., 64:723-737.
Palma, E. L., and Huang, A. S., 1974, Cyclic production of vesicular stomatitis virus caused by defective interfering particles, J. Infect. Dis., 129:402-410.
Patterson, W. C., Mott, L. O., and Jenney, E. W., 1958, A study of vesicular stomatitis in man, J. Am. Vet. Med. Assoc., 133:57-62.
Petri, W. A., and Wagner, R. R, 1979, Reconstitution into liposomes of the glycoprotein of vesicular stomatitis virus by detergent dialysis, J. Biol. Chem., 254:4313-4316.
Sellers, R. F., and Parker, J., 1969, Airborne excretion of foot-and-mouth disease virus, J. Hyg. (Camb.), 67:671-677.
Shelokov, A., and Peralta, P. H., 1967, Vesicular stomatitis virus, Indiana type: An arbovirus infection of tropical sandflies and humans, Am. J. Epidemiol., 86:149-157.
Sutmoller, P., and McVicar, J. W., 1976, Pathogenesis of foot-and-mouth disease: The lung as an additional portal of entry of the virus, J. Hyg. (Camb.), 77:235-243.
Stubbs, G. W., and Litman, B. J., 1978, Effect of alterations in the amphipathic microenvironment on the conformational stability of bovine opsin. 1. Mechanism of solubilization of disk membranes by the nonionic detergent, octyl glucoside, Biochemistry 17:215-219.
Sudia, W. D., Fields, B. N., and Calister, C. M., 1967, The isolation of vesicular stomatitis virus (Indiana strain) and other viruses from mosquitoes in New Mexico, Am. J. Epidemiol., 86:598-602.
Tesh, R. B., Chaniotis, B. N., and Johnson, K. M., 1972, Vesicular stomatitis virus (Indiana serotype): Transovarial transmission by phlebotomine sandflies, Science, 175:1477-1479.
Volk, W. A., Snyder, R. M., Benjamin, D. C., and Wagner, R. R., 1982, Monoclonal antibodies to the glycoprotein of vesicular stomatitis virus: Comparative neutralizing activity, J. Virol., 42:220-227.
Wagner, R. R., 1974, Pathogenicity and immunogenicity for mice of temperature-sensitive mutants of vesicular stomatitis virus, Infect. Immun., 10:309-315.
Wu, T.-Y., Yilma, T., and Huang, A. S., 1984, Correlation between virulence of vesicular stomatitis virus isolates and continued high titers during undiluted passage (submitted for publication).
Yilma, T., 1980, Morphogenesis of vesiculation of foot-and-mouth disease, Am. J. Vet. Res., 41:1537-1542.
Zinkernagel, R.M., Adler, B., and Holland, J. J., 1978, Cell-mediated immunity to vesicular stomatitis virus infection in mice, Exp. Cell. Biol., 46:53-70.

SURFACE ANTIGENS OF *BORDETELLA PERTUSSIS*

Charlotte D. Parker, Sandra K. Armstrong and
Dara W. Frank

Department of Microbiology; School of Medicine
University of Missouri
Columbia, Missouri 65212

ABSTRACT

Bordetella pertussis and other *Bordetella* species cause respiratory infections in humans and in a variety of animals. Clinical isolates of *B. pertussis* have multiple virulence factors, several of which have been reported to induce protective immunity. Using cell surface iodination techniques and monoclonal antibody immunoblots we have identified several proteins which are exposed on the surface of *B. pertussis* cells, including the filamentous hemagglutinin and outer membrane proteins 91, 18, and 15. Protein 91 is unique to virulent *B. pertussis* strains. Antibodies to protein 18 are found in convalescent serum of both humans and mice recovering from infection with *B. pertussis*.

INTRODUCTION

Bordetellae are small gram negative bacteria which cause respiratory infections in mammals. *B. pertussis* and *B. parapertussis* cause whooping cough in humans, while *B. bronchiseptica* causes animal diseases including kennel cough in dogs and atrophic rhinitis in swine (Pittman, 1970; Goodnow, 1980). These organisms produce a variety of substances which are toxic for animals or which may be implicated in virulence, including pertussis toxin (also called pertussigen, histamine sensitizing factor, islet activating protein and leucocytosis promoting factor); a heat-labile dermonecrotic toxin; endotoxin (see Wardlow and Parton, 1983, for a review of these three pertussis toxins); an extracytoplasmic adenylate cyclase (Hewlett and Wolff, 1976), and a filamentous hemagglutinin (Arai and Sato, 1976). Heat-labile surface antigens called agglutinogens are also produced (Pittman, 1970), and appear to consist of fimbriae-like filaments distinct from the filamentous hemagglutinin (Ashworth, et al., 1983). Only *B. pertussis* produces pertussis toxin, while the other substances are made by all three species.

Immunity to whooping cough results from natural infection or from immunization with suitably prepared whole cell vaccines. The exact nature of the protective immunogen is not clear, although both pertussis toxin and the filamentous hemagglutinin have been reported to induce passive immunity (Sato, et al., 1981).

Bordetellae undergo a phenotypic change in properties which has been called modulation (Lacey, 1960), and which is induced by growth at low temperature, growth at high magnesium concentration, or growth in the presence of certain pyridines (Pusztai and Joo, 1967; Schneider and Parker, 1982). Modulated cells vary from normal cells in several ways. Modulated cells of B. pertussis do not produce detectable levels of pertussis toxin, dermonecrotic toxin, adenylate cyclase, filamentous hemagglutinin or agglutinogens. Modulated cells also show a different pattern of membrane proteins (Parton and Wardlow, 1975). Mutants which resemble modulated strains and which are termed "degraded" or "phase IV" strains occur, and appear to be selected by in vitro growth (Standfast, 1951; Wardlow and Parton, 1975; Field and Parker, 1979). However, unlike the modulated phenotype, the mutant phenotype is not reversed by culture under normal conditions.

We have investigated the surface antigens of B. pertussis, reasoning that many of the proteins exposed on the bacterial cell are likely to be important in both virulence and immunity to whooping cough. We have used a variety of approaches, including isolation of outer membranes, surface labeling of cells with radioactive iodine, and immunologic analyses using monoclonal antibodies and immune sera. This paper summarizes our current knowledge of the surface antigens of B. pertussis.

EXPERIMENTAL

The general methods used were as follows: isolation and characterization of outer membrane fractions was as described by Schneider and Parker, 1982, using differential solubility in Triton X-100 to separate inner and outer membrane fractions, and sodium dodecyl sulfate polyacrylamide gel electrophoresis to characterize membrane components; surface radioiodination was by the lactoperoxidase method, or used a solid phase catalyst (Iodogen, a non-aqueous, solid phase catalyst used as described by Markwell and Fox, 1978; or Iodobeads, chloramine-T derivatized polystyrene beads, were both from Pierce Chemical Company); isolation of monoclonal antibodies was as described by Frank and Parker, 1984a, and immunoblotting followed the protocol in Frank and Parker, 1984b.

RESULTS

Isolation of B. pertussis Outer Membrane Fraction and its Characterization

We used sonication or a French pressure cell to break B. pertussis cells grown in a defined liquid medium which supported production of the toxins, the adenylate cyclase, the filamentous hemagglutinin, and the agglutinogens. The total envelope fraction was isolated by differential centrifugation, and treated with Triton X-100. The Triton-insoluble portion was recovered by centrifugation and subjected to sodium dodecyl sulfate polyacrylamide gradient gel electrophoresis in the presence of urea and mercaptoethanol to allow us to visualize the characteristic peptides. Modulated cells and mutant cells were grown similarly, except that for modulated cells, 4 mM nicotinic acid was added to the defined medium.

Figure 1 shows a typical Triton-insoluble envelope preparation stained with Coomassie blue. Normal cells (lane 3) show multiple peptide bands, ranging in apparent molecular weight from about 140,000 to less than 14,000. The starred doublet is characteristic of normal cells, and is missing from modulated (lane 1) or mutated strains. Gels stained with a silver stain show similar profiles, except that lipopolysaccharide is visualized as a single diffuse band at the bottom of the gel.

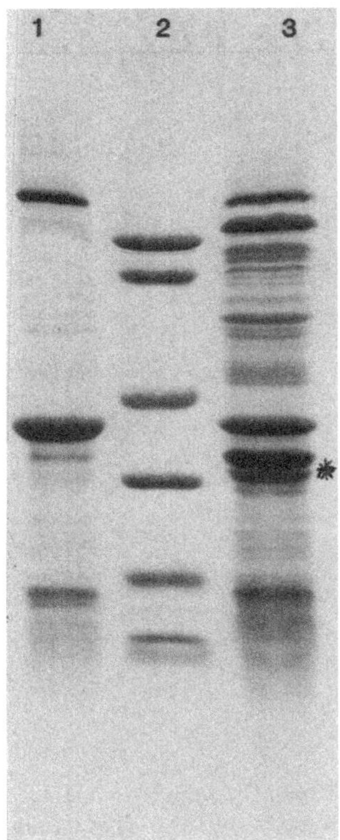

Fig. 1. Triton-insoluble envelope fraction of B. pertussis. Triton-insoluble proteins of the total envelope fraction of B. pertussis strain UT25 were analyzed by sodium dodecyl sulfate polyacrylamide gel electrophoresis in a 7.5-20% gradient gel containing 0.5M urea, and stained for protein with Coomassie blue R-250. Lane 1) modulated cells; Lane 2) molecular weight markers (phosphorylase B, 92,500; bovine serum albumin, 66,200; ovalbumin, 45,500; carbonic anhydrase, 31,000; soybean trypsin inhibitor, 21,500; and lysozyme, 14,400); Lane 3) normal cells.

Radioiodination of B. pertussis Whole Cells, Lysed Cells, and Triton-Insoluble Envelope Fraction

We analyzed intact cells, lysed cells, and the Triton-insoluble envelope fraction by autoradiography after labeling proteins with radioactive iodine. The results of such an experiment are shown in Figure 2. In this experiment, both lactoperoxidase and Iodogen methods were used to label peptides. The number of peptides radiolabeled with lysed cells (lanes 1 and 2) was much larger than the number labeled with whole cells (lanes 3 and 4), suggesting that intact cells were surface labeled. Peptides and proteins revealed by radioiodination of intact cells or Triton-insoluble envelope fraction were named for their calculated apparent molecular weights.

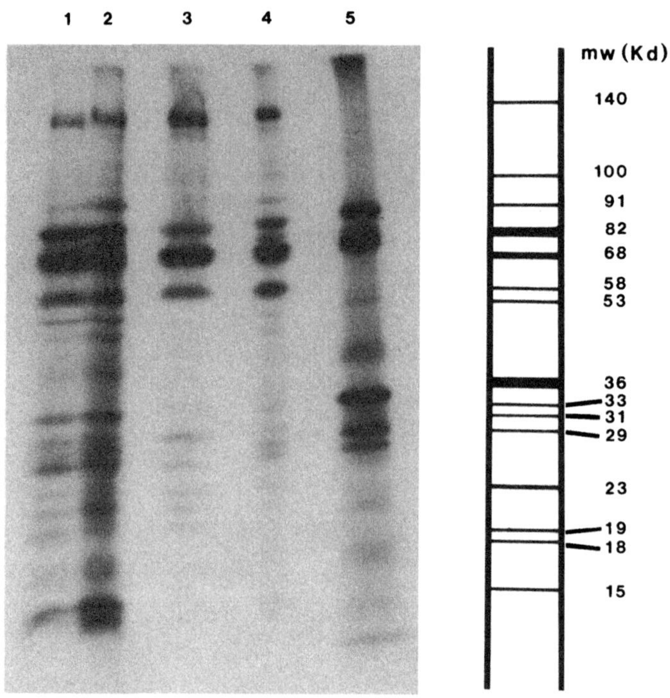

Fig. 2. Autoradiogram of radioiodinated cells and cell fractions of B. pertussis strain UT25. Cell fractions were radiolabeled, solubilized, electrophoresed into a sodium dodecyl sulfate polyacrylamide gradient gel, and the gel was placed on X-ray film to produce an image. Lane 1) cells lysed prior to labeling and iodinated using lactoperoxidase; Lane 2) cells lysed prior to labeling and iodinated using Iodogen; Lane 3) intact cells iodinated using lactoperoxidase; Lane 4) intact cells iodinated using Iodogen; Lane 5) Triton-insoluble envelope fraction iodinated using Iodogen; diagram, right) schematic summary of radiolabeled peptide pattern from intact cells, showing calculated apparent molecular weights in kilodaltons.

The peptide pattern revealed when whole cells were labeled was similar to the peptide pattern of Triton-insoluble envelope fraction (lane 5), except that a high molecular weight component was missing from the Triton-insoluble envelope fraction, and protein 36 was much reduced in the surface labeled cells in this experiment. In additional surface labeling experiments, protein 36 usually appeared (see lane A of figure 3, for example).

Both modulated cells and degraded mutants showed patterns of surface labeled proteins which were quite different from those of normal cells, and which appeared identical to each other (Figure 3, lanes B and C). There were changes in the peptide pattern in the area of the modulation doublet (*), but the most striking difference from normal cells was the reduction in the number of labelled components (especially higher molecular weight components) in modulated and mutant cells. After performing several experiments of this type, we constructed an idealized diagram of the polypeptides observed for surface labeled normal cells, surface labeled modulated cells or mutant cells, and Triton-insoluble envelope fraction (Figure 4).

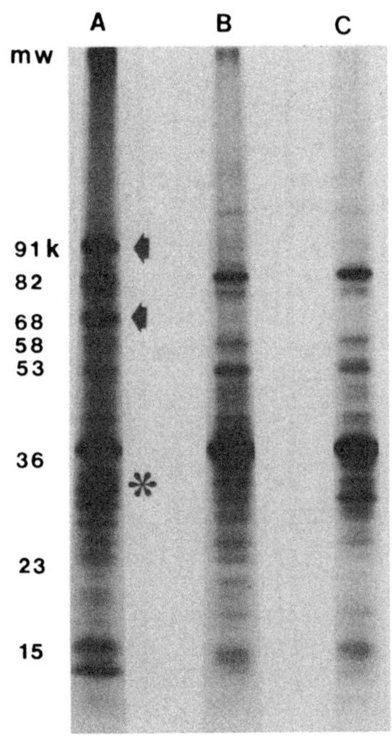

Fig. 3. Autoradiogram of radioiodinated intact cells of B. pertussis strain UT25. Intact cells were radioiodinated with Iodobeads, and analysed as described in Figure 2. Lane MW) calculated apparent molecular weights in kilodaltons for some of the proteins in lane A; Lane A) normal cells; Lane B) isogenic mutant cells; Lane C) modulated cells. The asterisk designates the area of the gel in which the modulation specific doublet is found in envelope fractions, while the arrows identify other modulation specific protein species.

Figure 4 summarizes the observations we made in repeated radio-iodination experiments. Protein 140, which was sometimes seen in normal cells was not seen in modulated or mutant cells or in Triton-insoluble envelope fractions. Proteins 91 and 68 (arrows, Figure 4) were present only in normal cells. Two peptides with apparent molecular weights of 40-45,000 were seen only in Triton-insoluble envelope fractions. Three peptides were seen in the area of the modulation specific doublet (*), but it was not clear which were missing from the modulated cells (which showed only one peptide in this region) and which was missing from Triton-insoluble envelope fractions (which showed two peptides in this region). Protein 91, like the filamentous hemagglutinin, was present only in normal cells of B. pertussis (see below).

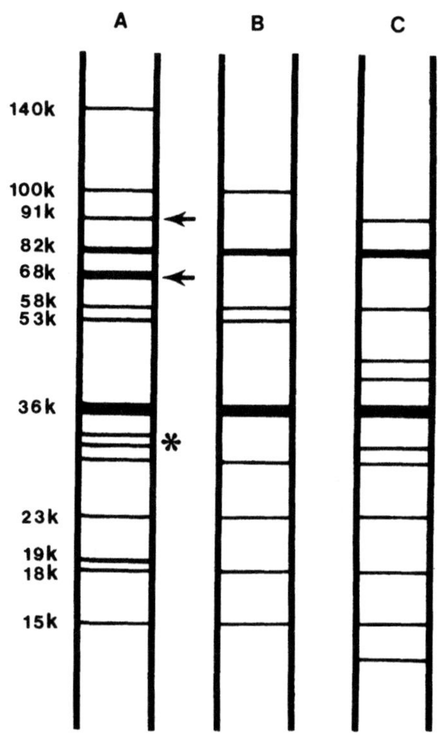

Fig. 4. Diagram of autoradiogram patterns of protein species detected by radiolabeling intact B. pertussis cells or Triton-insoluble envelope fraction. Lane A) normal intact cells; Lane B) mutant or modulated intact cells; Lane C) Triton-insoluble envelope fraction, iodinated following isolation. Calculated apparent molecular weights of some protein species are shown on the left in kilodaltons. The asterisk indicates the area of the gel in which the modulation specific doublet is found in envelope fractions, while the arrows identify other modulation specific protein species.

Convalescent Antibody Studies

We obtained blood samples from two individuals who were convalescing from culture-proven whooping cough, and we performed immunoblot studies using their serum. TF was a 30 year old male who had been fully immunized in childhood. He exhibited catarrhal symptoms for about 10 days, and was beginning to develop paroxysmal cough at the time of culture. Erythromycin was begun immediately following the culture, and the serum was collected 3 weeks later. AH was also a 30 year old male, but had never been immunized. He was symptomatic for 3 weeks prior to culture, with a typical paroxysmal cough of 2 weeks duration. Erythromycin was begun immediately following culture, and the serum was collected 3 weeks later. Figure 5, D and E, demonstrate the reactivity of these sera against Triton-insoluble envelope fraction. Similar results were obtained in immunoblots using whole cell lysate as antigen.

Both men showed antibody to lipopolysaccharide, which is seen as the heavily stained bands nearest the bottom of the paper. Both men also had antibody directed toward protein 18, as well as toward components which migrated with apparent molecular weights of 30,000 and 41,000 daltons. It is possible that antibody to other protein components was present; however, the volumes of serum available precluded further studies.

Figure 5, F and G, demonstrates the reactivity pattern of serum obtained from CFW mice challenged intranasally with B. pertussis, and bled 6 and 7 weeks later, respectively. Each serum is pooled from 3 mice, and represents the patterns which are typical of infected mice by six weeks following infection. Intranasally challenged mice developed little or no antibody directed to lipopolysaccharide, but showed antibodies to numerous protein species. The apparent molecular weights of the stained bands observed with mouse convalescent serum were 91,000, 77,000, 57,000, 49,000, 41,000, 34,000, 30,000, and 18,000 daltons. Antibody prepared by immunizing mice with Triton-insoluble envelope fraction (B and C) contained activity against lipopolysaccharide, as well as antibody to 66,000, 51,000, 25,000, and 16,000 dalton species, but not antibody to the 77,000 dalton protein.

Monoclonal Antibody Studies

We screened monoclonal antibodies for cell surface reactivity by an indirect immunofluorescence test, using B. pertussis cells harvested from Bordet-Gengou plates as antigen, individual monoclonal antibodies as primary antibody, and fluoresceinated goat anti-mouse IgG (specific for both light and heavy chains) as secondary antibody. Cells fluoresced brightly with all monoclonal antibodies directed against lipopolysaccharide, and fluoresced dully with some monoclonal antibodies directed to protein antigens, including monoclonal P12A3, specific for filamentous hemagglutinin (Frank and Parker, 1984a), and the three monoclonals specific for proteins 91, 18, and 15 (see Figure 5, lanes H-J). Antibodies to pertussis toxin and to several other proteins did not induce detectable cell fluorescense.

Immunoblotting studies were used to demonstrate that three of the monoclonal antibodies reacted against membrane proteins 91, 18, and 15. Lanes H, I and J of Figure 5 show the monoclonal reactivity patterns when Triton-insoluble envelope was used as antigen. Results were

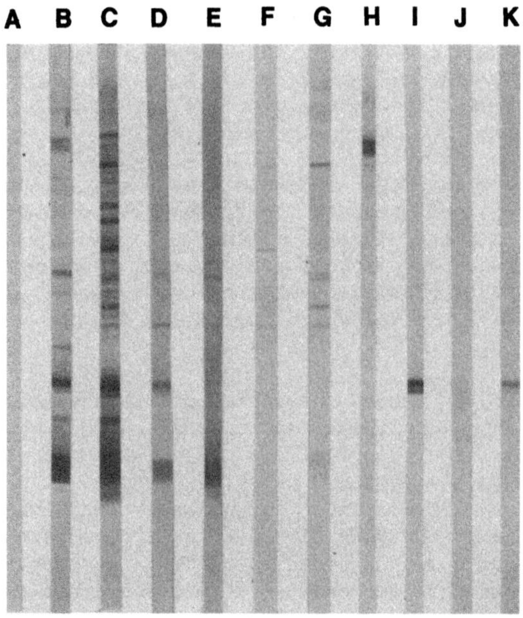

Fig. 5. Immunoblot reactivity of Triton-insoluble envelope fraction of B. pertussis UT25. Triton-insoluble envelope fraction was electrophoresed as in Figure 1, transferred electrophoretically to nitrocellulose paper, and strips of the paper cut from a single gel were reacted with various antisera. After washing, reactions were visualized by using a peroxidase conjugated secondary antibody and a suitable substrate. A) control, not treated with primary antiserum; B) polyclonal mouse antiserum to Triton-insoluble envelope fraction; C) polyclonal mouse antiserum to lysed whole cells; D) human convalescent serum, TF; E) human convalescent serum, AH; F) mouse convalescent serum, 6 weeks; G) mouse convalescent serum, 7 weeks; H) monoclonal P8E7 showing reaction to protein 91; I) monoclonal 1B6, showing reaction to protein 18; J) monoclonal DD3, showing reaction to protein 15; K) polyclonal antibody prepared against purified protein 15.

similar with total cell lysate and total envelope fraction as antigen, but no reactivity was observed against concentrated culture supernatant or cytoplasmic fractions. The monoclonals to proteins 15 and 18 reacted with modulated and mutant B. pertussis as well as with B. parapertussis and B. bronchiseptica.

The monoclonal directed against protein 91 was specific for normal B. pertussis cells. Figure 6 shows that it was unreactive with modulated or mutant B. pertussis. However, it differed in reactivity from the monoclonal directed to filamentous hemagglutinin. In experiments using cell fractions, such as concentrated culture supernatants (which contains both pertussis toxin and filamentous hemagglutinin), we found that both monoclonals reacted with proteins of about 91,000 daltons apparent molecular weight. However, monoclonal P12H3 reacted well only with concentrated culture supernatants, while monoclonal P8E7 reacted well only with fractions which contained cell envelope.

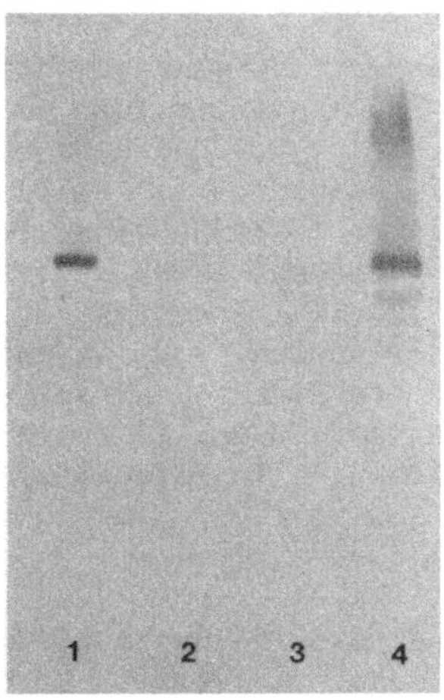

Fig. 6. Immunoblot showing specificity of monoclonal P8E7 for normal B. pertussis cells. Antigen preparations were electrophoresed into a gel as described in Figure 1, the gel was blotted onto paper and reacted with monoclonal P8E7 as primary antibody, and developed using peroxidase conjugated secondary antibody and a suitable substrate. Lane 1) normal whole cells; Lane 2) modulated whole cells; Lane 3) Triton-insoluble envelope fraction prepared from mutant cells; Lane 4) Triton-insoluble envelope fraction prepared from normal cells.

Further investigation of these two monoclonals revealed that they recognized different proteins. Authentic samples of filamentous hemagglutinin (which has a reported molecular weight of 89,000 daltons) were kindly provided to us by Dr. James Cowell, Food and Drug Administration, Bethesda, Maryland. We found that monoclonal P12H3 reacted specifically with filamentous hemagglutinin, while monoclonal P8E7 did not.

DISCUSSION

These studies have identified at least 4 cell surface proteins of B. pertussis strain UT25. Both the filamentous hemagglutinin and protein 91 were specific for normal B. pertussis, while proteins 15 and 18 were found in normal, mutant and modulated B. pertussis, as well as in B. parapertussis and B. bronchiseptica. Pertussis toxin was not detectable in membrane fractions analyzed by immunoblotting, nor did monoclonals induce cell fluorescense in indirect fluorescent antibody stains.

Human convalescent sera and mouse convalescent sera were both shown to react with surface protein 18. The responses of infected humans to protein antigens and lipopolysaccharide apparently differed from the responses of intranasally infected mice. The significance of this observation is not clear, and further work needs to be done to analyze antibodies to surface antigens produced during infections.

Because our monoclonal studies were limited to those antibodies which reacted to denatured antigens in immunoblots, we did not anticipate that this analysis would detect antibodies to agglutinogens. However, since many integral outer membrane proteins are rather heat stable in gram negative bacteria, we expected to find monoclonals to outer membrane proteins. These expectations were confirmed.

The use of monoclonal antibodies for analysis of bacterial surface antigens has been demonstrated to be a useful technique. Since bacteria have extremely complicated cell surfaces, analysis of surface antigens and structure has been difficult. The combination of cell fractionation techniques, surface radioiodination, and monoclonal antibody immunoblots has allowed us to begin to describe the array of surface proteins of B. pertussis, and to investigate which of the surface antigens induce the formation of serum antibodies in infected persons.

ACKNOWLEDGEMENTS

The research described here was supported by Department of Health and Human Services grant number AI 17682 to C. Parker.

REFERENCES

Arai, H., and Sato. Y., 1976, Separation and characterization of two distinct hemagglutinins contained in the purified leukocytosis-promoting factor from Bordetella pertussis, Biochem. Biophys. Acta, 444:765-782.

Ashworth, L., Irons, L., and Dowsett, A., 1982, Antigenic relationship between serotype-specific agglutinogen and fimbriae of Bordetella pertussis, Infect. Immun., 37:1278-1281.

Field, L., and Parker, C., 1979, Differences observed between fresh isolates of Bordetella pertussis and their laboratory passaged derivatives, in: "International Symposium on Pertussis," C. Manclark, and J. Hill, eds., pp. 124-132, DHEW Publication No. (NIH) 79-1830, Washington, D.C.

Frank, D., and Parker, C., 1984a, Isolation and characterization of monoclonal antibodies to Bordetella pertussis, J. Biologic. Standard., 12:(in press).

Frank, D., and Parker, C., 1984b, Interaction of monoclonal antibodies with pertussis toxin and its subunits, Infect. Immun., 46: (in press).
Goodnow, R., 1980, Biology of Bordetella bronchiseptica, Microbiol. Rev., 44:722-738.
Hewlett, E., and Wolff, J., 1976, Soluble adenylate cyclase from the culture medium of Bordetella pertussis, J. Bacteriol., 127: 890-898.
Lacey, B. W., 1960, Antigenic modulation of Bordetella pertussis, J. Hyg., 58:57-93.
Markwell, M., and Fox, C., 1978, Surface-specific iodination of membranes of viruses and eucaryotic cells using 1,3,4,6-tetrachloro-3,6-diphenylglycouril, Biochem., 17:4807-4817.
Parton, R., and Wardlow, A. C., 1975, Cell-envelope proteins of Bordetella pertussis, J. Med. Microbiol., 8:47-57.
Pittman, M., 1970, Bordetella pertussis: bacterial and host factors in the pathogenesis and prevention of whooping cough, in: "Infectious Agents and Host Reactions," S. Mudd, ed., pp. 239-270, W. B. Saunders Company, Philadelphia.
Pusztai, S., and Joo, I., 1967, Influence of nicotinic acid on the antigenic structure of Bordetella pertussis, Ann. Immunol. Hung., 10:63-67.
Sato, Y., Izumiya, K., Sato, H., Cowell, J., and Manclark, C., 1981, Role of antibody to leukocytosis-promoting factor hemagglutinin and to filamentous hemagglutinin in immunity to pertussis, Infect. Immun., 31:1223-1231.
Schneider, D., and Parker, C., 1982, Effect of pyridines on phenotypic properties of Bordetella pertussis, Infect. Immun., 38:548-553.
Standfast, A., 1951, The phase I of Haemophilus pertussis, J. Gen. Microbiol., 5:531-545.
Wardlow, A. C., and Parton, R., 1983, Bordetella pertussis toxins, Pharmac. Ther., 19:1-53.

SYNTHETIC HEPATITIS B SURFACE ANTIGEN PEPTIDE VACCINE

Gordon R. Dreesman,[1] James T. Sparrow,[2] Patrick J. Frenchick[1] and Ronald C. Kennedy[1]

[1]Virology and Immunology Department
Southwest Foundation for Biomedical Research
San Antonio, Texas 78284
and
[2]Department of Medicine
Baylor College of Medicine
Houston, Texas 77030

ABSTRACT

A synthetic hepatitis B surface antigen (HBsAg) peptide was prepared containing amino acid residues 122-137 of the major HBsAg polypeptide. This peptide was cyclized by the introduction of an intrachain disulfide bond between cysteine residues at positions 124 and 137 because previous studies had shown that intact disulfide bonds are critical for maintenance of HBsAg activity. An anti-HBs response was produced in mice by free peptide entrapped in liposomes. However, the immunogenicity was enhanced by aggregation into micelles, and by coupling to tetanus toxoid. Analysis of the peptide with a panel of monoclonal antibodies showed that peptide 122-137 contained a conformation (discontinuous) group _a_ epitope and a sequential (continuous) subgroup _y_ epitope. In addition, the cyclic peptide inhibited a human anti-HBs idiotype-anti-idiotype reaction with specificity for group _a_ determinant(s). The potential for synthetic peptides for hepatitis B virus vaccine development is discussed.

Abbreviations: HBsAg: hepatitis B surface antigen; HBcAg: hepatitis B core antigen; anti-HBs: antibodies to hepatitis B surface antigen; HBV: hepatitis B virus; SP1: synthetic hepatitis B surface antigen containing amino acid residues 122-137.

Acknowledgements: This work was supported in part by New Investigator Research Award AI-22307-01 (RCK) and grant no. 1RO1 AI-21966-01 (GRD) from the National Institutes of Allergy and Infectious Disease.

INTRODUCTION

Infections with hepatitis B virus (HBV) represent a major global health problem based on the fact that approximately 175 million new infections are recorded worldwide (Maupas and Melnick, 1981). The most serious aspect of this disease is that 5-10% of the infected individuals develop a chronic carrier state which may result in chronic liver disease with possible development of cirrhosis or hepatoma. The etiologic agent is visualized by electron microscopy as a double shelled particle with a diameter of 42 nm. A circular DNA genome is double stranded and contains a large gap in one strand which may represent approximately 50% of the genome. This single stranded gap can be repaired with DNA polymerase that has been identified within the nucleocapsid or core of the virus. The viral core with a diameter of 25-27 nm is antigenically distinct from the envelope or the surface of complete virion (Almeida, 1972). These two antigens have been termed hepatitis B core antigen (HBcAg) and hepatitis B surface antigen (HBsAg), respectively.

HBsAg is produced by infected hepatocytes and is observed in the blood stream of infected individuals as envelope material associated with complete virions (originally referred as Dane particles), spherical 22 nm particles, and filamentous forms with similar diameter of varying lengths. HBsAg is composed of lipid and a major disulfide crossed-linked complex of two subunits made up of a protein and a glycoprotein of molecular weights of 25,000 (P25) and 30,000 (gp30), respectively (Melnick et al., 1976). The protein moiety of these two subtypes appear to be identical, differing only by a carbohydrate side chain attached at asparagine residue 146.

Of particular importance to this study was the original observation that antibodies produced to HBsAg (anti-HBs) provide protection to the host (McAuliffe et al., 1980). Formaldehyde inactivated material of purified HBsAg 22 nm spheres derived from HBsAg positive plasma has provided an effective, safe vaccine. However, the high cost of producing vaccine derived from human plasma has been prohibitive for wide-scale distribution of this vaccine. The path for the development of an effective vaccine which can be produced on a reduced economic scale was facilitated by the cloning of the complete HBV genome (Tiollais et al., 1981). The gene for HBsAg P25 has been placed into a variety of expression vectors which has resulted in the production of a number of recombinant P25 products.

Our laboratory, along with a number of other researchers, has pursued the investigation of various immunogenic synthetic peptides which may have potential for HBV vaccine development (Dreesman, 1984). We based our selection of a synthetic peptide containing amino acid sequences analogous to P25 on a number of parameters. Computer analysis of the amino acid sequence for the ayw subtype of HBsAg (Charnay et al., 1979) showed that residue 122-137 represented a relatively hydrophilic region of P25, which is basically a hydrophobic molecule, and also contained a beta turn (Dreesman et al., 1982). The two dimensional structure of this region as visualized by a computer program developed by Corrigan and Hsiung (1982) is illustrated in Figure 1. A second major factor in selecting this sequence was based on the observation that amino acid residues 131 (Asn) and 134 (Phe) present in P25 subtype ayw were substituted with Thr and Tyr, respectively, in the adw polypeptide (Peterson, 1982). The final consideration in the basic design plan for our synthetic peptide was based on earlier observations that intact disulfide bonds are critical for maintenance of both antigenicity and immunogenicity of HBsAg (Dreesman et al., 1973). Therefore, we cyclized the 122-137 peptide via an intrachain disulfide bond between cysteine residues localized at positions 124 and 137.

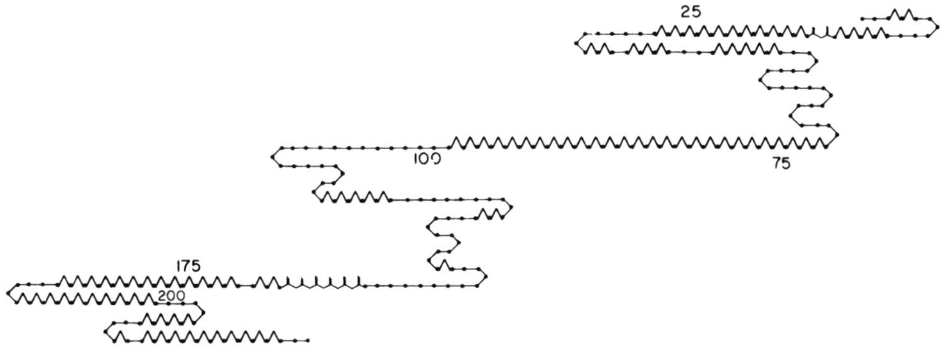

Figure 1. Two dimensional structure of P25 amino acid sequence of HBsAg/ayw (Charnay et al., 1979) as predicted from a computer program written by Corrigan and Huang (1982).

EXPERIMENTAL

Synthetic HBsAg peptides containing amino acid residues 122-137 and 117-137 associated with P25 derived from HBsAy/ayw were synthesized by solid phase methodology developed by Merrifield (1969) on a Schwarz Biosearch Synthesizer modified for computer control (Edelstein et al., 1981). The linear peptide was cleaved from a polystyrene resin at 0°C with anhydrous hydrogen fluoride containing 10% anisole and 1% ethanedithiol as scavengers. The peptides were then extracted from the resin with trifluoroacetic acid and precipitated with ether. The pellet was dissolved in 1 M Tris/6 M guanidine hydrochloride, adjusted to a pH of 8.0. At that point, this solution was desalted on a BioGel P-2 column in 0.1 M ammonium bicarbonate and the peptide containing fractions were pooled and subsequently lyophilized. The cyclic disulfide bond between cysteine residues 124 and 137 was accomplished by oxidation of a dilute solution of the peptide with potassium ferricyanide as described by Felix et al. (1980). The reaction products were again desalted by gel filtration and further purified by ion-exchange chromatography on a SP-Sephadex column equilibrated in 0.02 M NaH_2PO_4/5 M urea, pH 3.7 and eluted with a linear NaCl gradient utilizing the same buffer. The product was analyzed for amino acid composition by amino acid analysis and characterized for purity by C_{18}-reversed phase high-pressure liquid chromatography in a linear gradient of 1% triethyl ammonium phosphate, pH 3.2, and 2-propanol.

Several different adjuvant vehicles were tested to enhance the immunogenicity of the free cyclic peptide. These included emulsification in complete Freund's adjuvant (CFA), adsorption on aluminum potassium sulfate gels (alum) and incorporation into liposomes (Dreesman et al., 1982). In addition, peptide was prepared in a micelle form as previously described for solubilized HBsAg (Sanchez et al., 1983). Briefly, peptide was treated with 2% Triton X-100 in Tris buffer, pH 7.3, and 1 M NaCl at 37°C for 18 hr. The peptide aggregates were fractionated by linear sucrose gradient centrifugation. Lastly cyclic peptide was coupled to purified tetanus toxoid using water soluble carbodiimide (Sanchez et al., 1982).

Female Balb/c mice were immunized by the intraperitoneal route with 50 µg doses of the different peptide preparations described above. Blood was obtained from the tail vein and anti-HBs activity was ascertained using a commercial solid phase radioimmunoassay (Ausab, Abbott Laboratories, N. Chicago, IL) or with a micro-solid phase radioimmunoassay (micro-SPRIA) (Ionescu-Matiu et al., 1983a).

The generation and characteristics of a mouse monoclonal anti-HBs preparation have been described previously (Kennedy et al., 1983a). Inhibition analysis of monoclonal anti-HBs binding HBsAg was determined by preincubation of a limiting dilution of monoclonal anti-HBs with either peptide or intact HBsAg. This mixture was subsequently tested for residual antibody activity on HBsAg microtiter wells.

The ability of peptide to react with human anti-HBs was tested by peptide inhibition of idiotype-anti-idiotype reaction. The idiotype used was human anti-HBs directed at the cross-reactivity group a antigenic determinant(s) (Kennedy and Dreesman, 1983). The anti-idiotype reagent was prepared in rabbits as previously described (Kennedy and Dreesman, 1983).

RESULTS

Anti-HBs Response to Free Cyclic Synthetic Peptide

Both cyclic synthetic peptides 122-137 (SP1) and 117-137 were incorporated into three different adjuvant vehicles; CFA, alum and liposomes. A weak, but detectable, anti-HBs response was noted in a few mice within each group (Dreesman et al., 1982). Free SP1 adsorbed to liposomes appeared to induce the best immune response. The disappointing factor was that these anti-HBs titers either remained the same or decreased upon subsequent inoculation with a booster dose of the same preparation. The other significant observation was that linear SP1 failed to elicit a detectable immune response. That is, no antibody was noted in animals which were inoculated with linear SP1 incorporated into liposomes. Linear SP1 was produced by reduction of cyclic material with 2'-mercaptoethanol and alkylation of free thiol groups with iodoacetamide.

Anti-HBs Response to Aggregated or Conjugated Cyclic Synthetic Peptide

Either micelle forms of SP1 or SP1 coupled to tetanus toxoid were inoculated into mice. These two preparations were given as both aqueous solution and as alum adsorbed material. The primary antibody response to these four preparations was similar to that obtained in mice injected with one inoculation of liposome incorporated free SP1 (Sanchez et al., 1982). However, a brisk anti-HBs response was noted in 80-100% of the mice after a booster inoculation (Table 1). The soluble micelle preparation was the only exception to this response. This secondary response is even more impressive when compared to antibody responses in mice injected with intact 22 nm HBsAg preparations (Table 1). The mean antibody response to HBsAg was approximately 10-fold greater than to the SP1 tetanus toxoid preparation 14 days after the second inoculation.

However, it was of note that the anti-HBs titers in mice injected with HBsAg decreased significantly by day 35. However, the SP1-tetanus toxoid injected mice remained essentially unchanged on day 35 (Table 2).

Table 1. Production of Anti-HBs in Mice by Injection of SP1

SP1 Preparation	Primary Post-Day 32	Booster Post-Day 35
SP1-Micelles-saline	0/6[1]	1/6 (2.7)[2]
SP1-Micelle-alum	2/6 (2.7-4.0)	4/5 (7.5-76)
SP1-Tetanus toxoid-saline	3/6 (9)	4/5 (3.4-153)
SP1-Tetanus toxoid-saline	1/6 (5.3)	6/6 (2.1-123)

[1]The ratio of the number of mice responding with positive anti-HBs titers to the total number of mice inoculated. All sera were tested at a 1:6 dilution, by the AUSAB test (Abbott Laboratories, North Chicago, IL). Sera with S/N values <2.1 were considered negative.

[2]Numbers in parentheses represent the range of S/N in the responding animals.

Table 2. Production of Anti-HBs in Mice[a] by Injection of Alum-Adsorbed HBsAg or of SP1.

HBsAg Preparation	Days after Secondary Immunization 14	35
22 nm HBsAg particle	53,000[b]	6,250
SP1-tetanus toxoid conjugate	6,250	5,500

[a]Mean antibody titers in groups of six mice that were inoculated with each designated preparation on days 0 and 35.

[b]Reciprocal of the arithmetic mean titer of the 6 mice in each group determined by micro-SPRIA. Fivefold antisera dilutions were added to HBsAg *adw* coated wells. Affinity purified ^{125}I-goat IgG anti-mouse IgG was used as detecting probe.

Reactivity of Cyclic Synthetic Peptide with Human Anti-HBs.

A major question that has been asked was: Does SP1 react with human anti-HBs elicited by a natural HBV infection? This was monitored by the utilization of an idiotype-anti-idiotype inhibition assay. It has previously been demonstrated that the rabbit anti-idiotype reagents used in these studies reacts with a common interspecies anti-HBs idiotype which bears the internal image of a major cross-reactivity *a* epitope of HBsAg (Kennedy and Dreesman, 1983). It was noted that SP1 inhibited this reaction at the same level as either intact HBsAg or nonreduced native HBsAg polypeptides (Kennedy et al., 1983b). However, it should be noted that on a molar basis SP1 was less efficient as an inhibitor of this reaction. The other major observation noted in this series of experiments was that reduction of alkylation treatment of both native HBsAg polypeptides or of SP1 abolished its ability to react with the idiotype. This confirmed the importance of the SP1 intrachain and the native disulfide bonds in the maintenance of the cross-reactivity HBsAg epitope(s).

Group of Subgroup Epitopes Associated with Cyclic Synthetic Peptide

It has been demonstrated previously that antibody to group a epitope(s) affords protection against reinfection to the host (Szmuness et al., 1980). Therefore, it is critical that a vaccine candidate for HBV contain the group a epitope. We studied the specific epitopes associated with SP1 with a panel of monoclonal anti-HBs antibodies. This panel included 14 preparations with a reactivity, three preparations with anti-y subgroup reactivity and one which reacted with the w subgroup epitope. The results are summarized in Table 3 (Ionescu-Matiu et al., 1983b). Briefly, six of 14 anti-a monoclonals reacted with cyclic SP1. No reaction was noted with the other eight which reacted equally well with HBsAg subtypes ayw, adw, and adr. Again it was noteworthy that a linear SP1 produced by reduction of subsequent alkylation fails to react with the initial six anti-a monoclonal preparations. In addition, a positive reaction was noted with each of the three anti-y monoclonals reacted with either cyclic or linear peptide. No w reactivity was noted with SP1.

Table 3. Analysis of SP1 for HBsAg Group and Subtype Antigenic Determinants Utilizing a Panel of Monoclonal Murine Antibodies.

Monoclonal Antibody	Inhibition of Antibody with: Cyclic SP1	Inhibition of Antibody with: Linear SP1
Anti-a (6)[a]	+	-
Anti-a (8)	-	-
Anti-y (3)	+	+
Anti-w (1)	-	-

[a]Number in parentheses represents the number of individual monoclonal antibody preparations tested.

DISCUSSION

The development of a highly immunogenic synthetic viral vaccine has several advantages. First, theoretically it should be highly competitive in terms of cost effectiveness. Second, it should be free of any possible cross-reactivity with host or exogenous antigen. Thirdly, its preparation should be highly reproducible from lot to lot. Fourth, it should be free of infectious or transforming activity. With this method, we have actively continued our investigation of cyclic synthetic peptide

(SP1) which contains amino acid residues 122-137 of the P25 polypeptide derived from HBsAg, subtype ayw. We constructed a peptide containing an intrachain disulfide bond based on our previous observations that intact disulfide bonds are critical for maintenance of both HBsAg antigenicity and immunogenicity.

Our initial series of experiments established that a single inoculation of free cyclic SP1 into mice elicited an immune response in more than 50% of the animals (Dreesman et al., 1982). It was of note that a similar response also was noted in animals injected with SP1 aggregated into micelles (Sanchez et al., 1982). We conclude from these observations that free SP1 presented in a liposome or micelle preparation can induce an immune response.

The immunogenicity of SP1 was greatly enhanced when the cyclic peptide was presented in either alum adsorbed micelles or after coupling to a protein carrier such as tetanus toxoid (Sanchez et al., 1982). In comparison to immunization with intact HBsAg particles, two unique observations were made. 1) The primary response to alum adsorbed micelles or conjugates was considerably reduced. 2) The early response (14 days) to HBsAg was approximately 10-fold higher than to the peptide preparations; however, the anti-HBs activity in SP1 immunized mice remained relatively constant 35 days after the booster. Alternatively, the anti-HBs titer dropped significantly in the HBsAg immunized mice. The maintenance of anti-HBs activity for longer periods of time certainly would be an advantage in the construction of an effective HBV vaccine.

As noted above it has been established by several investigators that antibody to the cross-reaction a determinant(s) affords protection against HBV reinfection. Therefore, it was of great interest that a conformation a determinant was associated with the cyclic peptide as determined with a panel of murine monoclonal antibodies (Ionescu-Matiu et al., 1983b). Of some concern in these studies was that approximately one-half of our monoclonal antibody preparation did not react with SP1. Therefore, it became imperative to ascertain whether SP1 would react with antibody produced by humans naturally infected with HBV. The most instructive approach to test this was the utilization of an inhibition of human anti-HBs idiotype-anti-idiotype reaction (Kennedy and Dreesman, 1983). This assay was of value because we had previously shown that the common human idiotype was directed to the a determinant (Kennedy et al., 1982). The results of these experiments confirmed that SP1 contains a conformational a epitope which reacts with the human idiotype (Kennedy et al., 1983b). This adds additional evidence for the potential of the peptide for eventual vaccine development.

The experiment which we designed to detect group and subgroup epitopes with use of monoclonal antibodies indicated that SP1 contain a y subgroup determinant (Ionescu-Matiu et al., 1983b). Unlike the a epitope the y reactivity was not dependent in conformation, in that anti-y monoclonals reacted with both cyclic and linear forms of SP1. Gerin et al. (1983) confirmed these findings when they reported that chimpanzees inoculated with linear 122-137 peptide produced subtype-specific antibodies. This confirms the important role of maintaining conformation in an HBsAg synthetic peptide. It will be of interest whether intrachain disulfide bonds will enhance the immunogenicity of other, unrelated synthetic peptides.

Obviously a great deal of basic research lies ahead before an effective HBV vaccine will be used in a clinical field trial. Hopefully, the immunogenicity can be enhanced by use of other adjuvant or protein carriers and a more efficient method of chemical conjugation will be established. We will have to learn more about the role of T and B cell cooperation in response to synthetic peptides and the establishment of good immunologic memory. For instance, does SP1 contain both T and B cell epitopes? Lastly, the ability to induce protective antibody will have to be performed in a chimpanzee model. However, the present data does hold great promise that a synthetic peptide vaccine may have great potential for active immunization against HBV.

REFERENCES

Almeida, J. D., 1972, Individual morphological variations seen in Australia antigen-positive sera, Am. J. Dis. Child., 123:303.

Charnay, P., Mandart, E., Hampe, A., Fitoussi, F., Tiollais, P., and Galibert, F., 1979, Localization on the viral genome and nucleotide sequence of the gene coding for the two major polypeptides of the hepatitis B surface antigen (HBsAg), Nucl. Acids Res., 7:335.

Corrigan, A. J., and Huang, P. C., 1982, A BASIC microcomputer program for plotting the secondary structure of proteins, Comput. Programs Biomed., 15:163.

Dreesman, G. R., 1984, Polypeptide and synthetic peptide vaccines for hepatitis B virus, in: "Advances in Hepatitis Research," F. V. Chisari, ed., Masson Publishing Inc., New York, pp. 216-222.

Dreesman, G. R., Hollinger, F. B., McCombs, R. M., and Melnick, J. L., 1973, Alteration of hepatitis B antigen (HB Ag) determinants by reduction and alkylation, J. Gen. Virol., 19:129.

Dreesman, G. R., Sanchez, Y., Ionescu-Matiu, I., Sparrow, J. T., Six, H. R., Peterson, D. L., Hollinger, F. B., and Melnick, J. L., 1982, Antibodies to hepatitis B surface antigen after a single inoculation of uncoupled synthetic peptide, Nature, 295:158.

Edelstein, M. S., McNair, D. S., and Sparrow, J. T., 1981, The conversion of solid phase peptide synthesizers to computer control, in: "American Peptide Symposium, 7th, Madison, Univ. of Wisconsin, 1981; Peptides, synthesis, structure, function," D. H. Rich and E. Gross, eds., Pierce Chemical Co., Rockford, Illinois.

Felix, A. M., Jiminez, M. H., Wang, C.-T., and Meienhofer, J., 1980, Synthesis of somatostatin and [D-TRP^{8}]-somatostatin, Int. J. Pept. Protein Res., 15:342.

Gerin, J. L., Alexander, H., Shih, J. W.-K., Purcell, R. H., Dapolito, G., Engle, R., Green, N., Sutcliffe, J. G., Shinnick, T. M., and Lerner, R. A., 1983, Chemically synthesized peptides of hepatitis B surface antigen duplicate the d/y specificities and induce sub-type-specific antibodies in chimpanzees, Proc. Natl. Acad. Sci. USA, 80:2365.

Ionescu-Matiu, I., Sanchez, Y., Fields, H. A., and Dreesman, G. R., 1983a, Development of sensitive immunoassays for detection of antibodies against hepatitis B surface antigen, J. Virol. Methods, 6:41.

Ionescu-Matiu, I., Kennedy, R. C., Sparrow, J. T., Culwell, A. R., Sanchez, Y., Melnick, J. L., and Dreesman, G. R., 1983b, Epotopes associated with a synthetic hepatitis B surface antigen peptide. J. Immunol., 130:1947.

Kennedy, R. C., and Dreesman, G. R., 1983, Common idiotypic determinant associated with human antibodies to hepatitis B surface antigen, J. Immunol., 130:385.

Kennedy, R. C., Sanchez, Y., Ionescu-Matiu, I., Melnick, J. L., and Dreesman, G. R., 1982, A common human anti-hepatitis B surface antigen idiotype is associated with the group a conformation-dependent antigenic determinant, Virology, 122:219.

Kennedy, R. C., Ionescu-Matiu, I., Adler-Storthz, K., Henkel, R. D., Sanchez, Y., and Dreesman, G. R., 1983a, Characterization of anti-hepatitis B surface antigen monoclonal antibodies, Intervirology, 19:176.

Kennedy, R. C., Dreesman, G. R., Sparrow, J. T., Culwell, A. R., Sanchez, Y., Ionescu-Matiu, I., Hollinger, F. B., and Melnick, J. L., 1983b, Inhibition of a common human anti-hepatitis B surface antigen idiotype by a cyclic synthetic peptide, J. Virol., 46:653.

Maupas, P., and Melnick, J. L. (eds.), 1981, Hepatitis B virus and primary hepatocellular carcinoma, in: "Prog. Med. Virol.," Vol. 27, Karger, Basel.

McAuliffe, V. J., Purcell, R. H., and Gerin, J. L., 1980, Type B hepatitis: a review of current prospects for a safe and effective vaccine, Rev. Infect. Dis., 2:470.

Melnick, J. L., Dreesman, G. R., and Hollinger, F. B., 1976, Approaching the control of viral hepatitis type B, J. Infect. Dis., 133:210.

Merrifield, R. B., 1969, Solid-phase peptide analysis, Adv. Enzymol., 32:221.

Peterson, D. L., 1982, Structure of hepatitis B surface antigen: Correlation of amino acid sequence with subtype and location of the carbohydrate moiety, in: "Viral Hepatitis," W. Szmuness, H. J. Alter, and J. E. Maynard, eds., Franklin Institute Press, Philadelphia, pp. 707-708.

Robinson, W. S., 1977, The genome of hepatitis B virus, Annu. Rev. Microbiol., 31:357.

Sanchez Y., Ionescu-Matiu, I., Sparrow, J. T., Melnick, J. L., and Dreesman, G. R., 1982, Immunogenicity of conjugates and micelles of synthetic hepatitis B surface antigen peptides, Intervirology, 18:209.

Sanchez, Y., Ionescu-Matiu, I., Melnick, J. L., and Dreesman, G. R., 1983, Comparative studies of the immunogenic activity of hepatitis B surface antigen (HBsAg) and HBsAg polypeptides, J. Med. Virol., 11:115.

Szmuness, W., Stevens, C. E., Harley, E. J., Zang, E. A., Oleszko, W. R., William, D. C., Sadovsky, R., Morrison, J. M., and Kellner, A., 1980, Hepatitis B vaccine. Demonstration of efficacy in a controlled clinical trial in a high-risk population in the United States, N. Engl. J. Med., 303:833.

Tiollais, P., Charnay, P., Vyas, G. N. 1981, Biology of hepatitis B virus, Science, 213:406.

ANTIGEN MIMICRY BY ANTI-IDIOTYPE ANTIBODIES THAT RECOGNIZE A COMMON ANTI-HEPATITIS B SURFACE ANTIGEN IDIOTYPE

Ronald C. Kennedy and Gordon R. Dreesman

Virology and Immunology Department
Southwest Foundation for Biomedical Research
San Antonio, TX 78284

ABSTRACT

Anti-idiotype antibodies that recognize a common human idiotype present on antibodies to hepatitis B surface antigen (anti-HBs) from individuals naturally infected by hepatitis B virus (HBV) and an interspecies idiotype on anti-HBs produced by active immunization with hepatitis B surface antigen (HBsAg) were used _in vivo_ to modulate idiotype networks in mice. Injection of anti-idiotype antibodies without subsequent HBsAg stimulation induced an anti-HBs response. The anti-idiotype-induced anti-HBs recognized the group-specific _a_ determinant(s) of HBsAg, which has been shown to be responsible for induction of protective immunity to HBV in both chimpanzees and humans. In addition, the anti-HBs expressed an interspecies idiotype which is shared by humans naturally infected with HBV. Prior inoculation of mice with alum-adsorbed anti-idiotype potentiated the immune response to a subsequent inoculation with either native HBsAg particles or with a synthetic HBsAg peptide. These data suggest the potential use of anti-idiotype antibodies as a vaccine or vaccine potentiator for HBV.

Abbreviations: Anti-HBs: antibodies to hepatitis B surface antigen; Anti-Id: anti-idiotype antibodies; HBsAg: hepatitis B surface antigen; HBV: hepatitis B virus; PFC: plaque forming cells.

INTRODUCTION

Hepatitis B virus (HBV) infects 175 million people yearly and represents a major worldwide health problem. Hepatitis B surface antigen (HBsAg), the envelope material of HBV, is produced in an overwhelming excess by infected hepatocytes during the course of an HBV infection (Almeida, 1972; Robinson, 1977). The presence of HBsAg is often associated with the development of chronic liver disease and primary carcinoma of the liver (Maupas and Melnick, 1981). Serologically, HBsAg contains within its structure a group-specific determinant(s) termed _a_ and two sets of allelic subtype determinants, _d_ or _y_ and _w_ or _r_. Combinations of the _a_ determinant with the various allelic subtype determinants result in four major serotypes associated with HBsAg: adw, ayw, adr, and ayr (LeBouvier, 1971; Bancroft et al., 1972). Antibodies against HBsAg (anti-HBs) are protective against HBV infection (McAuliffe et al., 1980). Induction of this protective immunity is associated with the group-specific a determinant(s) since immunization with one HBsAg serotype confers protection against HBV infection with the other serotypes (Szmuness et al., 1980).

Immunoglobulin idiotypes have been studied extensively since they were first recognized by Kunkel and coworkers and Oudin and Michel in 1963 (Kunkel et al., 1963; Oudin and Michel, 1963). Idiotypes, along with homologous anti-idiotype antibodies (anti-Id), are thought to be components of a network of complex reactions that regulates a given immune response. This regulation of the level of idiotype recognition was postulated by Jerne (Jerne, 1974) to involve a series of idiotype-anti-Id reactions that are capable of controlling both the afferent and efferent limbs of the immune response. Numerous studies have implicated idiotype networks in regulating the immune response to a large variety of haptens and protein or carbohydrate antigens (Rodkey, 1980; Urbain et al., 1981).

Recent studies from this laboratory have characterized a common idiotype shared by human anti-HBs (Kennedy and Dreesman, 1983; Kennedy et al., 1982). The common idiotype was associated in part with the antibody-combining site, because HBsAg and a nondenatured HBsAg-derived viral polypeptide isolated in a micelle form partially inhibited the common idiotype-anti-Id reaction. Further characterization revealed that the common idiotype was induced by a conformation dependent group-specific a determinant because: (i) three serotypes of HBsAg (ayw, adw, and adr) inhibited the idiotype-anti-Id reaction equally well on a weight basis; (ii) reduction of the disulfide bonds in HBsAg and alkylation of free thiol groups abolished the ability to inhibit the idiotype-anti-Id reaction and; (iii) denatured HBsAg viral polypeptides had a decreased inhibitor capacity when compared with native polypeptides. This common idiotype was also expressed on anti-HBs produced in BALB/c mice and six other mammalian species indicating that an interspecies idiotype cross-reaction was detected (Kennedy et al., 1983a). The involvement of an idiotype-anti-idiotype network was indicated in that the numbers of anti-HBs plaque-forming cells (PFC) in the spleens of BALB/c mice were enhanced by prior injection of anti-Id (Kennedy et al., 1983b).

In this review, we summarize our observations of the effects of anti-Id as potentiators of the murine humoral anti-HBs response. The implications of these observations with regard to idiotype networks and anti-Id based vaccines for HBV are discussed.

EXPERIMENTAL

The generation and characterization of the anti-HBs idiotype (human) and the anti-idiotype (rabbit) reagents have been described in detail previously (Kennedy and Dreesman, 1983; Kennedy et al., 1982; 1983a,b,c). The purification of 22 nm HBsAg particles, subtypes adw, ayw, and adr, were performed by methods that have been described elsewhere (Dreesman et al., 1972). A cyclic synthetic HBsAg peptide analogous to amino acid residues 122-137 of HBsAg subtype ayw was synthesized by solid phase methodology (Merrifield, 1969) on a Schwarz Bioresearch Synthesizer modified for computer control (Edelstein et al., 1981). The cyclic disulfide bond between cysteines at positions 124 and 137 was formed by oxidation of a dilute solution of the peptide with potassium ferricyanide (Felix et al., 1980). After desalting on a BioGel P-2 column, the peptide was further purified by a combination of ion exchange (SP-Sephadex) and molecular sieve (BioGel P-10) chromatography. A single peak of the peptide was obtained by C_{18}-reverse phase high-pressure liquid chromatography.

For the in vivo modulation of the anti-HBs response, groups of adult female BALB/c mice received intraperitoneal injections of either affinity purified anti-Id or preimmune rabbit IgG which was purified from the serum of rabbits prior to injection of the idiotype. These two antibody preparations were given as an alum precipitate (Kennedy and Dreesman, 1984). The optimal dose of anti-Id, along with the optimal time interval between injections of anti-Id and either HBsAg or cyclic peptide 122-137 have been established previously (Kennedy and Dreesman, 1984; Kennedy et al., 1984a). Serum obtained from each mouse before any immunization served as a negative control for each animal.

The anti-HBs activity in the treated mice was determined by a solid phase radioimmunoassay using HBsAg coated onto the wells of microtiter plates (Ionescu-Matiu et al., 1983). The detection of the interspecies idiotype on mouse anti-HBs has been described in detail previously (Kennedy et al., 1983a).

RESULTS

Characteristics of the Common Human Anti-HBs Idiotype

The characterization of the common human anti-HBs idiotype-anti-Id reaction is summarized in Table 1. Inhibition of the idiotype-anti-idiotype reaction by three HBsAg subtypes indicated that the common idiotype was directed against the a determinant(s) of HBsAg. Previous studies had shown that the three HBsAg subtypes inhibited the reaction equally well on a molar basis (Kennedy et al., 1982). The importance of conformation on the anti-HBs idiotype was demonstrated by the inability of reduced and alkylated native HBsAg polypeptides to inhibit the idiotype-anti-Id reaction. In addition, SDS-denatured HBsAg viral polypeptides demonstrated a decreased ability to inhibit the idiotype-anti-Id reaction when compared with native polypeptides. These data indicated that the anti-HBs idiotype was induced by a conformation-dependent group specific a determinant of HBsAg. The idiotype-anti-Id reaction was also partially inhibited by a cyclic synthetic HBsAg peptide analogous to amino acid sequences 122-137. Reduction of the intrachain disulfide bond and subsequent alkylation destroyed its inhibitory activity, suggesting that a conformation-dependent group a determinant was also associated with the cyclic peptide (Table 1).

In Vivo Priming of the Anti-HBs Response by Anti-Id

The ability of the anti-Id to prime the anti-HBs response in mice is shown in Table 2. Injection of anti-Id prior to HBsAg stimulation increased the anti-HBs titer compared with that of groups of mice receiving preIgG prior to HBsAg (mean anti-HBs titer of 12,916 as compared with 13). These data indicated that anti-Id were capable of

Table 1. Percent Inhibition of the Anti-HBs Idiotype Anti-Id Reaction with Different Concentrations of Various Inhibitors

Inhibitor	Concentration (μg)	Inhibition (%)[a]
HBsAg[b]	25	33-35
	2.5	22-25
Native HBsAg derived polypeptide[c]	7.5	65
	0.75	22
Reduced and alkylated HBsAg	7.5	2
Native HBsAg polypeptide	0.75	0
SDS-denatured P25-6P30[d]	25	9-13
	2.5	2-5
Cyclic peptide 122-137	25	33
	2.5	30
Reduced and alkylated	25	1
peptide 122-137	2.5	0

[a]The inhibition values represent the mean of triplicate values. A solid phase radioimmunoassay was used to measure the percent inhibition of the idiotype-anti-idiotype reaction (Kennedy et al., 1982).
[b]The range of inhibition values obtained when each of the three HBsAg subtypes (ayw, adw, and adr) were individually used to inhibit the idiotype anti-Id reaction.
[c]Aggregate of P25-6P30 isolated as a micelle.
[d]Isolated by SDS-polyacrylamide gel electrophoresis under reducing conditions.

priming the anti-HBs response to HBsAg in mice. Similarly, anti-Id were found to enhance the anti-HBs response to a cyclic synthetic HBsAg peptide (Table 3). The injection of anti-Id prior to immunization with cyclic peptide 122-137 increased the anti-HBs response compared with that in groups of mice receiving preIgG and peptide 122-137 (mean anti-HBs titer of 38 as compared with 4). In addition, anti-Id in conjugation with peptide 122-137 induced anti-HBs titers comparable to that obtained with a single injection of intact HBsAg particles (Table 3).

Inducing Anti-HBs by Injecting Anti-Id Alone

In a previous study, we demonstrated that BALB/c mice receiving anti-Id alone generated IgG anti-HBs PFC (Kennedy et al., 1983b). These data indicated that anti-Id could induce an anti-HBs response without HBsAg stimulation. Injecting anti-Id alone in BALB/c mice produced a significant IgG anti-HBs response when compared with that of mice given similar injections or preIgG control antibodies ($p < 0.001$, two-tailed Student's t-test) (Table 4). The anti-Id induced anti-HBs was also found to express the interspecies idiotype since

Table 2. Enhancement of the Anti-HBs Response to HBsAg by Anti-ID Priming[a]

First Injection	Second Injection	Anti-HBs[b] Titer
Anti-Id	HBsAg	6,250
Anti-Id	HBsAg	6,250
Anti-Id	HBsAg	1,250
Anti-Id	HBsAg	31,250
Anti-Id	HBsAg	31,250
Anti-Id	HBsAg	1,250
PreIgG	HBsAg	5-50[c]

[a]Mice received either 50 μg of alum precipitated anti-Id or preIgG on day 0, followed by 6 μg of HBsAg on day 14. Mice were bled on day 26.
[b]Reciprocal of the anti-HBs endpoint titer which has been described elsewhere (Kennedy and Dreesman, 1984).
[c]Range of anti-HBs titers obtained from six mice.

Table 3. Priming the Anti-HBs Response to a Cyclic HBsAg Synthetic Peptide with Anti-Id[a]

First Injection	Second Injection	No. of Mice	Anti-HBs Response	
			Range	Mean
PreIgG	Peptide 122-137	6	0-10	4
Anti-Id	Peptide 122-137	7	10-50	38.6
PreIgG	HBsAg	5	10-50	34.0

[a]Mice received alum precipitated anti-Id or preIgG on day 0, following by 6 μg of HBsAg particles or 50 μg of peptide 122-137 on day 14. Serum was obtained on day 30 and the reciprocal dilution of the end-point titer was determined (Kennedy and Dreesman, 1984).

inhibition values obtained with a group of six individual mouse sera were from 27-54%. Conversely, less than 11% inhibition of the idiotype-anti-Id reaction was obtained with the nonanti-HBs-containing sera from six mice injected with control antibodies and the preimmune sera from the anti-Id induced anti-HBs-producing mice. The antibody specificity of the anti-HBs produced by anti-Id injection was determined by a radioimmunoassay using microtiter plates coated with HBsAg subtypes adw, adr, and ayw. Each of the six mouse antisera bound equally well the three different HBsAg subtypes indicating that the anti-Id anti-HBs was directed to the a determinant (Table 4). These data suggest that the anti-Id may act as an internal image of HBsAg and induce the production of anti-HBs without classical antigenic

Table 4. Production of Anti-HBs by Injection of Anti-Id[a]

First Injection	Second Injection	Anti-HBs Titer[b]		% Inhibition of Idiotype-Anti-Id[c]		S/N[d]		
		Range	Mean	Range	Mean	adw	ayw	adr
Anti-Id	anti-Id	250-1250	833	27-54	42	7.9-8.3	7.9-8.4	7.7-8.2
PreIgG	preIgG	< 5	< 5	0-11	4	< 1.8	< 1.6	< 1.8

[a]Each group of mice received two injections of anti-Id or preIgG, 14 days apart. Mice were bled 12 days after second injection.
[b]Reciprocal of the anti-HBs endpoint titer.
[c]All sera were tested for the ability to inhibit the idiotype-anti-Id reaction at a 1:10 dilution.
[d]The range of positive (S) to negative (N) cpm ratios obtained with the three HBsAg subtypes testing anti-Id treated sera at 1:100 dilution and preIgG sera at 1:10 dilution.

stimulation. It was noteworthy that a single injection of anti-Id failed to induce an anti-HBs response in mice (data not shown) and a second injection of anti-Id was required for the production of a detectable anti-HBs response.

DISCUSSION

This report characterizes a common idiotype associated with human anti-HBs and examines the use of anti-Id produced against the anti-HBs idiotype for priming the anti-HBs response in mice. The injection of anti-Id in conjugation with either HBsAg or a cyclic synthetic HBsAg peptide enhanced the *in vivo* anti-HBs when compared with that of groups of mice treated with control antibodies prior to immunization with HBsAg or synthetic peptide. Alternatively, two injections of anti-Id produced an anti-HBs response without subsequent HBsAg stimulation. The anti-Id induced anti-HBs was directed against the *a* determinant of HBsAg, which has been shown to be the epitope that induces protective immunity against HBV. In addition, the anti-HBs expressed an interspecies idiotypic determinant that is shared by humans naturally infected with HBV. These data lend experimental evidence that the anti-Id may represent homobodies (Lindenmann, 1979) or the internal image of HBsAg. Thus, the anti-Id can mimic the antigenic determinants of HBsAg and induce the immune system to produce anti-HBs without classical antigenic stimulation. This becomes particularly attractive as a potential vaccine when the antigen is an infectious viral agent. Anti-Id appear to have a potential use as vaccine primers or vaccine candidates whereby the susceptible host is not exposed to infectious components of the particular agent.

The theoretical implications for anti-Id vaccines against infectious agents have been discussed by others (Nisonoff and Lamoyi, 1981; Strosberg, 1983). Anti-Id have been used to induce protective immunity against African trypanosomiasis (Sacks et al., 1982) and Sendai virus (Ertl and Finberg, 1984). It is noteworthy that genetic restrictions were placed on the protection to trypanosomiasis (Sacks et al., 1983). This becomes important since humans represent a genetically outbred population, which may severely limit the effectiveness of anti-Id vaccines. Also, it is not known whether anti-Id would be able to boost preexisting antibody levels and how long the duration of anti-Id induced protection would be against infectious agents. Alternatively, anti-Id treatment may have deleterious effects and suppress the host's immune response to a particular agent as has been reported for herpes simplex virus in mice (Kennedy et al., 1984b). However, the problems associated with using anti-Id produced in a heterologous species for injection into humans may be circumvented by the use of human hybridoma antibodies.

The work presented here describes the potential use of anti-Id as vaccines or vaccine primers for HBV. This possiblity must await further testing since only chimpanzees and humans can be naturally infected with HBV.

ACKNOWLEDGEMENTS

This work was supported in part by grant A1-21966-01 from the National Institute of Allergy and Infectious Diseases.

REFERENCES

Almeida, J. D., 1972, Individual morphological variations seen in Australia antigen-positive sera, Am. J. Dis. Child., 123:303.

Bancroft, W. H., Mundon, F. K., and Russell, P. K., 1972, Detection of additional antigenic determinants of hepatitis B antigen, J. Immunol., 109:842.

Dreesman, G. R., Hollinger, F. B., McCombs, R. M., and Melnick, J. L., 1972, Production of potent anti-Australia antigen sera of high specificity and sensitivity in goats, Infect. Immun., 5:213.

Edelstein, M. S., McNair, D. S., and Sparrow, J. T., 1981, The conversion of solid phase peptide synthesizers to computer control, in: "American Peptide Symposium, 7th, Madison, Univ. of Wisconsin, 1981; Peptides, synthesis, structure, function," D. H. Rich and E. Gross, eds., Pierce Chemical Co., Rockford, Illinois.

Ertl, H.C.J., and Finberg, R. W., 1984, Sendai virus-specific T-cell clones: Induction of cytolytic T cells by an anti-idiotypic antibody directed against a helper T-cell clone, Proc. Natl. Acad. Sci., 81:2850.

Felix, A. M., Jiminez, M. H., Wang, C.-T., and Meienhofer, J., 1980, Synthesis of somatostatin and [D-TRP^8]-somatostatin, Int. J. Pept. Protein Res., 15:342.

Ionescu-Matiu, I., Sanchez, Y., Fields, H. A., and Dreesman, G. R., 1983, Development of sensitive immunoassays for detection of antibodies against hepatitis B surface antigen, J. Virol. Methods, 6:41.

Jerne, N. K., 1974, Towards a network theory of the immune system, Ann. Immunol. (Paris), 125c:373.

Kennedy, R. C., and Dreesman, G. R., 1983, Common idiotypic determinant associated with human antibodies to hepatitis B surface antigen, J. Immunol., 130:385.

Kennedy, R. C., and Dreesman, G. R., 1984, Enhancement of the immune response to hepatitis B surface antigen. In vivo administration of antiidiotype induces anti-HBs that expresses a similar idiotype, J. Exp. Med., 159:655.

Kennedy, R. C., Sanchez, Y., Ionescu-Matiu, I., Melnick, J. L., and Dreesman, G. R., 1982, A common human anti-hepatitis B surface antigen idiotype is associated with the group a conformation-dependent antigenic determinant, Virology, 122:219.

Kennedy, R. C., Ionescu-Matiu, I., Sanchez, Y., and Dreesman, G. R., 1983a, Detection of interspecies idiotypic cross-reactions associated with antibodies to hepatitis B surface antigen, Eur. J. Immunol., 13:232.

Kennedy, R. C., Adler-Storthz, K., Henkel, R. D., Sanchez, Y., Melnick, J. L., and Dreesman, G. R., 1983b, Immune response to hepatitis B surface antigen: Enhancement by prior injection of antibodies to the idiotype, Science, 221:853.

Kennedy, R. C., Dreesman, G. R., Sparrow, J. T., Culwell, A. R., Sanchez, Y., Ionescu-Matiu, I., Hollinger, F. B., and Melnick, J. L., 1983c, Inhibition of a common human anti-hepatitis B surface antigen idiotype by a cyclic synthetic peptide, J. Virol., 46:653.

Kennedy, R. C., Melnick, J. L., and Dreesman, G. R., 1984a, Antibody to hepatitis B virus induced by injecting antibodies to the idiotype, Science, 223:930.

Kennedy, R. C., Adler-Storthz, K., Burns, J. W., Sr., Henkel, R. D., and Dreesman, G. R., 1984b, Antiidiotype modulation of herpes simplex virus infection leading to increased pathogenicity, J. Virol., 50:951.

Kunkel, H. G., Mannik, M., and Williams, R. C., 1963, Individual antigenic specificity of isolated antibodies, Science, 140:1218

Le Bouvier, G. L., 1971, The heterogeneity of Australia antigen, J. Infect. Dis., 123:671.

Lindenmann, J., 1979, Homobodies do they exist?, Ann. Immunol. (Paris), 130C:311.

Maupas, P., and Melnick, J. L. (eds.), 1981, Hepatitis B virus and primary hepatocellular carcinoma, in: "Prog. Med. Virol.," Vol. 27, Karger, Basel.

McAuliffe, V. J., Purcell, R. H., and Gerin, J. L., 1980, Type B hepatitis: a review of current prospects for a safe and effective vaccine, Rev. Infect. Dis., 2:470.

Merrifield, R. B., 1969, Solid-phase peptide analysis, Adv. Enzymol., 32:221.

Nisonoff, A., and Lamoyi, E., 1981, Implications of the presence of an internal image of the antigen in anti-idiotypic antibodies: Possible application to vaccine production, Clin. Immunol. Immunopathol., 21:397.

Oudin, J. and Michel, M., 1963, Une nouvelle forme d'allotypie des globulines du serum de lapin apparemment liee a la fonction et a la specificite anticorps, C. R. Acad. Sci. Paris, 257:805.

Robinson, W. S., 1977, The genome of hepatitis B virus, Annu. Rev. Microbiol., 31:357.

Rodkey, L. S., 1980, Autoregulation of immune responses via idiotype network interactions, Microbiol. Rev., 44:631.

Sacks, D. L., and Sher, A., 1983, Evidence that anti-idiotype induced immunity to experimental African trypanosomiasis is genetically restricted and requires recognition of combining site-related idiotopes, J. Immunol., 131:1511.

Sacks, D. L., Esser, K. M., and Sher, A., 1982, Immunization of mice against African trypanosomiasis using anti-idiotypic antibodies, J. Exp. Med., 155:1108.

Strosberg, A. D., 1983, Anti-idiotype and anti-hormone receptor antibodies, Springer Semin. Immunopathol., 6:67.

Szmuness, W., Stevens, C. E., Harley, E. J., Zang, E. A., Oleszko, W. R., William, D. C., Sadovsky, R., Morrison, J. M., and Kellner, A., 1980, Hepatitis B vaccine. Demonstration of efficacy in a controlled clinical trial in a high-risk population in the United States, N. Engl. J. Med., 303:833.

Urbain, J., Wuilmart, C., and Cazenave, P. A., 1981, Idiotypic regulation in immune networks, Contemp. Top. Mol. Immunol., 8:113.

IMMUNOCHEMISTRY OF VARIANTS OF INFLUENZA VIRUS HEMAGGLUTININ AND NEURAMINIDASE

W.G. Laver

John Curtin School of Medical Research
P.O. Box 334
Canberra Act 2601 Australia

Abstract

Influenza is impossible to control by vaccination because of variation in the two surface antigens of the virus, the hemagglutinin (HA) and neuraminidase (NA). This variation is caused by sequence changes in the genes coding these antigens and this article summarizes recent information on the structure of the HA and NA, the way in which these glycoproteins vary and the effects of the changes on the antigenic properties of the virus.

Antigenic Structure of Influenza Viruses

Influenza viruses are classified into three types - A, B and C - on the basis of their type-specific nucleoprotein and matrix protein antigens. Type A influenza viruses are further classified into subtypes based on the antigenic characteristics of their surface antigens, hemagglutinin (HA) and neuraminidase (NA). Thirteen distinct HA subtypes and nine NA subtypes are now recognized in the nomenclature system for influenza A viruses recommended by the World Health Organization (WHO MEMO 1980).

Properties of Influenza A, B and C Viruses

The highly pleomorphic particles of Type A influenza virus (Figs. 1 and 2) are enclosed by a lipid envelope, derived from the plasma membrane of the host cell, to which the HA and NA antigens are attached by sequences of hydrophobic amino acids at one end of the molecules. Both surface antigens are glycosylated and some of the carbohydrate side chains possess antigenic activity characteristic of the host cell in which the virus grew.

Within the lipid envelope lies the matrix protein, which is believed to be structural in function. Within the matrix shell are eight single-stranded RNA molecules of negative sense (that is, the virion RNA is complementary to the messenger RNA) associated with a nucleocapsid protein and three large proteins, PA, PB1 and PB2, responsible for RNA replication and transcription (Krug, 1983). At least three virus-encoded non-structural proteins are also found in infected cells. Table 1 summarizes current information on the eight genes of the influenza virus and the molecular weights of the proteins encoded by them.

The properties of influenza Type B virus particles are similiar to those of type A (except that an additional non-structural protein, NB, is present Lamb, 1983); but Type C influenza differs considerably from the other two types.

Although influenza C is much less well characterized than influenzas A and B, it differs most significantly in possessing a unique surface glycoprotein which resembles the HA of influenzas A and B in many of its biochemical properties.

Influenza C glycoproteins form characteristic hexagonal arrays on the surface of both spherical and filamentous forms of the virus. Such arrays have not been observed on influenzas A or B, and it has been suggested (Herrler et al., 1981) that the presence of morphologically distinct trimeric HA and tetrameric neuraminidase of different symmetry may prevent lattice formation in these strains. Herrler et al., (1981) reported that the influenza C glycoprotein spikes viewed end-on in preparations negatively stained with sodium phosphotungstate (pH 7.2) have a rounded end, in contrast with the triangular shape observed for influenza A HA (Wrigley, 1979).

However, the influenza C glycoprotein is clearly seen to be a trimer in specimens prepared with uranyl stains. Three-dimensional reconstructions from naturally occurring hexagonal arrays show that at low resolution (~30 Å) the influenza C glycoprotein exhibits similar features to the hemagglutinin glycoprotein of influenza A (Hewat et al., (1984). Both have a triangular stalk

Table 1

Influenza Virus Genome RNA Segment and Coding Assignments

Segment	Length[a] (nucleotides)	Encoded[b] polypeptide	Nascent[c] polypeptide length (aa)	Mature polypeptides	Approx. No[d] molecules per virion	Remarks
1	2,341	PB2	759		30-60	Host cell RNA cap binding: component of RNA transcriptase.[e]
2	2,341	PB1	757		30-60	Initiation of transcription: possible endonuclease activity: component of RNA transcriptase.[e]
3	2,233	PA	716		30-60	Elongation mRNA chains?: component of RNA transcriptase.[e]
4	1,778	HA	566	HA1 326 HA2 222	500	Surface glycoprotein, trimer: major antigenic determinant.
5	1,565	NP	498		1000	Associated with RNA segments to form ribonucleoprotein: structural component of RNA transcriptase. Type-specific antigen.
6	1,413	NA	454	454	100	Surface glycoprotein; neuraminidase activity: Tetramer: Surface antigen.
7	1,027	M_1	252	252	3000	Major protein component of virus: underlies lipid bilayer: Type-specific antigen.
		M_2	96			Spliced mRNA, non-structural protein: function unknown.(f)
		?	?9			Spliced mRNA, peptide predicted by nucleotide sequence only.(f)
8	890	NS_1	230			Non-structural protein: function unknown.
		NS_2	121			Spliced mRNA, non-structural protein: function unknown.(f)
	13,588					

a For A/PR/8/34 strain: Winter et al., See Lamb (1983) for full references
b Determined by biochemical and genetic approaches (see text).
c Determined by nucleotide sequence analysis and protein sequencing (see text).
d Adapted from Compans and Choppin (1975).
e Krug (1983)
f Lamb (1983)

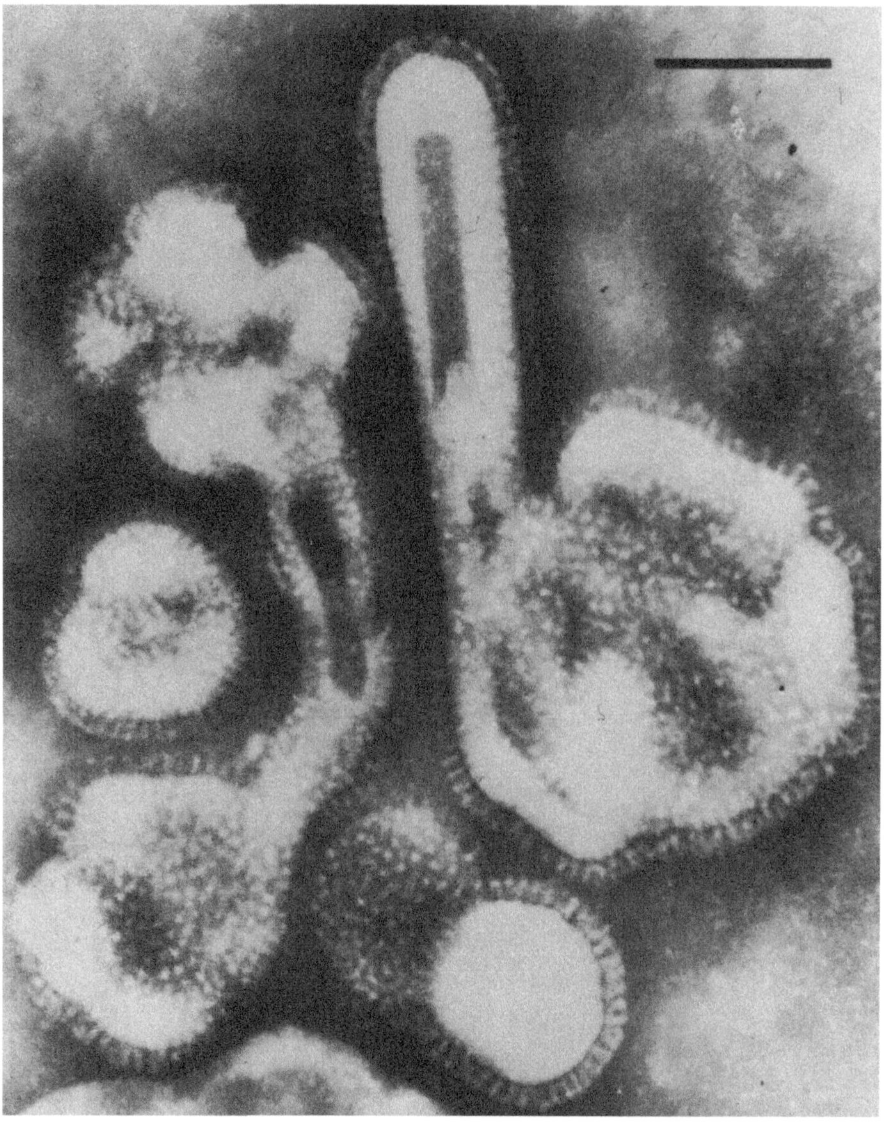

Figure I Electron micrograph showing particles of influenza type A virus.

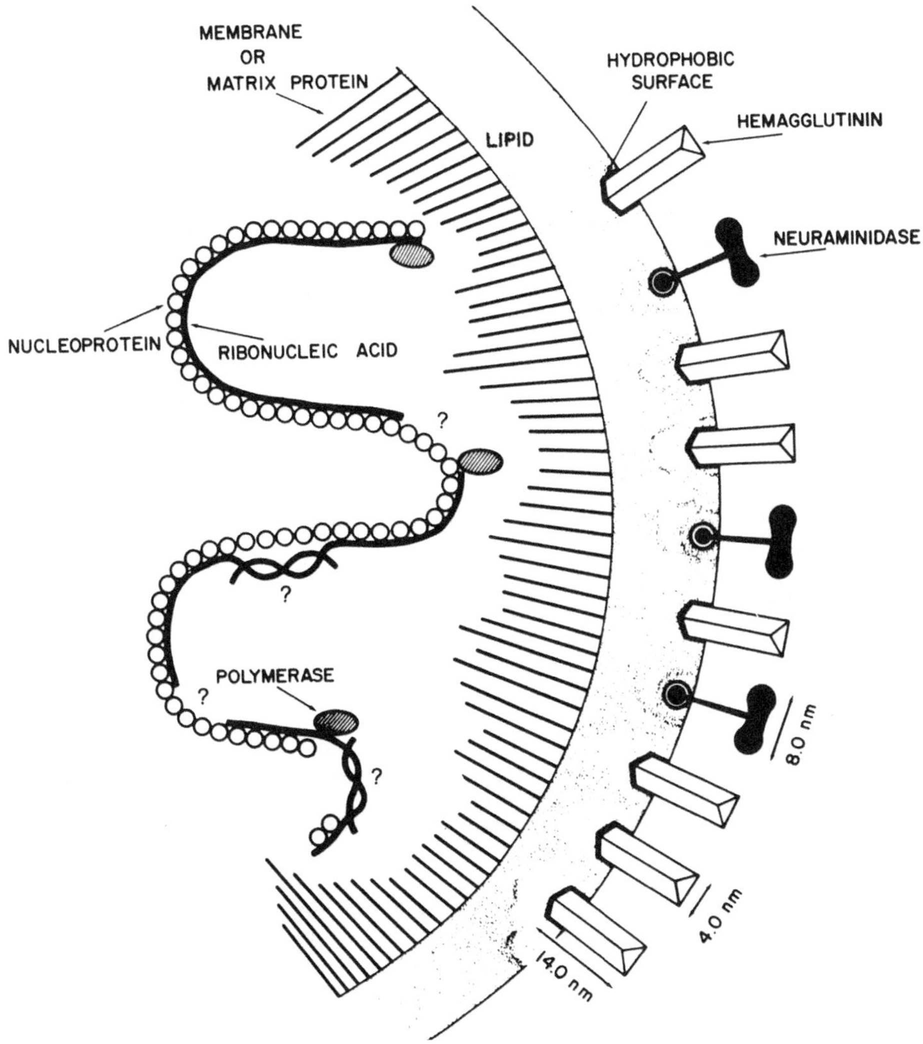

Figure 2 Diagram showing the surface structure of particles of influenza type A virus. There are eight RNA segments in the virus, each of which codes for one (in some cases two, using overlapping reading frames) of the virus proteins.

near the membrane. Further from the membrane, the stalk becomes broader and the monomers more separated, leaving an open centre. The molecule narrows at the top. The regions of greatest contact between adjacent trimers in the arrays are situated nearer the distal end of the molecule. These contact zones can be related to equivalent zones on the influenza A hemagglutinin. Differences between the structure of the influenza A hemagglutinin glycoprotein determined by X-ray analysis and reconstructions of the influenza C glycoprotein are greatest at either end of the molecule where the reconstructions are least reliable. Ordered glycoprotein arrays have not been observed on influenza C virions incubated at low pH. The staining patterns of glycoproteins on intact virions are essentially determined by the pH at which the virus is incubated, and the stain type, but not the pH of the stain.

Antigenic Variation

Two distinct kinds of antigenic variation have been demonstrated in influenza viruses, antigenic drift and major antigenic shifts. The first consists of relatively minor changes in the surface glycoproteins that occur gradually within a family of strains, all of which are clearly related to each other with respect to both internal and surface antigens. Among influenza A strains infecting man, each successive variant replaces pre-existing ones. This may be due to a selective advantage possessed by antigenic variants in overcoming immunological host resistance. Although antigenic variation occurs in type B influenza viruses it does not follow a sequential pattern as in type A.

No evidence has been obtained for any antigenic changes in influenza C virus since the first isolation of these agents in 1947. This antigenic stability resembles that of other respiratory agents such as the parainfluenza viruses and respiratory syncytial virus, rather than influenza A or B virus. Nucleotide sequences of the glycoprotein genes of two strains of influenza C (C/Johannesberg/66, Pfeifer and Compans, 1984 and C/California/78; Nakada et al., 1984) have been determined, and show 25 differences in amino acid sequence of the proteins.

This finding is consistent with the observation that influenza C viruses have not undergone significant antigenic variation. The reason for this difference in stability of the surface antigen is not understood, particularly since influenza C virus appears to share with the influenza A and B viruses most features of virus structure as well as virus replication.

The second kind of antigenic variation, which has been described only for influenza type A, involves much more dramatic antigenic changes. These are referred to as major antigenic shifts. These major shifts, which occur at intervals of 10-15 years, are marked by the sudden appearance of "new" subtypes in the population and it is these viruses which cause the major pandemics of influenza.

Structure of the hemagglutinin (HA)

In Type A influenza virus, the HA accounts for about 25% of the virus protein. It is responsible for the attachment of virus to cells and for the penetration of the virus into the cell during the initial stages of infection. Antibodies to the HA neutralize the infectivity of the virus and it is variation in the HA glycoprotein which is mainly responsible for the continually occurring outbreaks of influenza and for our inability to control these by vaccination.

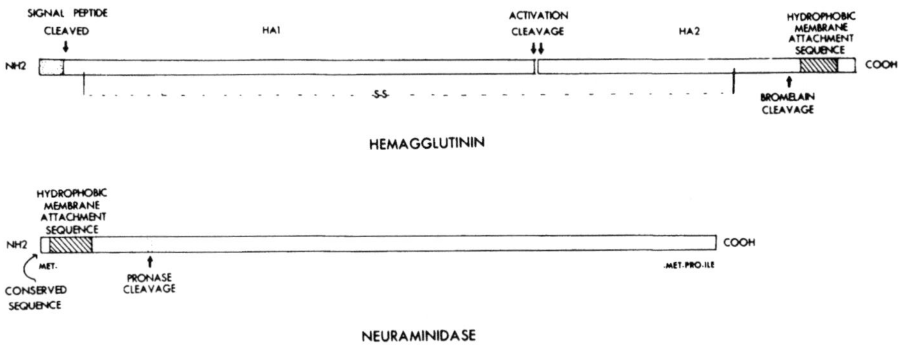

Figure 3 Diagrammatic representation of certain features of the hemagglutinin and neuraminidase polypeptides. Hemagglutinin: The HA is synthesized as a single polypeptide. Following its synthesis an N-terminal signal peptide is cleaved off and the molecule is cleaved further into HA1 and HA2 with the removal of one or more intervening amino acids. The latter cleavage is necesary for the virus to be infectious. HA1 and HA2 remain linked by a single disulfide bond and each HA spike contains three of these dimers. A sequence of hydrophobic amino acids near the C terminus of HA2 serves to anchor the HA in the lipid of the virus membrane. Treatment with bromelain removes this hydrophobic region without damaging the rest of the molecule, which, in some cases, can be crystallized. Neuraminidase: The neuraminidase is oriented in the virus membrane in the opposite way to the hemagglutinin. No post-translational cleavage of the NA polypeptide occurs, no signal peptide is split off, and even the initiating methionine is retained. No processing at the C terminus takes place; the C-terminal sequence -Met-Pro-Ile predicted from the gene sequence is found in intact NA molecules of the N2 subtype isolated from virus and in the pronase-released NA heads. A sequence of six polar amino acids at the N-terminus of the NA polypeptide, which are totally conserved in the nine different NA subtypes, is followed by a sequence of hydrophobic amino acids which probably represent the transmembrane region of the NA stalk. This sequence is not conserved at all between subtypes (apart from conservation of hydrophobicity). Pronase cleaves the polypeptide in the positions shown, removing the stalk and releasing the enzymically and antigenically active head of the NA, which, in some cases, can be crystallized.

The HA monomer is coded by the fourth largest RNA segment and is synthesized as a single polypeptide chain which undergoes post-translational cleavage in at least three places (Fig 3). An N-terminal signal sequence is removed and, depending on the host cell and virus strain, the molecule is cleaved, with the removal of one or more intervening residues, to give two polypeptide chains called HA1 and HA2, with molecular weights of 36,000 and 27,000 respectively. HA1 and HA2 remain joined by a single disulpide bond and each HA "spike" contains three of these HA1 and HA2 chains. A sequence of 25-32 hydrophobic amino acids near the C-terminus of HA2 serves to anchor the HA in the virus membrane. Cleavage of the HA polypeptide into HA1 and HA2 is necessary for the virus particles to be infectious. Intact HA molecules can be isolated from detergent-disrupted virus particles. Soluble HA, lacking the hydrophobic membrane - attachment sequence can be obtained by digesting virus particles with bromelain. This HA (BHA) can be crystallized and the three-dimensional structure of Hong Kong (H3N2) BHA has been determined (Wilson et al., 1981). It seems likely that HA molecules from other subtypes will have similar structures. The HA glycoprotein of influenza virus is a trimer built of two structurally distinct regions: a triple-stranded coiled-coil of α - helices extend 76Å from the membrane and a globular region of antiparalled β -sheet that contains the receptor binding site; the variable antigenic determinants are located on top of this stem (Wiley, et al., 1981). Each subunit has an unusual loop-like topology; it begins at the membrane, extends 135Å distally and folds back to enter the membrane.

Examination of the amino acids of the amino-terminal sequence of HA2 of several influenza A viruses of human and avian origin has revealed a highly conserved region of 10 residues. The sequence is homologous to the amino-terminus region of the fusion glycoprotein (F) of the paramyxovirus, Sendai virus (Gething et al., 1978) The F protein mediates fusion of the Sendai virus envelope with the plasma membrane of host cells, a function which is considered essential for virus infectivity. It has been postulated, therefore, that the infectivity of influenza virus is dependent on a fusion function of HA analogous to that of the Sendai fusion protein. Virus with uncleaved HA is not infectious, presumably because the N-terminus of HA2 is not accessible in uncleaved HA. Although the three-dimensional structure of the influenza HA shows that after cleavage the N-terminal HA2 peptide is buried in the molecule, it has recently been shown that, after incubation in conditions required for membrane fusion, (pH 5.0 Maeda and Ohnishi, 1980) a conformational change occurs in the influenza HA glycoprotein exposing the N-terminus of HA2 and the molecule acquires the ability to hemolyse erythrocytes and the bromelain-released HA can bind to lipid vesicles (Skehel et al., 1982).

Antigenic shift

An antigenic shift in type A influenza virus occurs when a virus suddenly appears in the human population with a "new" HA and sometimes also a "new" NA. Since the first human influenza virus was isolated in 1933, antigenic shifts have occurred in 1957 when the H2N2 subtype (Asian influenza) replaced the H1N1 subtype, in 1968 when the Hong Kong (H3N2) virus appeared and in 1977 when the H1N1 virus reappeared. All these major antigenic shifts in the virus occurred in China and anecdotal records suggest that previous epidemics (probably caused by the sudden emergence of 'new' viruses) also had their origin in China. Although we are still uncertain how these new subtypes appear or reappear in the human population, some clues are emerging from the plethora of recent sequence data.

Sequences up to 350 nucleotides from the 3' end of the HA gene and the predicted amino acid sequences from 32 type A influenza viruses, including representatives of each of the 13 known HA subtypes, have been determined (Air, 1981; Hinshaw et al., 1982). Cysteine residues and certain other amino acids are conserved in all sequences, indicating that the 13 HA subtypes evolved from a common ancestor and share a common basic structure. When the partial amino acid sequences are compared pairwise, the most distant subtypes are H1 and H3 (25% homology). Other subtypes are more similar to each other, the highest homology (80%) being between H2 and H5. Relationships between the partial sequences can be illustrated as a dendrogram (Fig. 4).

Within strains of a single subtype, the differences in deduced amino acid sequence are generally less than 10%, but recent extensions of the data have revealed differences in the N-terminal region of HA1 from two strains each of H7 and H10 viruses close to 20%, almost equal to the difference between the amino acid sequences of the two closest subtypes (H2 and H5).

Whereas it is possible to imagine that the H2 viruses could eventually mutate into H5, the change to H3 must involve a more drastic mechanism. How, then, could the H3 hemagglutinin have replaced the H2 strain in the human population in 1968? First, the 'new' virus may have caused an epidemic in man many years previously and have remained hidden and unchanged in some unknown place ever since. Evidence for this kind of event has been obtained. The strain of 'Russian 'flu' (H1N1) which reappeared in Anshan in northern China on 4 May 1977 and subsequently spread to the rest of the world, seems to be identical, in all genes, to the virus which caused an influenza epidemic in 1950 (Nakajima et al., 1978; Scholtissek et al., 1978). Where was this virus for 27 years? As yet we have no answer.

Second, some of the 'new' viruses may be derived from animal or avian viruses. One strain of human influenza has been shown to be such a recombinant (reassortant). The Hong Kong (H3N2) virus contains the NA (and other) genes from an Asian (H2N2) strain of human influenza and a HA which is antigenically related to that of A/Duck/Ukraine/63 (H3N8) and A/equine/2/Miami/63 (H3N8) viruses. The amino acid sequence homology between the HAs of A/Duck/Ukraine/63 and A/Aichi/2/68 viruses (both of subtype H3) is 96% (Fang et al., 1981; Ward & Dopheide 1981). We do not know if it was an animal or bird virus which donated the HA gene during the recombination event that led to the formation of the Hong Kong strain. The donating virus could have been one which contained Hong Kong HA that had persisted from a much earlier human influenza epidemic and maintained unchanged in the same way as the Russian 'flu. Antibodies in the sera of people who were born around 1900 suggest that a virus with a HA similar to that of the Hong Kong virus caused influenza at the time, but the NA of this virus was N8, not N2. It is possible that viruses such as equine (H3N8) and duck/Ukraine have evolved from the 1900 epidemic strain.

The third way in which new viruses could appear in the human population is if an animal or bird virus became infectious for man. This could be brought about by mutation, but the number of mutations required and the genes in which these would need to occur are unknown.

Perhaps each of these mechanisms has operated at one time or another to produce a 'major shift' in influenza viruses infecting man. We do not know how

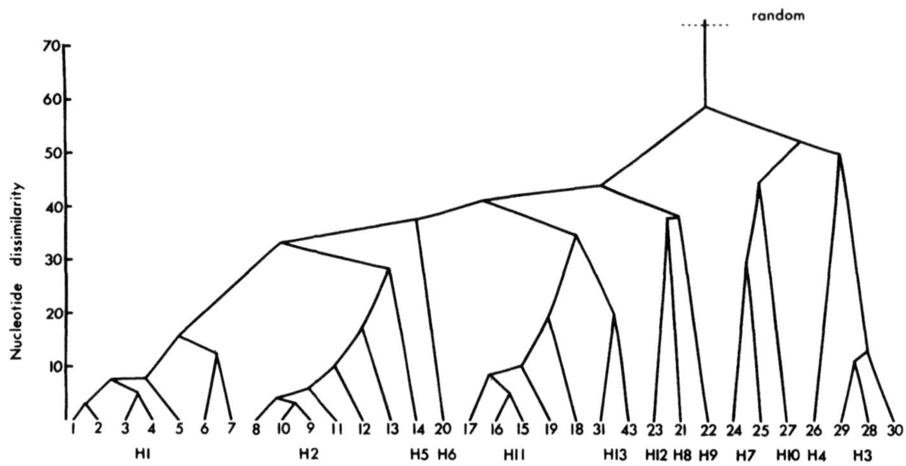

Figure 4 Dendrogram showing the relationships between the HA1 N-terminal amino acid sequences deduced from genomic sequences of viruses representing the 13 known subtypes. The sequences were aligned using amino acids that are invariable in certain positions of all sequences (amino acid position from N terminus of most subtypes: Cys 4, Gly 6, Thr 18, Val 26, Cys 42, Cys 55, Gly 63, Pro 65, Cys 67, Glu 81). The percentage sequence differences of all pairwise comparisons of the aligned sequences starting from the N-terminal Asp or corresponding amino acid were used to calculate the dendrogram. Thus, the positioins of each bifurcation in the dendrogram indicate the mean sequence difference of the sequences connected through that point.[2] The sequences used in the analysis are: 1, A/NWS/33 (H1N1); 2, A/PR/8/34 (H1N1); 3, A/BH/35 (H1N1); 4, A/Bel/42 (H1N1); 5, A/Loyang/4/57 (H1N1); 6, A/USSR/90/77(H1N1); 7, A/Fort Warren/1/50 (H1N1); 8, A/duck/Alberta/35/76 (H1N1); 9, A/swine/Wisconsin/15/30 (H1N1); 10, A/New Jersey/11/76 (H1N1); 11, A/RI/5-/57 (H2N2); 12, A/Tokyo/3/67 (H2N2); 13, A/Berkeley/68 (H2N2); 14, A/Netherlands/68 (H2N2); 15, A/duck/GDR/72 (H2N9); 16, A/duck/New York/6750/78 (H2N2); 17, A/duck/Ontario/77 (H2N1); 18, A/duck/Alberta/77/77 (H2N3); 19, A/pintail/Alberta/293/77 (H2N9); 20, A/shearwater/Australia/75 (H5N3); 21, A/shearwater/Australia/72 (H6N5); 22, A/duck/Alberta/1298/79 (H6N5); 23, A/duck/Alberta/1297/79 (H6N9); 24, A/tern/Australia/75 (H11N9); 25, A/duck/Ukraine/1/60 (H11N9); 26, A/duck/England/56 (H11N6); 27, A/duck/New York/12/78 (H11N6); 28, A/duck/Memphis/576/76 (H11N9); 29, A/gull/Mass/26/80 (H13N6); 30, A/gull/Md/704/77 (H13N6); 31, A/turkey/Ontario/6118/68 (H8N4); 32, A/duck/Alberta/827/78 (H8N4); 33, A/duck/Alberta/283/77 (H8N4); 34, A/turkey/Wisconsin/1/66 (H9N2); 35, A/duck/Alberta/60/76 (H12N5); 36, A/turkey/Oregan/71 (H7 N3); 37, A/equine/Prague/1/56 (H7N7); 38, A/duck/Manitoba/53 (H10N7); 39, A/quail/Italy/1117/65 (H10N8); 40, A/duck/Alberta/28/76 (H4N6); 41, A/black duck/Australia/702/78 (H3N8); 42, A/Memphis/1/71 (H3N2); 43, A/duck/Ukraine/1/63 (H3N8).

the different subtypes of type A influenza arose, but it is reasonable to suppose that 'shuffling' of genes between influenza A strains is occurring all the time, most frequently in birds, but also in mammals, including man.

Antigenic drift

Antigenic drift in the HA of influenza type A occurs by mutations in the gene, leading to an accumulation of amino acid sequence changes that alter the antigenic sites in such a way that they are no longer recognized by the body's immune system.

The HA from a large number of strains of influenza virus of types A and B have been completely sequenced (see Ward, 1981) but it is not possible to tell from this data precisely which of the sequence changes were responsible for changes in the antigenic properties of the HA.

Variants selected with monoclonal antibodies

Antigenic variants of several strains of HA have been selected by growing virus in the presence of monoclonal antibody to the HA (Gerhard et al., 1981; Laver et al., 1979, Caton et al., 1982, Nakajima et al., 1983). The variants occurred at a frequency of 10^{-5} in virus stock and did not bind at all to the antibody used for their selection. This dramatic change in antigenicity was found, in most cases, to be associated with single changes in the amino acid sequence of HA1 (Table 2) and these sequence changes were certainly related to antigenic changes.

The changes are scattered along the primary sequence of HA1 (Fig 5) but in the case of H3 HA cluster into four distinct sites (A-D) when the HA1 polypeptide folds into its three-dimensional structure (Fig 6). Since the sequence changes affect antibody binding, it is clear that the sites must be involved in the antigenicity of the HA. Many of the sequence changes found in natural variants of Hong Kong influenza virus (field strains) also are within these sites (Both and Sleigh 1981).

Drift in Influenza Type B Virus

Sequences of the HA gene of several field isolates of influenza B have been determined (Krystal et al., 1983; Verhoeyen et al., 1983, Berton et al., 1984), and it appears that variable areas in the B HA structure correspond to antigenic regions A, B and D of influenza A viruses. No changes were found in the B HAs in areas corresponding to antigenic region C of influenza A viruses.

The available sequence data for HA1 of influenza B HAs is shown in Fig. 7. Of 40 amino acid positions where variation is seen, 17 are conserved in all strains except B/Lee/40, and it has previously been noted that residue 90 of B/Lee/40 from another laboratory is valine, as in all other strains (Air and Compans, 1983). Otherwise substitutions tend to come and go over time, in contrast to the steady accumulation of amino acid substitutions with time found in HA1 of influenza A viruses of the H3 subtype (Both and Sleigh, 1981). Neither antigenic analyses (Lu et al., 1982) nor the sequence analyses indicate that antigenic drift occurs over time in the same way as in influenza A viruses.

The three-dimensional structure of the B HA is not known but comparison of sequence data and computer assisted model building suggested a structural similarity between influenza A and B HAs (Krystal et al., 1982).

Table 2

Amino Acid Sequences Changes in the HA of Variants Selected with Monoclonal Antibodies

H1[a]

Position in HA 1 (equivalent in H3 HA1)	Sequence change	Antigenic region
70 (78)	Leu to Pro	Cb
71 (79)	Leu to Pro	Cb
73 (81)	Val to Glu	Cb
74 (82)	Arg to Gly	Cb
75 (83)	Ser to Pro	Cb
115 (122)	Glu to Lys	Cb
124 (128)	Pro to Ser	Sa
125 (129)	Asn to Lys	Sa
125 (129)	Asn to Tyr	Sa
136 (140)	Ser to Pro	Ca2
139 (143)	Gly to Arg	Ca2
141 (145)	Ser to Ala	Ca2
152 (156)	Glu to Gly	Sb
154 (158)	Glu to Val	Sa
155 (159)	Gly to Asp	Sb
156 (160)	Ser to Leu	Sa
158 (162)	Pro to Leu	Sa
159 (163)	Lys to Thr	Sa
161 (165)	Lys to Glu	Sa
161 (165)	insert Lys	Sa
162 (166)	Asn to Lys	Sa
163 (167)	Ser to Tyr	Sa
163 (167)	Ser to Phe	Sa
165 (169)	Val to Ala	Cal
169 (173)	Gly to Val	Cal
169 (173)	Gly to Arg	Cal
178 (182)	Ile to Met	Cal
188 (192)	Gln to Leu	Sb
188 (192)	Gln to Arg	Sb
189 (193)	Asn to Ser	Sb
189 (193	Asn to Lys	Sb
192 (196)	Gln to Arg	Sb
194 (198)	Glu to Gly	Sb
203 (207)	Ser to Leu	Cal
220 (224)	Arg to Ile	Ca2
221 (225)	Asp to Gly	Ca2
236 (240)	Gly to Arg	Cal
236 (240)	Gly to Glu	Cal
270 (273)	Ser to Thr	Cal

H3[b]

Position in HA1	Sequence change	Antigenic region
53	Asn to Lys	C
63	Asp to Asn	E
133	Asn to Lys	A
135	Gly to Arg	A
143	Pro to Leu	A
143	Pro to Ser	A
143	Pro to Thr	A
143	Pro to His	A
144	Gly to Asp	A
144	Gly to Ser	A
145	Ser to Lys	A
145	Ser to Asn	A
146	Gly to Ser	A
156	Lys to Glu	B
157	Ser to Leu	B
158	Gly to Glu	B
188	Asn to Asp	B
189	Gln to His	B
193	Ser to Arg	B
198	Ala to Glu	B
198	Ala to Val	B
205	Ser to Tyr	D
218	Gly to Trp	D

B[c]

Position in HA1 (equivalent in H3 HA1)	Sequence change	Antibody used for selection
141 (137A)	Gly to Val	HK 21/6
147 (142)	Thr to Ala	HK 74/1
150 (145)	Asn to Ser	HK 21/6
162 (157)	Lys to Thr	HK 282/1
163 (158)	insert Asn	HK 16/1, 313/2, 430/1
166 (160A)	Lys to Ile	Ore 680/1, 710/1
166 (160A)	Lys to Asn	HK 74/1
196 (187)	Asn to Asp	Ore 401/2
196 (187)	Asn to Lys	Ore 146/1
197 (188)	Glu to Lys	Ore 280/2
199 (190)	Gln to Lys	Ore 21/2
201 (192)	Val to Leu	Ore 206/2
201 (192)	Val to Glu	Ore 113/2
201 (192)	Val to Ala	Ore 152/2
202 (193)	Lys to Thr	Ore 238/4
202 (193)	Lys to Asn	Ore 10/4 HK 74/1
202 (193)	Lys to Arg	HK 365/1
202 (193)	Lys to Gly	HK 74/1
207 (196 B)	Ser to Pro	Ore 128/2
207 (196 B)	Ser to Leu	Ore 391/1
240 (225)	Pro to Thr	HK 74/1
240 (225)	Pro to Gln	Ore 114/4

(a) Data from Caton et al., 1982
(b) Data from Daniels et al., 1983, Laver et al (1979, 1981) and Newton et al., (1983)
(c) Hovanec and Air (1984) Berton, Naeve and Webster (1984)

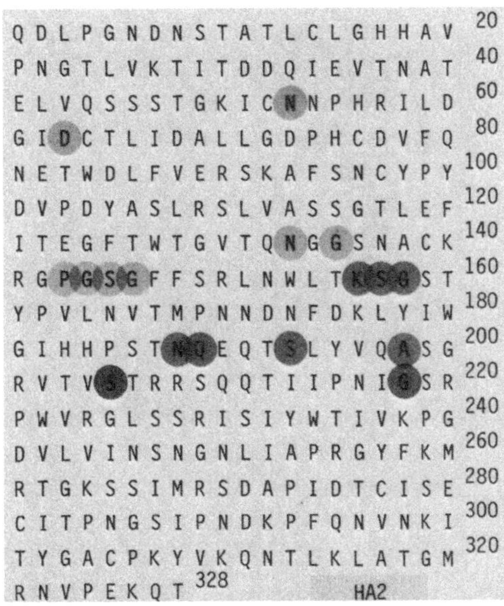

Figure 5 This illustration shows the amino-acid sequence of the HA1 polypeptide of A/Hong Kong/68 hemagglutinin. Residues which changed in variants selected with monoclonal antibodies are shown. The sequence changes which occurred in monoclonal variants of other influenza viruses are shown in Table 2.

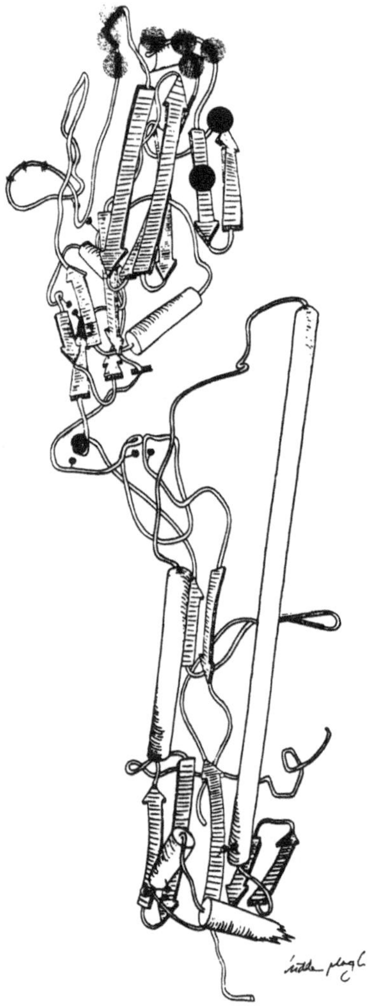

Figure 6 This illustration (by Hidde Pleogh) of the Hong Kong/68 (H3) hemagglutinin monomer shows folding of the HA1 and HA2 polypeptides and the location of amino-acid sequence changes in variants selected with monoclonal antibodies which delineate four distinct antigenic sites. These variants, which occur at a frequency of 1 in 10^5 of wild-type virus, have totally lost the ability to bind the particular monoclonal antibody used in the selection procedure.

To further investigate antigenic variation in influenza B viruses, variants have been selected with monoclonal antibodies and the sequence changes determined. Two strains have been used; B/Hong Kong/8/73 (Webster and Berton 1981; Hovanec and Air, 1984) and B/Oregon/5/80 (Berton et al., 1984).

For comparison, amino acid positions in the HA1 peptide can be described using the A/Aichi/2/68 numbers (Wiley et al., 1981 as aligned by Krystal et al., 1983).

In the B/Hong Kong/8/73 HA1 polypeptide there are 2 areas of antigenic variability. These areas would correspond to antigenic regions A and B of the A/Aichi/2/68 HA molecule (Fig 6). Amino acid substitutions found in these areas include residues 137A and 142 (region A) and residues 157, 158 and 193 (region B). However, two variants have a further nucleotide substitution at position 1742 which changes amino acid 211 of the HA2 peptide. This residue is located within the transmembrane region of the B/Hong Kong/8/73 HA and is therefore not even present in the bromelain-released HA crystal structure of A/Aichi/2/68 (Wilson et al., 1981), and is unlikely to be involved in antigenicity.

A noticeable feature of the sequences of HA genes of monoclonal variants of B/Hong Kong/8/73 is the relatively high incidence of multiple nucleotide changes; 4 of the 8 monoclonal variants sequenced had more than one nucleotide change.

The nucleotide sequences of the HA1 portions of the HA genes of the parent strain and of 18 laboratory-selected antigenic variants of B/Oregon/5/80 have been determined (Berton et al., 1984). Comparison of the deduced amino acid sequences of the parental and variant HA1 polypeptides revealed 16 different amino acid substitutions at 9 positions. All amino acid substitutions resulted from single point mutations and no double mutants were detected. The amino acid changes cluster in 4 regions in the primary structure of the influenza B virus HA and appear to identify at least 2 immunodominant sites which correspond to proposed antigenic sites A and B on the influenza A virus H3 HA. The positions of change are indicated in Fig. 7. However, in the influenza A studies only single amino acid changes were found and monoclonal anitbodies which recognized one region did not recognize any of the other antigenic sites. In both the influenza B studies monoclonal antibodies which belong to the same antigenic group recognize amino acid changes in regions corresponding to both antigenic sites A and B of the H3 HA (Berton et al., 1984, Hovanec and Air, 1984).

One interpretation of these results is that, although the basic three-dimensional structure of the B HA may be similar to that of A/Aichi/2/68, the loop regions of antigenic sites A and B are in closer proximity to each other than in the A structure. Although the distance between antigenic sites A and B of the Aichi HA (Wilson et al., 1981) could possibly be spanned by an antibody combining region, results with H3 and H1 variants have not indicated that it is.

The sequence data of the variants suggest that there is a considerable overlap between the antigenic regions of the influenza B HA molecule, and distinct antigenic sites, as found in the HA of A/Aichi/2/68 may not exist as such in the HA of influenza B viruses. The HA1 peptide in B viruses may be folded into a more compact structure, such that a relatively larger proportion of the sequence is recognized by a single monoclonal antibody.

```
                              10                  20                  30                  40                  50                  60
B/LEE/40    D R I C T G I T S S N S P H V V K T A T Q G E V N V T G V I P L T T T P T K S H F A N L K G T Q T R G K L C P N C F N C
B/MD/59                                                                                                                       L
B/HK/73                                                                                                                       L
B/SING/79                                                                                                 K                   L
B/ORE/80                                                                                                  K                   L

                              70                  80                  90                 100                 110                 120
B/LEE/40    T D L D V A L G R P K C M G N T P S A K V S I L H E V K P A T S G C F P I M H D R T K I R Q L P N L L R G Y E N I R L S
B/MD/59         M                       T I                           V
B/HK/73                                 T I         A                 V
B/SING/79                               T I         A                 V
B/ORE/80                                T I         A                 V

                             130                 140                 150                 160                 170                 180
                                                    *           *     #                       * *     #
B/LEE/40    T S N V I N T E T A P G G P Y K V G T S G S C P N V A N G N G F F N T M A W V I P K D N N K T A I N P V T V E V P Y I C
B/MD/59       R         A                 T                     T     K       E         A V     N K         T     L
B/HK/73     A R     T   A                 I                     T             A         A V     - -         T     L
B/SING/79     R         A   R             I I                   T             A         A V       -         T     L
B/ORE/80      R         A   R             I I                   T             A         A V       -         T     L

                             190                 200                 210                 220                 230                 240
                                                      #
                                          # #   #   # *         #                                                                 #
B/LEE/40    S E G E D Q I T V W G F H S D D K T Q M E R L Y G D S N P Q K F T S S A N G V T T H Y V S Q I G G F P N Q T E D E G L K
B/MD/59     T K                             E   L   V I           K
B/HK/73     T K                             E       V K           K                                                   A           P
B/SING/79   T K                           T E       V K           K                                                         G     P
B/ORE/80    T K                           N E A     V K           K                                                         G     P

                             250                 260                 270                 280                 290                 300
B/LEE/40    Q S G R I V V D Y M V Q K P G K T G T I V Y Q R G I L L P Q K V W C A S G R S K V I K G S L P L I G E A D C L H E K Y G
B/MD/59                       I                               V
B/HK/73                                         A         V                         R       E
B/SING/79                                                 V
B/ORE/80                                                  V

                             310                 320                 330                 340                 350
B/LEE/40    G L N K S K P Y Y T G E H A K A I G N C P I W V K T P L K L A N G T K Y R P P A K L L K E R
B/MD/59
B/HK/73
B/SING/79
B/ORE/80
```

Figure 7 Sequences of the HA1 polypeptide of influenza B strains. Shown is the sequence of B/Lee/40 (Krystal et al., 1982), and differences in B/Maryland/59 (Krystal et al., 1983), B/Hong Kong/8/73 (Hovanec and Air, 1984) B/Singapore/222/79 (Verhoeyer et al., 1983) and B/Oregon/5/80 (Berton et al., 1984) are shown. Amino acids which change in variants of B/Oregon/5/80 (#) and B/Hong Kong/8/73 (*) are indicated.

Structure of the Neuraminidase (NA)

The NA accounts for about 5-10% of the virus protein and exists as a mushroom-shaped spike on the surface of the virion. It is a tetramer with a box-shaped head, 100 x 100 x 60Å, made out of four coplanar and roughly spherical subunits and a centrally attached stalk containing a hydrophobic region by which it is embedded in the viral membrane (Fig 8). Several roles have been suggested for the neuraminidase (acylneuraminyl hydrolase, EC 3.2.1.18). The enzyme catalyses cleavage of the α-ketosidic linkage between terminal sialic acid and an adjacent sugar residue. This reaction permits transport of the virus through mucin and destroys the hemagglutinin receptor on the host cell, thus allowing elution of progeny virus particles from infected cells. The removal of sialic acid from the carbohydrate moiety of newly synthesized hemagglutinin and neuraminidase is also necessary to prevent self-aggregation of the virus. In general, then, the role of neuraminidase may be to facilitate mobility of the virus both to and from the site of infection.

The neuraminidase molecule is composed of a single polypeptide chain, coded by RNA segment 6 and it is oriented in the virus membrane in the opposite way to hemagglutinin (Fig. 3). No post-translational cleavage of the NA polypeptide occurs, no signal peptide is split off and even the initiating methionine is retained (Blok et al., 1982). No processing at the C-terminus takes place - the C-terminal sequence, -MET-PRO-ILE predicted from the gene sequence for N2 NA is found in intact NA molecules isolated from virus, and in the pronase-released NA heads. A sequence of six polar amino acids at the N-terminus of the NA polypeptide, which are totally conserved in each of the 9 different NA subtypes, is followed by a sequence of hydrophobic amino acids which must represent the transmembrane region of the NA stalk. This sequence is not conserved at all between subtypes (apart from conservation of hydrophobicity). Pronase cleaves the polypeptide in the positions shown (Fig. 3), removing the stalk and releasing the enzymically and antigenically active head of the NA which, in some cases, can be crystallized (Fig. 9).

The three-dimensional structure of N2 neuraminidase heads determined from an electron density map at 2.9Å resolution, shows that each monomer is composed of six topologically identical β-sheets arranged in a propeller formation (Varghese et al., 1983). The tetrameric enzyme has circular 4-fold symmetry stabilized in part by metal ions bound on the symmetry axis. The catalytic sites are located in deep clefts which occur on the upper corners of the box-shaped tetramer. Sugar residues are attached to four of the five potential glycosylation sequences, and in one case contribute to the interaction between subunits in the tetramer (Fig. 11).

Type B influenza virus RNA segment 6 codes two distinct glycoproteins, NA and NB, using overlapping reading frames. The amino acid sequence of polypeptide NB deduced from the nueclotide sequence of the B/Lee/40 strain consists of 100 amino acids with a molecular weight of 11,242. The sequence contains four potential glycosylation sites, and the protein has been found to be glycosylated in infected cells. NB has not been found in virions. A protein analogous to NB has not been found in influenza A virus and this represents a major difference between the two virus types.

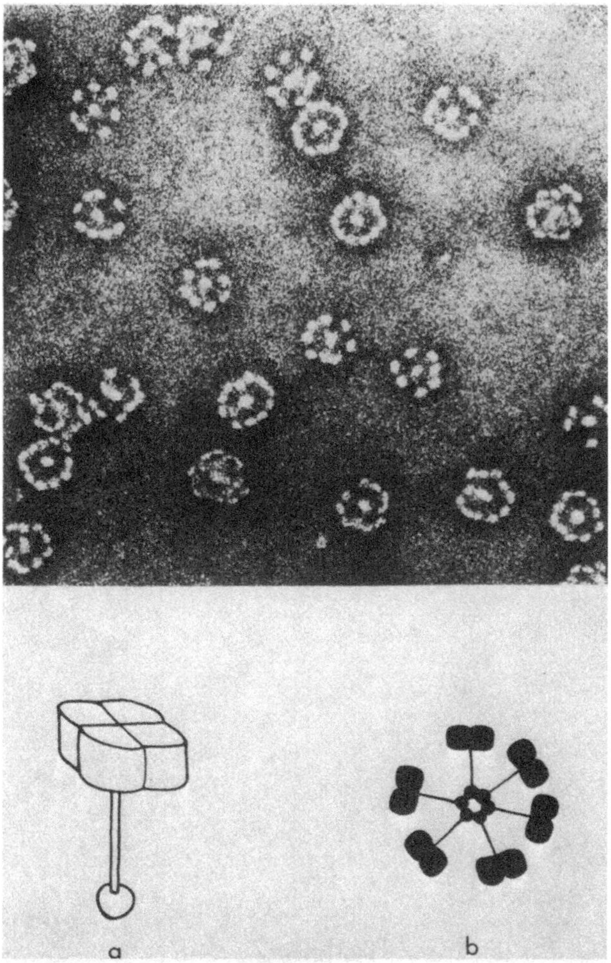

Figure 8 Electron micrograph and diagram showing detergent-released neuraminidase molecules from which the detergent has been removed. The NA molecules have aggregated by the hydrophobic region near the end of the stalk, which served to attach the NA to the lipid of the virus envelope. Treatment of virus particles with pronase cleaves the stalk and releases the head of the NA, which carries the enzymic and antigenic activities of the molecule and, in some cases, can be crystallized. (Electron micrograph courtesy of Nick Wrigley).

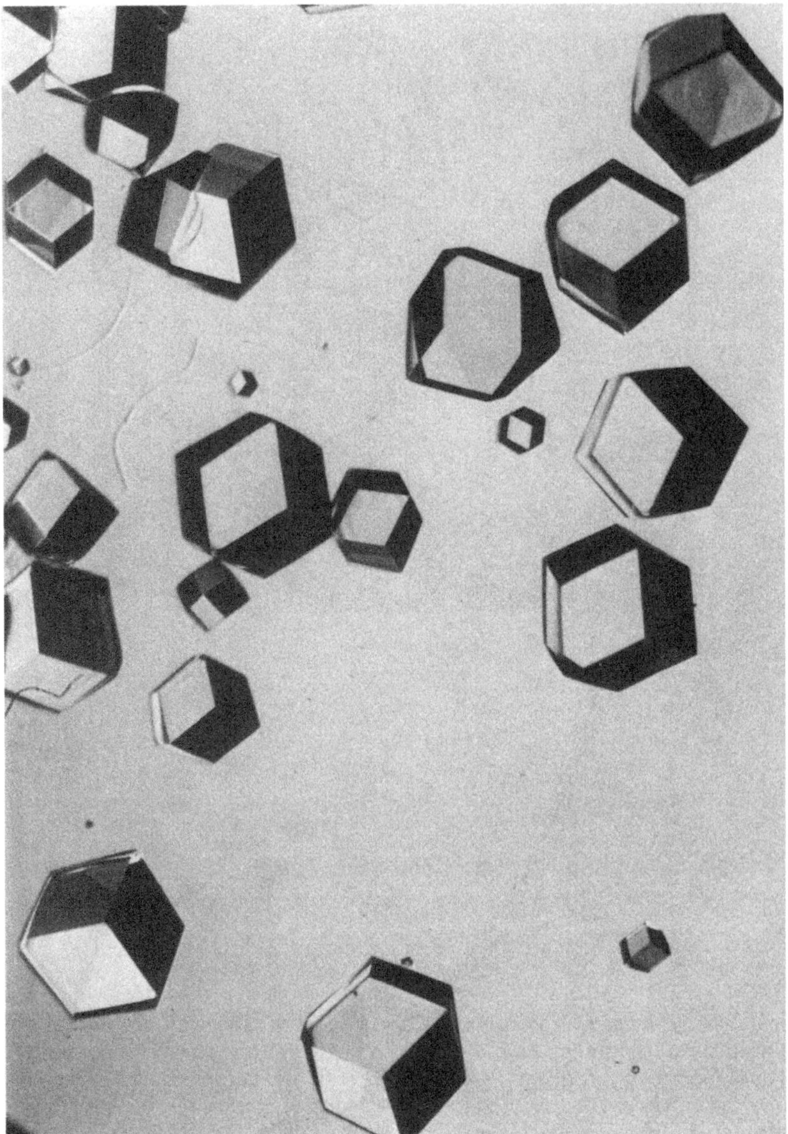

Figure 9 Crystals of pronase-released neuraminidase heads of the N9 subtype suitable for X-ray diffraction studies. The crystals were grown from phosphate buffer solution by vapour diffusion.

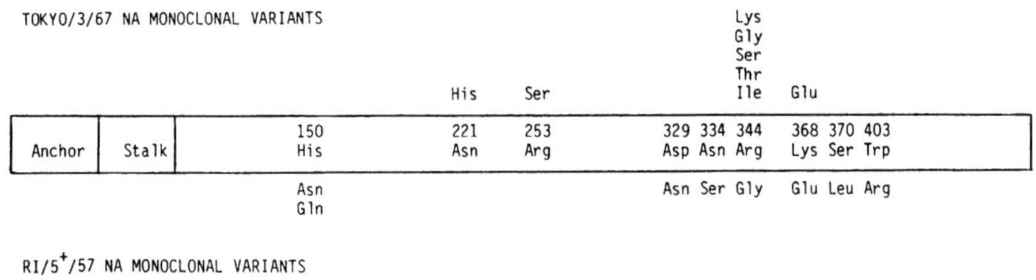

Figure 10 Amino acid substitutions found in variants of NA selected with monoclonal antibodies of Tokyo/3/67 (Lentz et al., 1984) and RI/5^+/57 (Air and Els, unpublished results).

Antigenic variants of neuraminidase

A number of monoclonal antibodies to neuraminidase of the N2 subtype have been obtained and competitive radioimmunoassays with these antibodies showed that the NA possessed at least 4 overlapping antigenic regions. Some of the antibodies were found to effectively neutralize the infectivity of the virus, but careful experiments showed that these antibodies to the NA did not prevent the virus from infecting cells. However, low concentration of the antibodies in the fluid surrounding the cells prevented these from yielding virus and these monoclonal antibodies could be used to select antigenic variants of the NA (Webster et al., 1982, Webster et al., 1984).

These variants were found to have single amino acid sequence changes in the NA polypeptide (Fig. 10).

The regions in which these changes occurred were certainly involved in the antigenicity of the NA and the location of the sequence changes on the 3-dimensional structure of the NA (Colman et al., 1983) is shown in Fig. 11.

It is not known however, if the amino acids which changed formed part of the antigenic determinants recognized by the monoclonal antibodies used to select the variants, or whether the sequence alterations caused conformational changes in the NA which in turn altered an antigenic site some distance away. Attempts have been made to answer this question by examining the effects of chemical modification of certain amino acid residues on the antigenic properties of the NA. N2 Neuraminidase was reacted with succinic anhydride until 70% of the lysine was modified. Despite the drastic alteration in charge, the succinylated NA reacted with polyclonal antisera in immunodiffusion tests to almost the same extent as untreated NA (Fig. 12). This surprising finding suggests that lysines do not exist in the antigenic sites of the NA (Laver et al., 1983). When the succinylated NA was tested with a panel of monoclonal antibodies, some failed to bind, indicating that a small number of antigenic sites were destroyed, but why the majority are unaffected by succinylation remains a mystery. Similar results were obtained when other reagents were used, and when similar experiments were done using the HA (Laver et al., 1981).

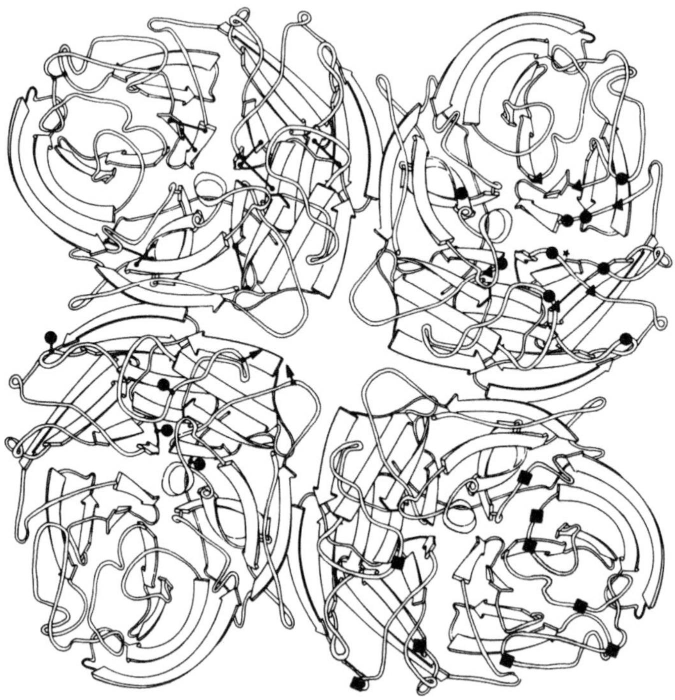

Figure 11 Schematic of the neuraminidase tetramer viewed from above down the symmetry axis. The four subunits high-light different features of the structure.

Top left: disulfide bonds.

Bottom left: carbohydrate attachment sites at 86, 146, 200 and 234 and metal ligands ASP 113 and ASP 114

Bottom right: Residues which changed in variants selected with monoclonal antibodies.

Top right: conserved acidic and basic residues in influenza A and B neuraminidase. *Marks the sialic acid binding site.

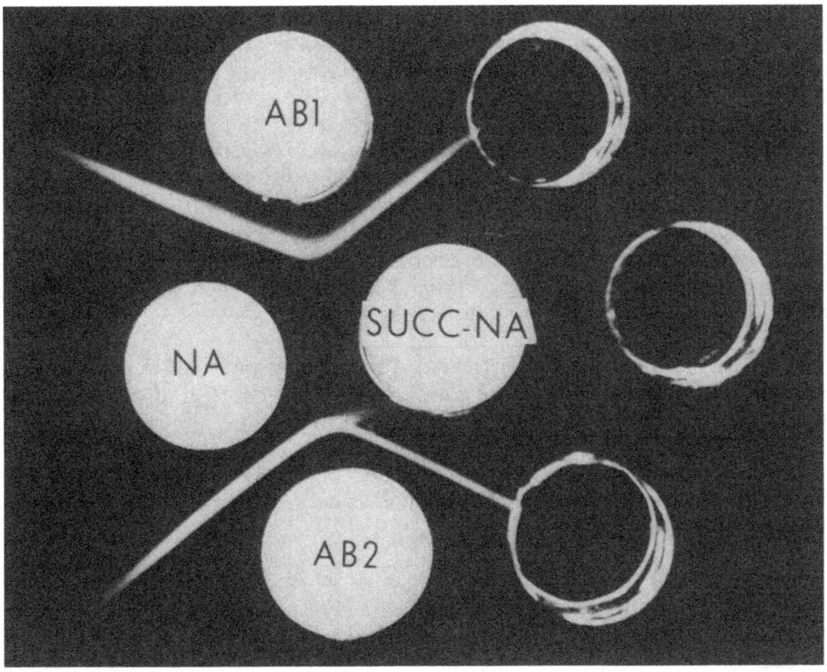

Figure 12. Double immunodiffusion tests showing: Top: line of identity, bottom: small spur between Tokyo NA and succinylated Tokyo NA when tested with polyclonal antisera from two individual rabbits.

Despite the extensive studies using monoclonal antibodies and sequence analysis of variants, the precise nature of the interaction between influenza antigens and antibody molecules is not known. Precise and unambiguous information about the structure of individual antigenic sites on the HA and NA and the way in which these change during antigenic drift may come only from X-ray diffraction analyses of crystals of complexes between the Fab fragments of monoclonal antibodies and the antigens (Colman et al., 1982). This work is currently being undertaken.

Conclusions

We do not know how major shifts occur in type A influenza virus (several mechanisms have been proposed) and this topic is not covered here. The way in which drift occurs, on the other hand, is slowly being understood.

The 3-dimensional structures of both surface antigens of influenza virus (hemagglutinin and neuraminidase) are known, some of the antigenic sites have been located and the sequence changes which alter these sites during drift have been determined.

Many unsolved problems remain, however. We do not know why drift occurs in influenza virus when other viruses, such as measles, which yield variants with the same frequency as influenza, when grown in the presence of monoclonal antibodies, do not vary antigenically in nature. We do not know precisely how the single sequence changes in the HA and NA totally abolish binding by monoclonal antibodies. Binding is abolished even when the change is to a similar amino acid (e.g., ARG to LYS).

Why does drift in type A influenza follow a set path while drift in type B apparantly does not. Can drift be predicted? Why do few, if any, sequential changes occur at particular positions in the HA molecule?

Why do particular areas on the HA and NA induce neutralizing antibody while others do not? Does drift involve changes in the immunogenicity of different antigenic determinants?

Does drift continue indefinitely? Can it go backwards? It is not known if the same antigenic change occurs simultaneously in different areas of the world or whether a single mutant arises somewhere and all spread derives from a singel focus.

The answers to these questions may come from a deeper understanding of the structure of individual antigenic sites on the HA and NA and the complementary sites on the antibody molecules with which they interact.

ACKNOWLEDGEMENT

The writing of this article was greatly assisted by international telephone facilities donated by the Australian Overseas Telecommunications Commission.

REFERENCES

Air, G.M. (1981). Proc. Natl. Acad. Sci. 78, 7639-7643.

Air, G.M. and Compans, R.W. (1983). In "Genetics of Influenza Viruses" (P. Palese and D.W. Kingsbury, eds.), pp. 280-304. Springer-Verlag, New York.

Berton, M.T., Naeve, C.W., and Webster, R.G. (1984). Submitted.

Blok, J., Air, G.M., Laver, W.G., Ward, C.W., Lilley, G.G., Woods, E.F., Roxburgh, C.M., and Inglis, A.S. (1982). Virology, 119, 109 - 121.

Both, G.W. and Sleigh, M.J. (1981). J. Virol. 39, 663-672.

Caton, A.J., Brownlee, G.G., Yewdell, J.W., and Gerhard, W (1982). Cell 31, 417-427.

Colman, P.M., Varghese, J.N., and Laver W.G. (1983). Nature 303, 41-44.

Colman, P.M., Gough, K.H., Lilley, G.G., Blagrove, R.J., Webster, R.G., and Laver, W.G. (1982). J. Mol Biol 152, 609-614.

Compans, R.W. and Choppin, P.W. (1975). In Comprehensive Virology (H. Fraenkel-Conrat and R.R. Wagner, eds.), Vol IV, pp. 179-252 Plenum.

Daniels, R.S., Douglas, A.R., Gonsalves-Scarano, F., Palu, G., Skehel, J.J., Brown, E., Knossow, M., Wilson, I.A., and Wiley, D.C. (1983). In: The Origin of Pandemic Influenza Viruses, W.G. Laver (ed.), Elsevier Sci. Pub. Co., Inc., 9 -18.

Fang, R., Min Jou, W., Huylebroeck, D., Devos, R., and Fiers, W. (1981). Cell 25, 315-323.

Gerhard, W., Yewdell, J., Frankel, M., and Webster, R.G. (1981). Nature 290, 713-717.

Gething, M.J., White, J.M., and Waterfield, M.D. (1978). Proc. Natl. Acad. Sci. U.S.A. 75, 2737-2740.

Herrler, G., Nagele, A., Meier-Ewert, H., Bhown, A.S., and Compans, R.W. (1981). Virology 113, 439-451.

Hewat, E.A., Cusack, S., Ruigrok, R.W.H. and Verwey, C. (1984). J. Mol. Biol. 175, 175-193.

Hinshaw, V.S., Air, G.M., Gibbs, A.J., Graves, L., Prescott, B., and Karunakaran, D. (1982). J. Virol. 42, 865-872.

Hovanec, D.L. and Air, G.M. (1984). Submitted.

Krug, R.M. (1983). In Genetics of Influenza Viruses. (P. Palese and D.W. Kingsbury, eds.), pp. 70-98.

Krystal, M., Elliott, R.M., Benz, E.W., Young, J.F., and Palese, P. (1982). Proc. Natl. Acad. Sci. USA 79, 4800-4804.

Krystal, M., Young, J.F., Palese, P., Wilson, I.A., Skehel, J.J., and Wiley, D.C. (1983). Proc. Natl. Acad. Sci. USA 80, 4527 - 4531. Correction. Proc. Natl. Acad. Sci. USA 81, 1261.

Lamb, R.A. (1983). In Genetics of Influenza viruses (P. Palese and D.W. Kingsbury, eds.), pp. 21-94, Springer-Verlag.

Laver, W.G., Air, G.M., and Webster R.G. (1981). J. Mol. Biol. 145, 339-361.

Laver, W.G., Webster, R.G., and Air, G.M. (1983). In The Origin of Pandemic Influenza Viruses (W.G. Laver, ed.), 113-120 Elsevier.

Laver, W.G., Air, G.M., and Webster, R.G. (1981). Virology 111, 538-548.

Laver, W.G., Air, G.M., Webster, R.G., Gerhard, W., Ward, C.W., and Dopheide, T.A.A. (1979). Virology 98, 226-237.

Lentz, M.R., Air, G.M., Laver, W.G., and Webster, R.G. (1984). Virology 135, 257-265.

Lu, B.L., Webster, R.G., Brown, L.E., and Nerome, K. (1983). Bull. WHO 61, 681-687.

Maeda, T. and Ohnishi, S. (1980). FEBS Lett. 122, 283-287.
Nakada, S., Creager, R.S., Krystal, M., Aaronson, R.P., and Palese, P. (1984) J. Virol 50, 118-124.
Nakajima, K., Desselberger, U., and Palese, P. (1978). Nature 274, 334-339.
Nakajima, S., Nakajima, K., and Kendal, A.P. (1983). Virology 131, 116-127.
Newton, S.E., Air, G.M., Webster, R.G., and Laver, W.G. (1983). Virology, 128, 495-501.
Palese, P., Tobita, K., Veda, M. and Compans, R.W. (1974). Virology 61, 397-410.
Pfeifer, J.B. and Compans, R.W. (1984). Virus Research in press.
Scholtissek, C., van Hoyningen, V., and Rott, R. (1978). Virology 89, 613-617.
Shaw, M.V., Lamb, R.A., Erickson, B.W., Briedis, D.J., and Choppin, P.W. (1982). Proc Natl Acad Sci 79, 6817 - 6821.
Skehel, J.J., Bayley, P.M., Brown, E.B., Martin S.R., Waterfield, M.D., White, J.M., Wilson, I.A., and Wiley, D.C. (1982). Proc. Natl. Acad. Sci. USA 79, 968-972.
Varghese, J.M., Laver, W.G., and Colman, P.M. (1983). Nature 303, 35-40.
Verhoeyen, M., Van Rompuy, L., Min Jou, W., Huylebroeck, D., and Fiers, W. (1983). Nucl. Acids Res. 11, 4703-4712.
Ward, C.W. (1981). Curr. Topics Microbiol. Immunol. 94/95, 1-74.
Ward, C. and Dopheide, T. (1982). In: Genetic Variation Among Influenza Viruses (Nayak, D.P., Fox, C.F., eds.), (ICN-UCLA Symposia on Molecular and Cellular Biology, Vol XXI), 323-340, New York: Academic Press.
Webster, R.G. and Berton, M.T. (1981). J. Gen. Virol. 54, 243-251.
Webster, R.G., Brown, L.E., and Laver, W.G. (1984). Virology 135, 30-42.
Webster, R.G., Laver, W.G., and Air G.M. (1983). In Genetics of Influenza Viruses (P. Palese and D.W. Kingsbury, eds.), pp. 127-168.
Webster, R.G., Hinshaw, V.S., and Laver, W.G. (1982). Virology 117, 93-104.
W.H.O. Memorandum: (1980). Bull. Wld. Hlth. Org. 58, 585-591.
Wiley, D.C., Wilson, I.A., and Skehel, J.J. (1981). Nature 289, 373-378.
Wilson, I.A., Skehel, J.J., and Wiley, D.C. (1981). Nature 289, 366-373.
Wrigley, N.G. (1979). Brit. Med. Bull, 35, 35-38.

IMMUNE INDUCTION BY A PROTEIN ANTIGEN AND BY A PEPTIDE SEGMENT OF THE PROTEIN

Anne Marie Wan, Charles D. Estin, Beatrice C. Langton, Matthew L. Andria, and Eli Benjamini

Department of Medical Microbiology and Immunology
School of Medicine, University of California
Davis, California

ABSTRACT

The immune induction by a protein (the tobacco mosaic virus protein - TMVP) was compared to the immune induction by the free, non-conjugated eicosa tryptic peptide fragment of the protein (tryptic peptide 8 representing residues 93-112 of the protein). The results demonstrated that like TMVP, peptide 8 was immunogenic in A/J mice. TMVP and peptide 8 do not cross react on the T cell level. However, immunization with TMVP or with peptide 8 induces antibodies which react with both TMVP and peptide 8. Characterization of the antibodies produced by both immunogens revealed that: 1) their isotope composition is similar with IgG1 and IgG2 being the predominant isotypes; this composition indicates that both immunogens are T cell dependent antigens, 2) the antibodies induced by TMVP and by peptide 8 are directed against the C-terminal decapeptide portion of peptide 8 (residues 103-112 of the protein), 3) the fine specificity of these antibodies is the same. These results, and results of adoptive transfer experiments, indicate that antigen specific T cells had no effect on the expression of the fine antibody specificity. The results demonstrate the feasibility of immunizing with a portion of a protein for the purpose of inducing antibodies with the same isotype composition and specificity towards a protein epitope as those induced by immunization with the whole protein.

INTRODUCTION

There is currently great interest in the production of synthetic vaccines specific for protein antigens of etiologic agents. These vaccines utilize peptides of the protein antigen, either free or in the form of peptide-protein conjugates, as the immunogen. The rational behind the use of these vaccines is to elicit an effective immune response against the etiologic agent through the use of simplified immunogens. Peptides consisting of portions of the influenza virus hemagglutinin protein (Green et al., 1982; Muller, 1982), hepatitis B virus surface antigen (Prince et al., 1982), foot and mouth disease virus protein (Bittle et al., 1982), or rabies virus protein (Lerner, 1982), serve as examples. This current interest in synthetic vaccines and optimization of their efficacy has refocused many questions regarding the nature of antigenic determinants of proteins, the presentation of their epitopes to T and B lymphocytes and the activation of these lymphocytes leading to the induction of an immune response.

Considerations in the Design of Synthetic Vaccines

From work performed during the last twenty years it is apparent that some epitopes of protein antigens are highly dependent upon conformation while others are less dependent. Thus, some isolated fragments from protein antigens do not react with antibodies which do react with the same amino acid sequences, but in the conformation they assume as a part of the whole protein (Brown et al., 1959; Crumpton and Sonall, 1967; Gerwing and Thompson, 1968; Arnon and Sela, 1969; Sela, 1969; Arnon et al., 1971; Schechter, 1971; Scibienski et al., 1978). In contrast, it has been demonstrated that other small isolated fragments of protein antigens are capable of reacting with antibodies elicited by immunization with the whole protein (Benjamini et al., 1964; Atassi, 1975; Hurrell et al., 1978; Chavez and Scheraga, 1979; Sakata and Atassi, 1980) Moreover, immunization with small peptides conjugated to protein carriers can induce antibodies capable of reacting not only with the peptide but also with the native protein. Thus, Anderer and coworkers demonstrated as early as 1963, that immunization of rabbits with the tobacco mosaic virus led to the production of antibodies which recognized the C-terminal hexapeptide of the viral protein (Anderer, 1963). Subsequently they showed that immunization with the C-terminal hexapeptide or even tripeptide conjugated to a protein carrier induced antibodies capable of reacting with the peptide as well as with the whole virus (Anderer and Schlumberger, 1965). In fact, these antibodies were capable of neutralizing viral infectivity.

Similarly, work from our laboratory performed over fifteen years ago, showed that immunization of rabbits with the isolated tobacco mosaic virus protein (TMVP) led to the induction of antibodies capable of reacting with an isolated pentapeptide of the TMVP representing residues 107-112 of the protein (Young et al., 1967). Immunization with this pentapeptide conjugated to a protein carrier led to the induction of antibodies capable of binding with the peptide as well as with the parent globular protein (Fearney et.al., 1971). It was also shown in our laboratory that immunization of rabbits with a synthetic octapeptide related to collagen conjugated to a protein carrier induced antibodies capable of reacting with the peptide as well as with collagen derivatives of many species (Kettman et al., 1967; Benjamini et al., 1973). Subsequent to these early observations several investigators demonstrated that an immune response to various proteins can be induced by immunization with peptide fragments of the proteins conjugated to heterologous carrier proteins (Langbeheim et al., 1976; Green et al., 1982; Muller et al., 1982; Prince et al., 1982; Lerner, 1982). This cross reactivity at the antibody level between an epitope in a native protein and that epitope contained within a peptide fragment of the protein has also been demonstrated for epitopes which are highly conformation dependent. For example, the "loop peptide" of lysozyme, representing residues 64-83, constitutes a conformation dependent epitope of this protein (Aron et al., 1971; Aron and Sela, 1969). The tertiary structure of this isolated peptide can be maintained by a disulfide bond between residues 64 and 90.

In this form, the peptide reacts with antibodies to lysozyme. Moreover, immunization with the loop peptide conjugated to a protein carrier, induces the production of antibodies which can react with the loop peptide as well as with lysozyme. In contrast, the reduced peptide does not react with antilysozyme and its conjugate is not capable of inducing antibodies which exhibit binding with lysozyme, thus demonstrating the conformational dependence of this determinant.

These achievements not withstanding, there is an important immunological disadvantage in the use of immunogens which consist of a peptide hapten conjugated to a heterologous carrier protein. In the classical view of immune induction by a hapten-carrier conjugate (Mitchison, 1971), the peptide represents a hapten. The carrier protein provides T cell help for the induction of the various anti-hapten antibodies. Invoking the hapten-carrier concept of immune induction it is easy to visualize how repeated immunization with the conjugate will induce primary and secondary responses with activation of hapten specific memory B cells and carrier specific T helper cells. However, subsequent exposure to the native protein from which the peptide was derived would not stimulate memory B cells. This is because this native protein is unable to activate the expanded T helper cells which are specific for the original carrier protein. Thus, the expanded carrier specific T helper cells are unable to provide T help for the hapten specific B cells in the generation of an anamnestic response to the desired protein. Consequently, in order to create a synthetic vaccine capable of activating an anamnestic response upon encounter with the native protein it is imperative that an immunogen be constructed from areas of the native protein which are recognized by T cells and by B cells. Such an immunogen should then induce both T and B responses which would be of similar specificities to those induced by the native protein.

It has been shown for several protein antigens that while some epitopes are recognized by and are capable of activating both T and B lymphocytes (Spitler et al., 1970; Waterfield et al., 1972; Bixler and Atassi, 1983), there are other epitopes which activate exclusively T lymphocytes or exclusively B lymphocytes (Salvin and Smith, 1960; Kaattari et.al., 1980; Bixler and Atassi, 1983; Senyk et al., 1983). Thus, for optimal T-B cell collaboration and immunogenicity, it is desirable to construct an immunogen which contains distinct T and B cell recognizable areas of the whole protein. Indeed, large protein fragments (Hashim, 1978) as well as free small peptides (Young and Atassi, 1982), have been shown to induce immune responses to the whole protein suggesting that these fragments contain epitopes capable of activating both T and B lymphocytes. Such synthetic peptide vaccines would therefore, be expected to induce an immune response to the native protein which would result in an anamnestic response upon further encounter with the protein.

The observation that an immune response to a protein antigen can be induced by peptide vaccines consisting of free, non-conjugated peptides raises several questions of biological significance regarding the nature of the response and how it compares to the response induced by immunization with the native protein. First, how does the fine specificity of the antibodies induced by these two immunogens compare? Second what role do antigen specific T cells play in generating the antibody response to these two immunogens? Specifically how do the immunoglobulin isotype profiles compare, and what influence, if any, does the specificity of the T helper cell have on the fine specificity of the elicited antibodies? The latter question is of particular interest since it has been suggested by several workers that the expression of antibody specificity is influenced by antigen specific T cells (Cecka et al., 1976; Sercarz et al., 1977; Berzofsky, 1983a, Berzofsky, 1983b).

In this communication we present our finding germane to the above questions, utilizing as model antigens the tobacco mosaic virus protein and a peptide segment of this protein.

IMMUNE RESPONSE TO THE TOBACCO MOSAIC VIRUS PROTEIN AND PEPTIDE SEGMENTS OF THIS PROTEIN

For the past twenty years the immune response to the tobacco mosaic virus protein (TMVP) has been the subject of extensive investigations in our laboratory (Benjamini, 1979). Because of its known amino acid sequence (Anderer et al., 1960; Tsugita et al., 1960) and three dimensional structure (Champness et al., 1976). This protein has proved to be very useful in eludicating the specificity and the regulation of the immune response to a defined region of a protein antigen. TMVP is a globular protein consisting of 158 amino acids. It is a single polypeptide chain containing several alpha-helical regions, but no disulfide bonds. TMVP can be cleaved by trypsin to yield twelve tryptic peptides (Woody and Knight, 1959). We have demonstrated that tryptic peptide number 8, representing residues 93-112 of the protein, contains a major antigenic determinant of the protein, recognized by anti-TMVP sera from many species (Benjamini, 1977). For rabbits and many strains of mice, immunological specificity and activity have been localized in the C-terminal decapeptide of peptide 8 (Benjamini, 1977; Morrow et.al., 1983). The ability of this antigenic area to induce antibodies has been investigated in a variety of animals. These studies showed that the non-conjugated C-terminal pentapeptide or even C-terminal decapeptide portions of peptide 8 are not immunogenic (Spitler et al., 1970). However, a conjugate consisting of a protein carrier to which the C-terminal pentapeptide is coupled through its N-terminus was immunogenic in rabbits (Fearney et.al., 1971). Immunization with the conjugate induces antibodies capable of binding with C-terminal peptides of tryptic peptide 8. In addition, these antibodies were shown to bind with the whole native protein, TMVP.

The findings that the free C-terminal decapeptide is nonimmunogenic, but that it is capable of inducing in rabbits anti-TMVP antibodies when conjugated to a protein carrier indicate that the peptide contains areas recognized only by B cells. T helper cells have to be provided by the protein carrier. In the generation of anti-decapeptide antibodies by the native protein TMVP, anti-TMVP T and B cell populations are expanded. Subsequent encounter with TMVP leads to activation of decapeptide specific memory B cells resulting in an anamnestic response to the decapeptide. However, when the immunogen consists of a decapeptide conjugated to a heterologous carrier protein, immunization expands a decapeptide specific B cell population, but, of course, would not expand TMVP specific T helper cells. As such, the conjugate would be ineffective in generating a functional memory B cell response upon encounter with the parent protein TMVP.

The inability of immunogens consisting of peptide haptens conjugated to protein carriers to induce a state of immunity, such that subsequent encounter with the protein antigen would result in an anamnestic response, must be overcome in the development of effective synthetic peptide vaccines. It should be possible to circumvent this problem by immunizing with a fragment of the parent protein containing both T and B cell determinants. We approached this goal by first identifying a T cell determinant in tryptic peptide 8 of TMVP, in addition to the B cell determinant already discussed. We then assessed the immunogenicity of this entire peptide with respect to the induction of T cells and antibodies. Furthermore, we compared the immune induction by peptide 8 to the immune induction by the whole protein TMVP. The results of these studies are discussed below.

Identification of a T cell Determinant in Tryptic Peptide 8

The T cell responses of TMVP immune mice to various tryptic peptides of TMVP have been extensively investigated in our laboratory. T cell responsiveness was assessed by measuring the *in vitro* proliferation of lymph node T cells in the presence of antigen. The results revealed that differences exist in the T cell responses of various strains of mice to different regions of the protein. Germane to the present discussion are results shown in Table 1 which indicate that T cells from all TMVP immune mouse strains respond to TMVP, but not all respond to peptide 8. The responsiveness to peptide 8 appears

Table 1. The stimulation by TMVP and by peptide 8 of T cells obtained from lymph nodes 14 days following immunization of various strains of mice with 10 μg TMVP. Culture conditions were essentially as described (Corradin et al., 1977). A strain is considered a responder strain (+) to an antigen when the incorporation of tritiated thymidine in the presence of 5 to 30 μM antigen is >1.5 times incorporation of tritiated thymidine in the absence of antigen.

STRAIN	H-2	MHC						RESPONSE	
		K	I-A	I-J	I-E	S	D	TMVP	PEPTIDE 8
A/J	a	k	kkk	k	k	d	d	+	-
CSW	b	b	bbb	b	b	b	b	+	+
CWB	b	b	bbb	b	b	b	b	+	+
C57BL/10	b	b	bbb	b	b	b	b	+	+
B10.BR	k	k	kkk	k	k	k	k	+	-
B10.S	s	s	sss	s	s	s	s	+	+
B10.D_2	d	d	ddd	d	d	d	d	+	+
B10.A	a	k	kkk	k	k	d	d	+	-
B10.A (2R)	h2	k	kkk	k	k	d	b	+	-
B10.A (3R)	i3	b	bbb	b	k	d	b	+	+
B10.A (4R)	h4	k	kkk	b	b	b	b	+	-
B10.A (5R)	i5	b	bbb	k	k	d	d	+	+
B10.A (18R)	i18	b	bbb	b	b	b	d	+	+

Table 2. The stimulation by TMVP and by peptide 8 of T cells obtained from lymph nodes 14 days following immunization of the various strains of mice with 10 μg peptide 8. Culture conditions were essentially as described (Corradin et al., 1977).

STRAIN	H-2	MHC K I-A I-J I-E S D	RESPONSE TMVP	RESPONSE PEPTIDE 8
A/J	a	k kkk k k d d	-	+
CSW	b	b bbb b b b b	+	+
C57BL/10	b	b bbb b b b b	+	+
B10.BR	k	k kkk k k k k	-	+
B10.A (5R)	i5	b bbb k k d d	-	+

to be under genetic control as evident by data obtained from recombinant strains of B10.A. Of particular significance to the present discussion are the findings that immunization of A/J mice with TMVP induced T cells capable of being stimulated with TMVP but not with peptide 8. It is interesting that this nonresponsiveness to peptide 8 is a characteristic of strains which are k in the IA region, irrespective of their genetic background. Thus, A/J mice, which are IA of k but of the H-2^a haplotype and B10.BR mice which are also IA of K but of the H-2^k haplotype are both nonresponders to peptide 8 with respect to T cell, following immunization with TMVP.

Immunogenicity of Tryptic Peptide 8

The immunogenicity of free, non-conjugated peptide 8 in various strains of mice was assessed. The results of these studies (Table 2) indicate that peptide 8 is immunogenic to T cells: it induces, in all strains tested, T cells capable of proliferating *in vitro* in the presence of peptide 8. However, the results indicate that the *in vitro* response to TMVP varies between the strains. Again, the results in A/J mice are of particular interest. On the one hand, as indicated in Table 1, immunization of A/J mice with TMVP leads to the induction of TMVP specific T cells but not peptide 8 specific T cells. On the other hand, the findings summarized in Table 2 show that immunization of A/J mice with peptide 8 leads to the induction of T cells capable of reacting with peptide 8 but not with TMVP.

Although no cross reactivity on the T cell level between TMVP and peptide 8 is evident in A/J mice, repeated immunization of A/J mice with 50 μg of peptide 8 in Freund's complete adjuvant, leads to the production of antibodies capable of binding with peptide 8 as well as with TMVP. Results in Figure 1 demonstrate the reactivity with peptide 8 of individual sera from A/J mice immunized with peptide 8.

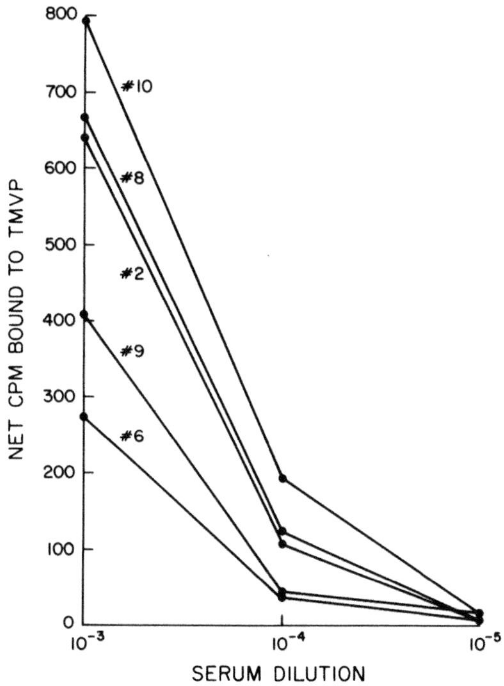

Figure 1. Solid phase radioimmunoassay (SPRIA) for peptide 8 specific antibodies in individual antisera (numbered) from A/J mice immunized intraperitoneally with 100 μg peptide 8 in Freund's complete adjuvant (FCA) followed by several biweekly injections of 50 μg peptide 8 in FCA. SPRIA was performed as described (Mishell and Shiigi, 1980) using peptide 8 conjugated to succinylated BSA (Goodfried et al., 1964) as antigen.

Data in Figure 2 show the reactivity of these same sera with TMVP. The results demonstrate that the antisera exhibit significant binding with both peptide 8 and TMVP. Thus, while immunization with TMVP induces the production of antibodies capable of binding with peptide 8 (Benjamini, 1977), immunization with peptide 8 induces antibodies capable of binding with TMVP.

The findings that in this strain of mice, peptide 8 and TMVP do not cross react on the T cell level yet they both induce antibodies capable of reacting with TMVP and with peptide 8, allow an in depth analysis of the effect of T helper cell populations of different specificities on the fine specificity of antibodies.

Comparison of Antibodies Induced by TMVP and by Peptide 8

The distribution of anti-peptide 8 immunoglobulin isotypes present in the antisera from peptide 8 and TMVP immune mice is presented in Figure 3. The data demonstrate that the isotype composition of antibodies from both groups is essentially the same, with the

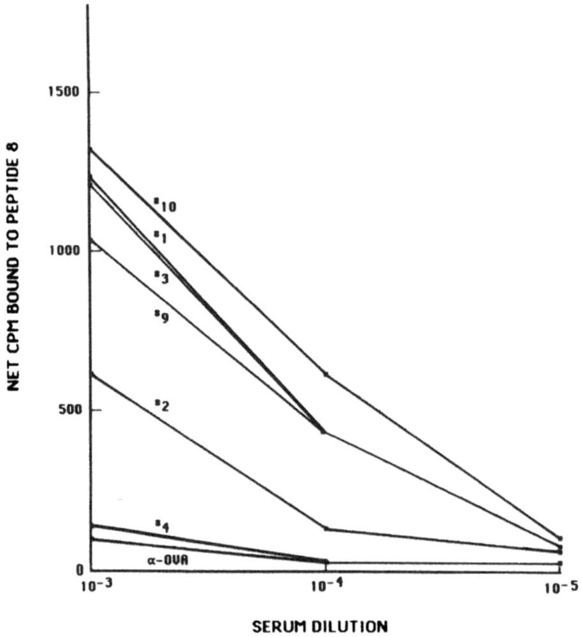

Figure 2. Solid phase radioimmunoassay (SPRIA) for TMVP binding antibodies in individual antisera (numbered) from A/J mice immunized with peptide 8 as described for Fig 1.

predominant isotypes being IgG1, and IgG2a and IgG2b. Since the switch from IgM to IgG involves T cells, this isotype profile indicates that peptide 8, like TMVP (Rennick et al., 1983) is a T dependent antigen.

The specificity of the antibodies induced by both immunogens was assessed. First, the precise area on peptide 8 to which antibodies are directed was determined. This was performed by measuring the ability of the C-terminal decapeptide to inhibit the reaction between labelled peptide 8 and antibodies induced by TMVP or by peptide 8. Results in Figure 4 demonstrate that essentially complete inhibition is achieved by the C-terminal decapeptide of peptide 8. This is true when antisera produced by either immunogen is used. Thus, both TMVP and peptide 8 induce, in A/J mice, antibodies which recognize exclusively the C-terminal decapeptide portion of peptide 8.

Having localized the determinant on peptide 8, the fine specificity of the antibodies to this determinant was assessed. This was achieved by determining the ability of anti-peptide 8 and anti-TMVP antibodies to bind to a panel of several synthetic peptide analogues of the C-terminal decapeptide portion of peptide 8. The results of the direct binding of various labelled peptides with antibodies are summarized in Table 3. These results corroborate the conclusion reached from the inhibition studies, namely that antibodies induced by immunization with TMVP or with peptide 8 are capable of binding with the C-terminal decapeptide portion of peptide 8. For comparision purposes, a value of 1 was assigned to the binding of the decapeptide

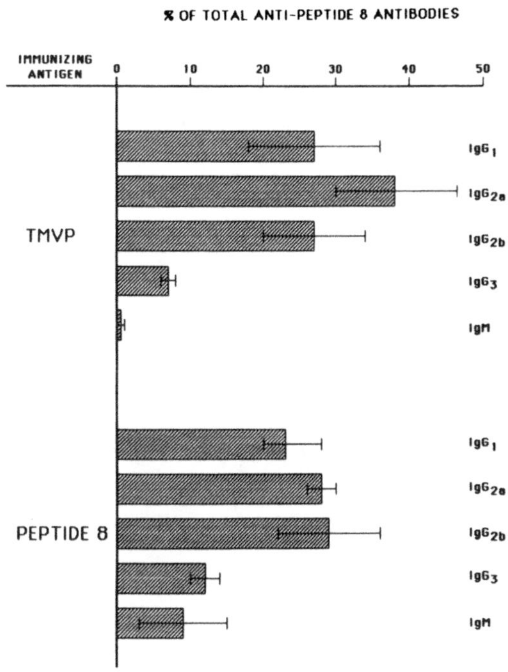

Figure 3. The relative immunoglobulin isotype composition of peptide 8 specific antibodies from A/J mice immunized either with peptide 8 (as described for Figure 1) or with TMVP (100 μg TMVP in Freund's complete adjuvant followed by several biweekly aqueous boosts with 50 μg TMVP). The isotope composition was assessed as described (Rennick et al., 1983).

with antibodies induced by TMVP or by peptide 8. The relative binding with the various analogues of the decapeptide of both sets of antisera is represented as a ratio of the binding with the native decapeptide. The results in Table 3 indicate that antibodies produced by immunization with TMVP or with peptide 8 were incapable of binding with analogues in which the N-terminal portions of the decapeptide were substituted by alanines. In addition, neither of these sera exhibited binding with the synthetic analogue in which alanine at position 110 was replaced by leucine. However, sera induced by both TMVP and peptide 8 exhibited strong heteroclytic binding with an analogue in which the arginine at position 112 was replaced by cysteine. Thus, within the resolution of the peptide panel, the fine specificity pattern of the antibodies induced by both imunogens appears identical. This is in spite of the fact that both immunogens utilize T cells of different specificities. These findings indicate that the antigenic specificity of the T helper cells does not influence the fine specificity of the expressed antibodies.

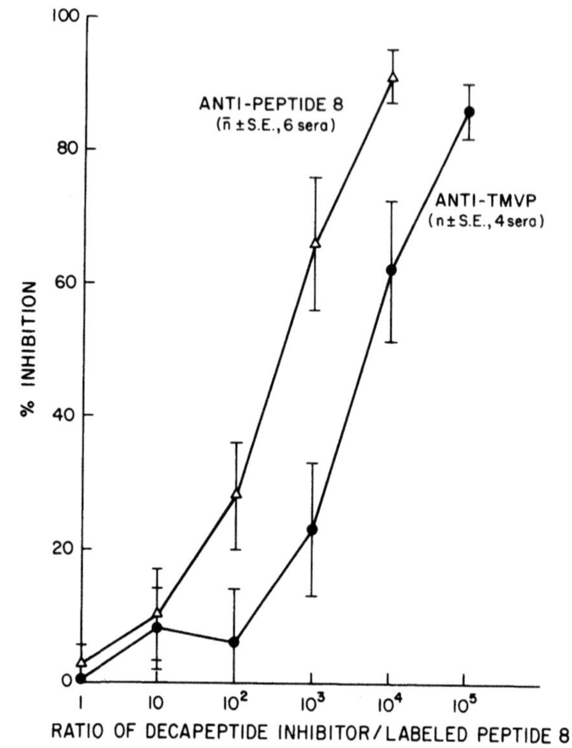

Figure 4. Inhibition of binding between [^{3}H]-N-Acetyl-peptide 8 and antisera from either peptide 8 immune or TMVP immune A/J mice by the C-terminal decapeptide of peptide 8 assessed as described (Rennick et al., 1983).

Table 3. Relative binding of anti-TMVP and anti-peptide 8 sera to a decapeptide (residues 103-112) of TMVP and to synthetic analogues of this decapeptide. Peptides were synthesized as described (Merrifield, 1964).

ANALOGUE ABBREVIATED DESIGNATION	AMINO ACID RESIDUE OF TMVP 103	104	105	106	107	108	109	110	111	112	IMMUNOGEN	RELATIVE BINDING (AVE±SE of 4 Sera)
DECAPEPTIDE	Thr	Thr	Ala	Glu	Thr	Leu	Asp	Ala	Thr	Arg	TMVP	1
											PEPTIDE 8	1
$(Ala)_5$-108-112	Ala	Ala	Ala	Ala	Ala	—	—	—	—	—	TMVP	0.02±0.01
											PEPTIDE 8	0.01±0.004
$(Ala)_4$-107-112	Ala	Ala	Ala	Ala	—	—	—	—	—	—	TMVP	0
											PEPTIDE 8	0.02±0.02
Leu_{110}	—	—	—	—	—	—	—	Leu	—	—	TMVP	0
											PEPTIDE 8	0
Cys_{112}	—	—	—	—	—	—	—	—	—	Cys	TMVP	3.70±0.15
											PEPTIDE 8	2.19±0.49

The Role of T Helper Cells in the Expression of Fine Specificity of Antibodies

The putative role of T cells in regulating the fine specificity of the antibody response was further investigated using another strain of mice, namely the C3H.SW (CSW) strain. This strain was chosen because previous work has shown that anti-TMVP sera from individual mice of this strain exhibit varied patterns of fine specificity to the decapeptide determinant of peptide 8 (Morrow et al., 1983). This was demonstrated by testing the reactivity of individual sera with an enlarged panel of synthetic peptide analogues of the C-terminal decapeptide of peptide 8. This panel is shown in Table 4. The reactivity of anti-TMVP sera produced by individual CSW mice with these synthetic analogues is summarized in Table 5. The response of the mice can be grouped into several major specificity profiles depending upon their pattern of binding to the panel of analogues. The pattern exhibited by serum from a given CSW mouse represents a polyclonal summation of its decapeptide binding antibodies. Long term experiments conducted in our laboratory have shown that once the pattern of binding is set following TMVP immunization of a CSW mouse, the binding specificity remains locked in this pattern throughout the life of the mouse (manuscript in preparation). This suggests that the same B cell clones which are activated during the first time that a given mouse is immunized with TMVP become activated in the same frequency ratios following subsequent exposure to the TMVP antigen. The question may be asked whether a mouse expressing a unique antibody fine specificity to the decapeptide would demonstrate a different fine specificity if a T cell source, from a mouse expressing a different fine specificity pattern, were to supply the T help function.

To elucidate this question, a large number of CSW mice (approximately 100) were immunized with TMVP. It has been repeatedly demonstrated in our laboratory that 70-90% of immune TMVP CSW mice make

Table 4. Panel of synthetic peptides related to the antigenic decapeptide of the tobacco mosaic virus protein (TMVP). Peptides were synthesized as described (Merrifield, 1964).

ABBREVIATED DESIGNATION	AMINO ACID RESIDUE OF TMVP									
	103	104	105	106	107	108	109	110	111	112
DECAPEPTIDE	Thr	Thr	Ala	Glu	Thr	Leu	Asp	Ala	Thr	Arg
Ala_{103}	Ala	—	—	—	—	—	—	—	—	—
$(Ala)_2$-105-112	Ala	Ala	—	—	—	—	—	—	—	—
$(Ala)_4$-107-112	Ala	Ala	Ala	Ala	—	—	—	—	—	—
$(Ala)_5$-108-112	Ala	Ala	Ala	Ala	Ala	—	—	—	—	—
Tyr_{106}	—	—	—	Tyr	—	—	—	—	—	—
Met_{107}	—	—	—	—	Met	—	—	—	—	—
Leu_{110}	—	—	—	—	—	—	—	Leu	—	—
Cys_{112}	—	—	—	—	—	—	—	—	—	Cys

Table 5. Fine specificities of decapeptide specific CSW anti-TMVP antibodies (Taken from Morrow et al., 1983).

BINDING PATTERN	# OF ANIMALS	103-112	ALA_{103}	$(ALA)_2$-107-112	$(ALA)_5$-108-112	TYR_{106}	MET_{107}	LEU_{110}	CYS_{112}
Monoclonal C10		+++	+++	+++	++	+++	++++	+++	-
Monoclonal B2		+++	++	++	-	+	-	-	++++
1	3	+++	+++	++++	++	+++	+++	+++	-
2	4	+++	++	++	+	+	++	-	+
3	2	+++	++	++	++	++	++++	±	±
4	5	+++	+++	++	+	+	++	∓	-
5	2	+++	++++	++++	++	++	++++	++	++
6	2	+++	++	++	++	+	+++	±	++
7	1	+++	+	+	-	++	+	-	++++
8	2	+++	+	∓	±	±	+++	-	++++
9	1	+++	+++	-	-	-	++++	-	++++

anti-decapeptide antibodies (Morrow et al., 1983). The fine specificity pattern of antibodies produced by each mouse was determined using the panel of synthetic peptides. The mice were rested for seven months to allow the anti-decapeptide titers to drop to undetectable levels. At this point half of the mice were irradiated with 600 RADS (this procedure inactivates B cells but leaves T helper function intact (Hamaoka et.al., 1972). These mice were used as recipients of B cells. The B cells were obtained from the spleens of the remaining decapeptide positive mice. The spleen cells were pooled and treated with anti-theta serum plus complement to remove T cells. These B cells were injected intravenously into the recipient mice at $1x10^8$ per mouse. Following two aqueous boosts with TMVP, the reconstituted animals were bled, their titers to TMVP and to the decapeptide were assessed and the fine specificity of the anti-decapeptide antibodies was determined by the panel of synthetic peptides. The preliminary results obtained from this experiment are summarized in Table 6. The results show that mice which originally did not produce anti-decapeptide antibodies produced such antibodies when they were given TMVP primed B cells from animals producing anti-decapeptide antibodies. This indicates that the lack of anti-decapeptide responses in the recipients prior to transfer was not due to a lack of the proper T helper cells but was rather inherent in the B cell population: either they lacked decapeptide positive B cells in their repertoire or, if such cells were present they were not activated. However, these mice did possess TMVP primed T cells capable of providing help for the transferred TMVP primed decapeptide specific B cells. Such help was similar in magnitude to that provided by T cells of animals which were originally decapeptide positive.

The results in Table 6 also indicate that the fine specificity of the resulting antibodies in all the recipients was the same, regardless of whether the animals were originally decapeptide negative or decapeptide positive. More interesting, the fine specificity of the decapeptide positive recipients was the same in all recipients regardless of the fine specificity originally exhibited by the recipient

Table 6. Decapeptide specific antibody response of TMVP immune CSW mice following irradiation (600rads) and reconstitution with B cells from TMVP immune, decapeptide positive CSW mice. Adoptive transfers were performed as described (Rennick et al., 1983).

RESPONSES BEFORE RECONSTITUTION		RESPONSE AFTER RECONSTITUTION	
B CELL SOURCE	T CELL SOURCE	DECAPEPTIDE BOUND (P MOLES) BY 40 µL OF SERUM	FINE SPECIFICITY
POOL OF ALL FINE SPECIFICITIES	NON-IMMUNE	0	NO RESPONSE
	TMVP POSITIVE DECAPEPTIDE NEGATIVE	4.2 ± 1.6	ALL DECAPEPTIDE POSITIVE OF THE SAME FINE SPECIFICITY
	TMVP POSITIVE, DECAPEPTIDE POSITIVE, EACH INDIVIDUAL WITH A UNIQUE FINE SPECIFICITY	4.8 ± 1.6	

prior to transfer. These findings indicate that the fine specificity of the antibodies produced by B memory cells is not influenced by the T helper cells. In addition, preliminary results from a similar experiment indicate that T cells also had no effect on the fine antibody specificity of the primary anti-decapeptide response.

It has been suggested by several workers that idiotype specific T helper cells influence the expression of the antibody specificity (Janeway et.al., 1977; Bottomly, 1981; Sercarz et al., 1981). Results presented herein tend to minimize this possibility. It would be expected that if present, idiotype specific T helper cells which determine the original response of the recipient would influence the expression of the fine antibody specificity of the transferred cells. The results indicate that this was not the case. Thus, in addition to the lack of influence of antigen specific T cells on the fine antibody specificity shown earlier in the present communication, there appears to be no influence of idiotype specific T helper cells on the expression of fine antibody specificity.

In conclusion, using TMVP as a model protein antigen, results presented herein demonstrate that:

1. An immune response to a protein antigen may be induced by a small fragment of the protein, provided that it contains regions capable of activating T and B lymphocytes.

2. The isotype composition of antibodies induced by the peptide immunogen appears to be similar to that induced by immunization with the whole protein.

3. The fine specificity of antibodies induced by the peptide immunogen appears to be similar to that induced by immunization with the whole protein.

4. The specificity of antigen specific T helper cells does not influence the fine specificity of the antibodies induced to a given epitope.

5. Idiotype specific T helper cells do not appear to influence the antibody fine specificity to a given epitope.

Finally, an important conceptual aspect of utilizing synthetic vaccines is that while immunization with peptide conjugated to heterlogous carrier proteins should not lead to a memory response upon encounter with the native protein, immunization with an immunogenic peptide representing a portion of a protein antigen should lead to an anamnestic response following an encounter with the native protein.

ACKNOWLEDGEMENT: The authors wish to acknowledge the competent assistance of Mrs. C.Y. Leung throughout these studies. The work was supported in part by a National Science Foundaton Grant PCM 8103264 and by the Ann B. Ashworth memorial Grant for Cancer Research IM 291 from the American Cancer Society.

REFERENCES

Anderer, F.A. (1963) Preparation and properties of an artificial antigen immunologically related to tobacco mosaic virus. Biochem. Biophys. Acta, 71:246.

Anderer, F.A., and Schlumberger, H.D. (1965) Properties of different artificial antigens immunologically related to tobacco mosaic virus. Biochem. Biophys. Acta, 97:503.

Anderer, F.A., Uhlig, H., Weber, E., and Schramm, G. (1960) Primary structure of the protein of tobacco mosaic virus. Nature, 186:922.

Arnon, R., Maron, E., Sela, M., and Anfinsen, C.B. (1971) Antibodies reactive with lysozyme elicited by a completely synthetic antigen Proc. Natl. Acad. Sci., 68:1450.

Arnon, R., and Sela, M. (1969) Antibodies to a unique region in lysozyme provoked by a synthetic antigen conjugate. Proc. Natl. Acad. Sci., 62:163.

Atassi, M.Z. (1975) Antigenic structure of myoglobin: the complete immunochemical anatomy of a protein and conclusions relating to antigenic structures of proteins. Immunochemistry, 12:423.

Benjamini, E. (1977) Immunochemistry of the tobacco mosiac virus protein. in: immunochemistry of proteins (M.Z. Atassi, ed.), Plenum Press, New York, 2:265.

Benjamini, E., Michaeli, D., Leung, C.Y., Wong, K., and Scheuenstuhl, H. (1973) Immunochemical studies with synthetic peptides related to collagen. Immunochemistry, 10:629.

Benjamini, E., Young, J.D., Shimizu, M., and Leung, C.Y. (1964) Immunochemical studies on the tobacco mosaic virus protein. I. The immunological relationship of the tryptic peptides of the tobacco mosaic virus protein to the whole protein. Biochemistry, 3:1115.

Berzofsky, J. (1983a) T cell activation by antigen: promising clues to receptor genes and molecules. Immunology Today, 4:299.

Berzofsky, J.A. (1983b) T-B reciprocity: an Ia-restricted epitope-specific curcuit regulating T cell-B cell interaction and antibody specificity. Surv. Immunol. Res., 2:223.

Bittle, J.L., Houghten, R.A., Alexander, H., Shinnick, T.M., Sutcliffe, J.G., Lerner, R.A., Rowlands, D.J., and Brown, F. (1982) Protection against foot-and-mouth disease by immunization with a chemically synthesized peptide predicted from the viral nucleotide sequence. Nature, 298:30.

Bixler, G.S., Jr., and Atassi, M.Z. (1983) Molecular localization of the full profile of the continuous regions recognized by myoglobin primed T cells using synthetic overlapping peptides encompassing the entire molecule. Immunological Commun., 12:593.

Bottomly, K. (1981) Activation of the idiotypic network: environmental and regulatory influences. in: Immunoglobin idiotypes (C.A. Janeway, E.E. Sercarz, H. and Wigzell, eds.), Academic Press, New York, pp. 517.

Brown, R.K., Delaney, R., Levine, L., and van Vanakis, H. (1959) Studies on the antigenic structure of ribonuclease. J. Biol. Chem., 2334:2343.

Cecka, J.M., Stratton, J.A., Miller, A. and Sercarz, E. (1976) Structural aspects of immune recognition of lysozyme. III. T cell specificity restriction and its consequences for antibody antibody specificity. Eur. J. Immunol., 6:639.

Champness, J.N., Bloomer, A.C., Bricogne, G., Butler, P.J.G., and Klug, A. (1976) The structure of the protein disc of the tobacco mosaic virus to 5 A resolution. Nature, Lond., 259:20.

Chavez, L.G., and Scheraga, H.H. (1979) Location of antigenic determinants of bovine pancreatic ribonuclease. Biochemistry, 18:4386.

Corradin, G., Etlinger, H.M., and Chiller, J.M. (1977) Lymphocyte specificity to protein antigens. I. Characterization of the antigen-induced in vitro T cell-dependent proliferative response with lymph node cells from primed mice. J. Immunol.,119:1048.

Crumpton, M.J., and Sonall, P.A. Jr. (1967) Conformation of immunologically active fragments of sperm whale myoglobin in solution. J. Mol. Biol., 26:143.

Fearney, F.J., Leung, C.Y., Young, J.D., and Benjamini, E. (1971) The specificity of antibodies to a peptide determinant of the tobacco mosaic virus proeinm induced by immunization with the peptide conjugate. Biochim. Biophys. Acta, 243-509.

Gerwing, J., and Thompson, K. (1968) Studies on antigenic properties of eggwhite lysozyme. I. Isolation and characterization of a tryptic peptide from reduced and alkylated lysozyme exhibiting hapten activity. Biochemistry, 7:3888.
Goodfried, T.L., Levine, L., and Fasman, G.D. (1964) Antibodies to bradykinin and angiotensin: a use of a carbodiimide in immunology. Science, 14:1344.
Green, N., Alexander, H., Olson, A., Alexander, S., Shinnick, T.M., Sutcliffe, J.B., and Lerner, R.A. (1982) Immunogenic structure of the influenza virus hemagglutinin. Cell, 28:477.
Hamaoka, T., Katz, D.H., and Benaceraf, B. (1972) Radio-resistance of carrier-specific helper thymus-derived lymphocytes in mice. Proc. Natl. Acad. Sci., (U.S.A.) 69:3453.
Hashim, G.A. (1978) Myelin basic protein: structure function and antigenic determinants. Immunol. Review, 39:60.
Hurrell, J.G.R., Smith, J.A., and Leach, S.J. (1978) The detection of five antigenically reactive regions in the soybean leghemoglobin "a" molecule. Immunochemistry, 15:297.
Janeway, C.A. Jr., Murgita, R.A., Weinbaum, F.I., Asofsky, R., and Wigzell, H. (1977) Evidence for an immunoglobulin-dependent antigen-specific helper cell. Proc. Natl. Acad. Sci., 74:4582.
Kaattari, S., Scibienski, R.J., and Benjamini, E. (1980) The immunoregulatory role of antigen-antibody complexes. Immunology, 40:9.
Kettman, J.R., Benjamini, E., Michaeli, D., and Leung, C.Y., (1967) The synthesis and immunological activity of a peptide related to collagen, Biochem. Biophys. Res. Commun., 29:623.
Langbeheim, H.N., Arnoon, R., and Sela, M. (1976) Antiviral effect on MS-2 coliophage obtained from a synthetic antigen. Proc. Natl. Acad. Sci., 73:4636.
Lerner, R.A. (1982) Tapping the immunological repertoire to produce antibodies of predetermined specificity. Nature, 299:592.
Mishell, B.B., and Shiigi, S.M. (1980) "Selected Methods in Cellular Immunology," W.H. Freeman and Co., San Francisco.
Mitchison, N.A. (1971) The carrier effect in the secondary response to hapten protein conjugates II: Cell cooperation. Eur. J. Immunology, 1:18.
Morrow, P.R., Rennick, D.M., and Benjamini, E. (1983) The antibody response to a single antigenic determinant of the tobacco mosaic virus protein (TMVP): effects of allotype-linked genes and restricted heterogeneity of the response. J. Immunol. 131:2875.
Muller, G.M., Shapira, M., and Arnon, R. (1982) Anti-influenza response achieved by immunization with a synthetic conjugate. Proc. Natl. Acad. Sci., 79:569.
Prince, A.M., Ikram, H.I., and Hopp, T.P. (1982) Hepatitis B virus vaccine: identification of HBsAg/a and HBsAg/d but not HBsAg/y subtype antigenic determinants on a synthetic immunogenic peptide Proc. Natl. Acad. Sci .,79:579.
Rennick, D.M., Morrow, P.R., and Benjamini, E. (1983) Functional heterogeneity of memory B lymphocytes: in vivo analysis of TD-primed B cells responsive to secondary stimulation with TD and TI antigens. J. Immunol., 131:561.
Sakata, S., and Atassi, M.Z. (1980) Immunochemistry of serum albumin. X. Five major antigenic sites of human serum albumin are extrapolated from bovine albumin and confirmed by synthetic peptides Molecular Immunology., 17:139.
Salvin, S.B., and Smith, R.G. (1960) Delayed hypersensitivity and an anmnastic response. J. Immunol., 84:44.

Schechter, B., Schechter, I., Ramachandran, J., Conway-Jacobs, A., Sela, M., Benjamini, E., and Shimizu, M. (1971) Synthetic containing the same L-tyrosyl-alanyl-L-glutamyl sequences. Eur. J. Biochem., 20:309.

Scibienski, R.J., Klingman, V., Leung, C., Thompson, K., and Benjamini, E.(1978) Recognition of lysozyme by lymphocyte subunits. in: Immunobiology of Proteins and Peptides 1 (M.Z. Medicine and Biology 98:305.

Sela, M. (1969) Antigenicity: some molecular aspects. Science., 166:1365.

Senyk, G., Williams, E.B., Nitecki, D.E., and Goodman, J.W. (1983) The functional dissection of an antigen molecule: specificity of humoral and cellular immune responses to glucagon J. Exp. Med., 133:1294.

Sercarz, E., Cecka, J.M., Kipp, D., and Miller, A. (1977) The steering function of T cells in expression of the antibody repertoire directed against multideterminant protein antigen. Ann. Immunol., Paris. 128c:599.

Sercarz, E., Wicher, L., Stratton, J., Miller, A., Metzyer, D., Maizel, R., Katz, M., Harvey, M., and Benjamini, C. (1981) Regulation of antibody specificity and idiotypy by independent T cells. in: Immunoglobin Idiotypes (C.A. Janeway, E.E. Sercarz, H. Wigzell, eds.) Acad. Press, New York, pp. 533.

Spitler, L., Benjamini, E. Young, J.D., Kaplan, H., and Fudenberg, H H. (1970) Studies on the immune response to a characterized antigen determinant of the tobacco mosaic virus protein. J. Exp. Med.,, 131:133.

Tsugita, A., Gish, D.T., HYoung, J. Fraenkel-Conrat, H., Knight, C.A., and Stanley, W.M. (1960) The complete amino acid sequence of the protein of the tobacco mosaic virus. Proc. Natl. Acad. Sci., 46:1460.

Woody, B,.R., and Knight, C.A. (1959) Peptide maps obtained with tryptic digests of the proteins of some strains of tobacco mosaic virus. Virology, 9:359.

Young, C.R., and Atassi, M.Z. (1982) Antibodies to sperm-whale myoglobin evoked by free synthetic peptides fo an antigenic site. Immunological Communications 11:9.

Young, J.D., Benjamini, E., Stewart, J.M., and Leung, C.Y. (1967) Immunochemical studies on the tobacco mosaic virus protein. V. The solid phase synthesis of peptides of an antigenically active decapeptide of tobacco mosaic virus protein and the reaction of these peptides with antibodies to the whole protein. Biochemistry, 6:1455.

Waterfield, D., Levy, J.G., Kilburn, D.G. and Teather, R.M. (1972) The effect of haptenic peptides from performic acid oxidized ferredoxin from Clostridium pasteurionium and protein carrier-hapten conjugates on the immune response of macrophages and lymphoid cells from animals immunized against oxidized ferredoxin. Cell. Immunol., 113:58.

PROTECTIVE IMMUNITY EVOKED BY SYNTHETIC PEPTIDES OF STREPTOCOCCAL M PROTEINS

Edwin H. Beachey

VA Medical Center and
University of Tennessee Center for the Health Sciences
Memphis, Tennessee 38104

INTRODUCTION

Protective immunity against group A streptococci is directed against the M protein fibrils on the surface of virulent organisms (1,2). Since this protein is often tenaciously associated with antigens that evoke toxic reactions or immunological cross reactions with host tissues (3), efforts have been directed toward the isolation and purification of protective epitopes of the M protein molecule that are free of harmful products. Recent studies of the immunogenicity of polypeptide fragments of M protein extracted from type 24 group A streptococci by limited peptic digestion (4) have been promising. The purified extracts lacked toxicity or detectable heart cross-reactivity but retained protective immunogenicity as demonstrated by vaccination studies in laboratory animals (4,5) and by preliminary vaccine trials in human volunteers (6).

Similar peptic extracts of type 5 M protein (pep M5), however, evoked heart cross-reactive antibodies in rabbits immunized with the purified protein (7). Detailed immunochemical analyses revealed that the cross-reactive determinant resided within the covalent structure of the M protein molecule. Furthermore, antibody directed toward the heart cross-reactive epitope was shown to be opsonic and protective against types 5 and 19 streptococci, both serotypes of which had been shown to contain type specific heart cross-reactive antigens (3,7). Inhibition studies with purified preparations of M protein showed that heart cross-reactive epitope(s) represented only a limited immunogenic region(s) of the type 5 M protein molecule because many other protective epitopes clearly were not heart cross-reactive (7). This raised the hope that one might be able to cleave M protein in such a way as to eliminate potentially harmful regions while retaining protective regions of the molecule.

In this paper, we shall review studies of the chemistry and immunology of natural and chemically synthesized peptide fragments of streptococcal M proteins. Evidence is presented that selected regions of the M protein molecule consisting of as few as 12 amino acid residues contain epitopes that inhibit opsonic anti-streptococcal antibodies. Chemically synthesized peptide fragments as short as 13 amino acid residues long, when covalently linked to tetanus toxoid are capable of evoking protective antibodies against the related streptococci. Moreover, a synthetic copy of the NH_2-terminal 20-residue peptide of an M protein molecule (type 5) that contains heart cross-reactive epitopes within its covalent structure is capable of eliciting protective anti-streptococcal antibodies without eliciting heart crossreactive immunity.

PROTECTIVE IMMUNOGENICITY OF NATIVE PEPTIDE FRAGMENTS OF STREPTOCOCCAL M PROTEIN

The idea that one might be able to evoke protective immune responses against group A streptococci by immunization with peptide fragments of M protein is based on studies of the immunogenicity of polypeptide fragments extracted from streptococcal cells with various agents. Thus, most M protein preparations extracted by a number of different methods (2,5, 9-12) have consisted only of large polypeptide fragments of the M protein molecule. These polypeptide fragments contain many antigenic determinants, most of which are type specific (9,11,13-16) but some of which are cross-reactive with other M serotypes (17-21). Moreover, site specific cleavage of M protein with cyanogen bromide has yielded multiple subpeptides each of which evoke protective antibodies against the homologous serotype of streptococci (22). The final proof that subpeptides contain protective epitopes was obtained by the chemical synthesis of M protein peptide fragments which are able to evoke similar type specific protective antibodies (23,24).

PRIMARY STRUCTURE AND CONFORMATION OF M PROTEIN

Recent studies of the primary structures of three different serotypes of M protein have revealed several remarkable features. First, although the amino acid sequences of each of three different M serotypes (types 5, 6 and 24) are unique, certain amino acids are conserved among the sequences of all three (20,25-27). Second, the M protein molecules are composed in varying degrees of internally repeating covalent structures (15,26,28). Third, the M proteins contain a seven-residue periodicity in their amino acid sequences which is reminiscent of α tropomyosin of muscle tissue (26,28,29). The latter characteristic accounts for an α helical coiled-coil structure of the M protein molecule (8); it is the first such conformation to be demonstrated for any surface appendage of bacteria. Electron microscopy of the rotary shadowed coiled-coils of pepsin-, detergent-, and phage lysin-extracted M protein of type 6 streptococci suggested that the pepsin- and detergent-extracted M proteins represented the distal ends of these surface fibrillar structures, whereas the lysin-extracted M protein represented most, if not all, of these structures (8).

Thus far, most studies of primary structure have employed pepsin or detergent extracts of M protein. The elucidation of the covalent structures of M protein has enabled the chemical synthesis of peptide fragments and testing of the protective immunogenicity of selected regions of the M protein molecules.

SYNTHETIC PEPTIDE FRAGMENTS OF M PROTEIN

The synthesis of M protein peptides was based on the covalent structures determined for the cyanogen bromide peptides of type 24 M protein (pep M24) and on the NH_2-terminal sequence analysis of type 5 M protein.

Since we had shown that at least one protective determinant resided in each of the seven purified repeating covalent structures of pep M24 (14,21), and had established the immunogenicity of the 35-residue peptides whose sequences had been completed (22-24), we decided to chemically synthesize two of these, CB3 and CB7; the synthetic peptides being designated S-CB3 and S-CB7.

In addition, we synthesized subpeptides of the 35-residue S-CB7, including S-CB7(13-35), S-CB7(18-35), S-CB7(23-35)C and S-CB7(18-29) (Table 1). The latter peptide overlaps two subpeptides derived by digesting lysyl-blocked CB7 (see below). A 20-residue peptide of pep M5, S-M5(1-20) also was synthesized. Each of the peptides was synthesized by the solid phase method of Merrifield (30) as previously described (23,24,31,32). The identity of the synthetic peptides was confirmed by amino acid analyses and automated Edman degradation (22).

When emulsified in complete Freund's adjuvant and injected as a single 25 nmol dose, the 35-residue S-CB7 evoked antibody titers at 6 weeks of 1:400, 1:1,280 and 1:25,600, respectively in each of three rabbits as determined by enzyme linked immunosorbent assays (ELISA). Only the serum of the rabbit showing the highest ELISA titer was able to opsonize type 24 streptococci. In contrast, the same dose of S-CB7 covalently linked to polylysine and emulsified in complete Freund's adjuvant evoked strong ELISA as well as opsonic antibody titers in each of three rabbits (Table 2). The results obtained by using S-CB3 instead of S-CB7 were virtually identical (24). In opsono-bactericidal assays using types 5, 6 and 24 streptococci, the immune sera were capable of promoting phagocytosis and killing only of the type 24 streptococci indicating that the humoral immune responses to S-CB7 were type specific (22). None of the S-CB7 immune sera reacted with the sarcolemmal membranes of frozen sections of human heart as determined by immunofluorscence (6).

Table 1. Covalent Structures of Synthetic Peptides of Streptococcal M Protein

Synthetic Peptides	Amino Acid Sequences
<u>Type 5 M Protein</u>	5 10 15 20 25 30 35 40
S-M5-(1-20)	AVTKGTINDPQAAKEALDKY
S-M5-(20-40)	YELENHDLKTNNEGLKTENTG
<u>Type 24 M Protein</u>	5 10 15 20 25 30 35
S-CB7-(1-35)	NFSTADSAKIKTLEAEKAALAARKADLEKALEGAM
S-CB7-(13-35)	LEAEKAALAARKADLEFALEGAM
S-CB7-(18-35)	AALAARKADLEKALEGAM
S-CB7-(23-35)C	RKADLEKALEGAMC
S-CB7-(23-32)KAMC	RKADLEKALEKAMC
S-CB7-(18-29)	AALAARKADLEK

Each amino acid sequence was confirmed by sequential Edman degradations. The single-letter code for amino acid residues is used to conserve space and for ease of comparison. Italicized residues indicate substitutions or additions with respect to native CB7. Data from (32-33).

Table 2. Immune Response in Rabbits After Immunization with S-CB7 Conjugated to Polylysine[a]

Rabbit #	ELISA Titer[b]	Opsonic Antibody Titer[b]
1	6,400	2
2	25,600	8
3	102,400	32

[a]Each rabbit was injected with 25 nmol of S-CB7 conjugated to polylysine and emulsified in complete Freund's adjuvant (CFA) followed a week later by the same dose in incomplete FA.
[b]Titers are expressed as the reciprocal of the highest dilution giving a positive test by ELISA (enzyme linked immunosorbent assay) or by opsonophagocytic tests of type 24 streptococci in whole heparinized human blood as previously described (23). Data from (23).

The protective activities of the S-CB7 immune sera were tested by passive mouse protection tests. Mice were injected with 0.2 ml of a pool of the immune sera and 24 h later challenged with type 24 or type 6 streptococci. The immune sera clearly protected against the homologous type 24 but not against the heterologous type 6 streptococci (Table 3).

Having demonstrated that a protective determinant of M protein resides in a synthetic peptide as small as 35 residues long, it was of interest to determine whether protective determinants were retained in even smaller peptide fragments. Therefore we cleaved CB7 with trypsin after blocking lysine residues with maleic anhydride. Since CB7 contains only one arginine residue at position 23, the molecule was cleaved into a COOH-terminal dodecapeptide and an NH_2-terminal 23-residue peptide (Fig.

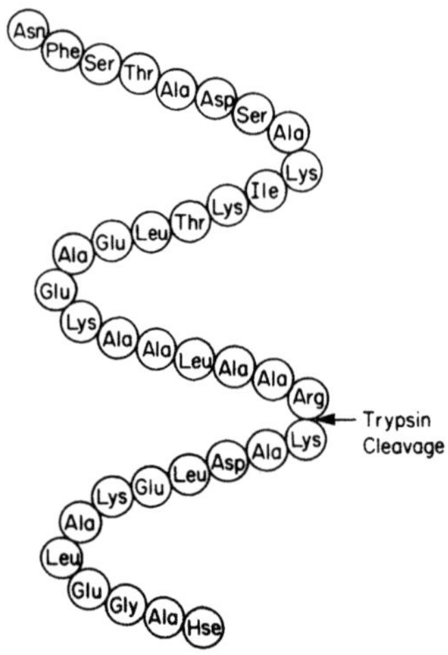

Fig. 1. Sequence of CB7 of type 24 M protein. After blocking the lysine residues with maleic anhydride, trypsin cleaves CB7 at the arginine residue as indicated into a COOH-terminal dodecapeptide and an NH^2-terminal 23-residue peptide.

Table 3. Mouse protection against challenge infections with type 24 streptococci by sera of rabbits immunized with S-CB7

Serum Used to Passively Immunize Mice	LD_{50} in Mice Challenged with: Type 6 Streptococci	Type 24 Streptococci
Preimmune serum	<500 (2/15)[a]	<500 (3/15)
Pooled (3 rabbits) immune anti-S-CB7 serum[b]	<500 (2/15)	3,500,000 (14/15)

[a]Figures in parenthesis indicate the number of survivors/number of challenged mice.

[b]Three rabbits were immunized intracutaneously with 25 nmol of S-CB7 conjugated to polylysine (m.w. ≈ 35,000) and emulsified in complete Freund's adjuvant as previously described. The initial immunizing dose was followed one week later by the same dose emulsified in incomplete Freund's adjuvant and injected subcutaneously. Preimmune and immune sera obtained before immunization and at 6 wks after the initial immunizing dose, respectively were pooled and white Swiss mice injected intraperitoneally with 0.2 ml of either preimmune or immune serum. 24 h later, the mice were challenged by the same route with various doses (ranging from 500 to 4 X 10^6 CFU) of type 6 or type 24 streptococi. The survivals were recorded over a 7 day period and are shown in parenthesis. These data are from (22).

1). After deblocking the lysine residues with pyridine/acetate, the subpeptides were separated and purified by HPLC reverse-phase chromatography (23). The purified peptides were then tested for their abilities to inhibit opsonization by anti-S-CB7. In addition, the synthetic overlapping peptide, S-CB7(18-29), was tested for inhibitory activity. The NH_2-terminal 23-residue peptide and the COOH-terminal 12-residue peptide both were found to be inhibitory. The overlapping 12-residue peptide, S-CB7(18-29), on the other hand, was without effect even at concentrations as high as 100 nmol (23).

These results indicated that a protective determinant of type 24 M protein resided in a peptide fragment containing only 12 residues. The inability of the overlapping synthetic peptide to block opsonization suggests that CB7 contains at least 2 distinct protective antigenic determinants neither one of which resides in the overlapping peptide.

That protective epitopes reside in the subpeptides of CB7 was confirmed by studies of the immunogenicity of the synthetic subpeptides S-CB7-(13-35), -(18-35) and -(23-35)C covalently linked to tetanus toxoid (32). Each of these peptides evoked opsonic antibodies in rabbits (Table 4). The substitution of a single amino acid residue at position 33 (lysine substituted for glycine) rendered the protective 13 residue peptide, S-CB7(23-35)C nonprotective (Table 4). These results indicate that a synthetic peptide containing as few as 13 residues of type 24 streptococcal M protein is capable of evoking protective immunity against the related streptococci.

These findings suggested the possibility that one might be able to

Table 4. Immune Response in Rabbits after Immunization with Synthetic Subpeptides of Type 24 M or Type 5 Protein

Immunizing Peptide[a] of:	ELISA Titers[b]	Opsonic Antibody Titers[b]
Type 24 M Protein		
S-CB7-(13-35)	200-102,400	<2-64
S-CB7-(18-35)	1,600-12,500	<2-8
S-CB7-(23-35)C	1,600-51,200	<2-8
S-CB7-(23-32)KAMC	200-800	<2
Type 5 M Protein		
S-M5-(1-20)	6,400-25,600	2-16
S-M5-(20-40)	<200	<2

[a]A set of three rabbits was immunized with 25 nmol of each of the peptides conjugated to lysylated tetanus toxoid as described (32,33). A booster of the same dose of vaccine in saline was administered subcutaneously 4 wks later.

[b]ELISA and opsonic antibody titers were determined as described in footnote to Table 2 by using the respective pep M 24 and Pep M 5 as the ELISA antigen and types 24 and 5 streptococci in the opsonophagocytic assays. The titers are given as ranges from lowest to highest at 6 wks.

delete tissue cross-reactive epitopes from M protein vaccines by selecting synthesizing peptide regions of M protein that contain only protective epitopes. With this in mind, we attacked the problem of type 5 M protein, a molecule demonstrated to contain cardiac tissue cross-reactive epitopes within its covalent structure (7). Having established the primary structure of the NH_2-terminal region of type 5 M protein, we synthesized peptides 1-20 and 20-40. When covalently linked to tetanus toxoid with glutaraldehyde, the synthetic peptide S-M5(1-20) but not S-M5(20-40) evoked antibodies that were protective against the related type 5 streptococci (Table 4). Even though each of the animals received five times the molar dose required to produce heart cross-reactive antibody with the intact pep M5 molecule, none of the immunized animals developed heart cross-reactive antibodies (33).

These studies are promising for the development of streptococcal vaccines composed of small synthetic peptide regions of M protein that contain protective, but not tissue cross-reactive, antigenic determinants. The knowledge of the complete covalent structures of several serotypes of M protein should allow us to select antigenic regions to synthesize based on the location of protective as opposed to tissue cross-reactive epitopes within these antigenically complex molecules.

CONCLUSION

Many issues remain to be investigated to define the optimal procedures to be used to immunize with synthetic peptides of streptococcal M protein. It is generally assumed that the administration of artificial vaccines in humans will require the use of a carrier such as tetanus toxoid along with an adjuvant such as aluminum hydroxide or MDP (NacMur-L-Ala-D-isoGln), a synthetic adjuvant currently undergoing clinical trials (34). The repeated use of carriers, however, may create other problems because the carriers often are, themselves, potent immunogens and sometimes may even suppress the immune reponse in an irreversible fashion (35). Adjuvants may give rise to local toxic reactions or unwanted immune responses to unrelated epitopes including self antigens.

Preliminary studies in our laboratories have shown that one might be able to circumvent the problems of carriers and adjuvants by immunizing with hompolymers of synthetic peptide fragments. Mice sensitized with the natural M protein in saline gave strong immune responses to a secondary injection in saline of a 35-residue synthetic peptide (S-CB7) that had been homopolymerized with glutaraldehyde (31). Furthermore, the secondary response was epitope specific in that virtually all of the antibody raised was directed against the synthetic peptide. Thus, it is possible that one may be able to devise effective synthetic vaccines that can be administered without the need of a carrier or even an adjuvant.

It is now clear that the whole M protein molecule is not needed to effectively immunize against group A streptococcal infections. Chemically synthesized peptide fragments of carefully chosen regions of various M protein molecules should provide a vaccine that can be safely administered to humans without the fear of evoking host tissue cross-reactive antibodies. There is some evidence to suggest that certain protective determinants of M protein may be shared among different serotypes. The identification and synthesis of such common protective determinants should provide the basis for the formulatin of a broadly protective vaccine against many serotypes of rheumatogenic group A streptococci.

ACKNOWLEDGEMENT

I thank Drs. Jerome Seyer, James Dale, Louis Chedid, Francois Audibert, Michelle Jolivet, Andrew Tartar and Helene Gras-Masse for valuable contributions to some of the previous studies that are summarized in this review. The research studies of the authors are supported in part by Institutional Research Funds from US Veterans Administration and by NIH Research Grants AI-10085 and AI-13550. The expert secretarial assistance of Connie Carrier is gratefully acknowledged.

REFERENCES

1. G.H. Stollerman, in: "Rheumatic Fever and Streptococcal Infection," Grune and Stratton, New York (1975).
2. R.C. Lancefield,, J. Immunol. 89:307 (1962).
3. M.H. Kaplan, J. Immunol. 90:595 (1963).
4. E.H. Beachey, G.L. Campbell, and I. Ofek, I., Infect. Immun. 9:891 (1972).

5. E.H. Beachey, G.H. Stollerman, E.Y. Chiang, T.M. Chiang, J.M. Seyer, and A.H. Kang, J. Exp. Med. 145:1479 (1977).
6. E.H. Beachey, G.H. Stollerman, R.H. Johnson, I. Ofek, and A.L. Bisno, J. Exp. Med. 150:862 (1979).
7. J.B. Dale and E.H. Beachey, J. Exp. Med. 156:1165 (1982).
8. G.N. Phillips, P.F. Flicker, C. Cohen, B.N. Manjula, and V.A. Fischetti, Proc. Natl. Acad. Sci. USA 78:4689 (1982).
9. E.N. Fox and M.K. Wittner, Immunochemistry 6:11-24(1969).
10. H. Russel, R.R. Facklam, L.R. Edwards, Infect. Immun. 12:679 (1975).
11. V.A. Fischetti, E.C. Gotschlich, G. Siviglia, and J.B. Zabriskie, J. Exp. Med. 144:32 (1976).
12. K.L. Vosti and W.K. Williams, Infect. Immun. 21:546 (1978)..
13. E.H. Beachey and M.W. Cunningham, Infect. Immun. 8:19 (1973).
14. M.W. Cunningham and E.H. Beachey, J. Immunol., 115: 1002 (1975).
15. E.H. Beachey, J.W. Seyer, and A.H. Kang, Proc. Natl. Acad. Sci. USA, 75:3163 (1978).
16. D.L. Hasty, E.H. Beachey, W.A. Simpson, and J.B. Dale, J. Exp. Med. 155:1010 (1982).
17. E.N. Fox and M.K. Wittner, J. Immunol. 100:39 (1968).
18. M.K. Wittner and E.N. Fox, Infect. Immun. 15:104 (1977).
19. V.A. Fischetti, J. Exp. Med. 146:1108 (1977).
20. J.B. Dale, I. Ofek, and E.H. Beachey, J. Exp. Med. 151:1026, 1980.
21. E.H. Beachey and J.M. Seyer, in: "Bacterial Vaccines," J.B. Robbins, J.C. Hill, and J.C. Sadoff, eds. Thieme-Stratton, Inc., New York (1982).
22. E.H. Beachey, J.M. Seyer, and A.H. Kang, J. Biol. Chem. 255:6284 (1980).
23. E.H. Beachey, J.M. Seyer, J.B. Dale, W.A. Simpson, and A.H. Kang, Nature 292:457 (1981).
24. E.H. Beachey, J.M. Seyer, J.B. Dale, and D.L. Hasty, J. Biol. Chem. 258:13250 (1983).
25. B.N. Manjula and V.A. Fischetti, J. Immunol. 124:261 (1980).
26. B.N. Manjula and V.A. Fischetti, J. Exp. Med. 151:695 (1980).
27. V.A. Fischetti and B.N. Manjula. in: "Bacterial Vaccines," J.B. Robbins, J.C. Hills, and J.C. Sadoff, eds., Thieme-Stratton, Inc., New York (1982).
28. B.N. Manjula, A.S. Acharya, S.M. Mische, T. Fariwell, and V.A. Fischetti, J. Biol. Chem. 259:3686 (1984).
29. B. Hosein, M. McCarty, and V.A. Fischetti, Proc. Natl. Acad. Sci. USA 76:3765 (1979).
30. R.B. Merrifield, J. Am. Chem. Soc. 85:2149 (1963).
31. M. Jolivet, F. Audibert, E.H. Beachey, A. Tarter, H. Gras-Masse, and L. Chedid, Biochem. Biophys. Res. Comm. 117:359 (1984).
32. E.H. Beachey, A. Tarter, J.M. Seyer, and L. Chedid, Proc. Natl. Acad. Sci. USA 81:2203 (1984).
33. J.B. Dale, J.M. Seyer, and E.H. Beachey, J. Exp. Med. 158: 1727 (1983).
34. L.A. Chedid, M.A. Parant, F.M. Audibert, G.J. Riveau, F.J. Parant, E. Lederer, J.P. Choay, and P.L. Lefrancier, Infect. Immun. 35:417 (1982).
35. L.A. Herzenberg, T. Tokuhisa, and L.A. Herzanberg, Nature 285:664 (1980).

ROTAVIRUS ANTIGENS

Mary K. Estes and David Y. Graham

Departments of Virology and Epidemiology and of Medicine
Baylor College of Medicine
Houston, Texas 77030

ABSTRACT

The rotaviruses are important medical and veterinary pathogens that cause acute diarrheal disease in children and in the young of most mammalian and avian species. The successful development of a rotavirus vaccine should significantly reduce worldwide infant morbidity and mortality and economic losses and increase food-production. Recent successes in the cultivation and characterization of rotavirus strains from humans and animals have led to a better understanding of the major antigens of these viruses and the genes encoding these antigens. The products of three independent genes (1) the neutralization antigen (the outer capsid glycoprotein); (2) the subgroup antigen (the major inner capsid protein); and (3) the hemagglutinin (the outer capsid protease-sensitive protein) are the major viral antigens. The properties and structure of these antigens are discussed. Further knowledge should facilitate the synthesis and development of novel antigen or subunit vaccines.

INTRODUCTION

Rotaviruses are principal etiologic agents of acute viral gastroenteritis in both man and animals (Estes et al., 1983; Holmes, 1983). The significance of diarrheal disease as a worldwide cause of mortality and morbidity is now clearly established. It has been estimated that there are 220 diarrheal episodes and 1.4 deaths from diarrhea each year for every 100 children under five years of age in developing countries, resulting in nearly five million deaths per year (Snyder and Merson, 1982). In developed countries, gastroenteritis causes low mortality but significant morbidity, and it ranks second in frequency behind respiratory illness among diseases affecting families (Monto and Koopman, 1980).

Such data are staggering and they focus the need for the development of effective disease prevention and control. Although oral rehydration is effective in reducing diarrheal disease mortality, other interventions are needed to reduce morbidity and, if possible, to eradicate the disease (Feachem et al., 1983). Protection from disease should be possible if the global population can be specifically immunized with the pathogens themselves or with pathogen-specific antigenic proteins.

This paper presents our current knowledge of rotavirus antigens and of their antigenic determinants. The impact of this information on the practical problem of developing conventional and novel types of rotavirus vaccines is also discussed.

ROTAVIRUS STRUCTURE AND CLASSIFICATION

Rotaviruses are classified in a separate genus of the family Reoviridae. Rotaviruses share common morphologic and biochemical characteristics which include: (i) Mature virus particles are nonenveloped and they possess a double-shelled capsid with a distinctive "wheel-like" morphology. Infectious particles, approximately 70 nm in diameter, are composed of an outer capsid, an inner capsid, and an icosahedral core. Double- and single-shelled particles are the predominant forms found in infected cells and in diarrheal stools, suggesting that the proteins on the outer surfaces of these particles are those most likely to be exposed to the immune system. (ii) The viral genome contains eleven discrete segments of double-stranded RNA and each segment codes for one virus-specific protein. (iii) Rotaviruses are capable of genetic reassortment. (iv) Virus replication occurs in the cytoplasm of infected cells. (v) Virus particles are formed by budding into the endoplasmic reticulum, and enveloped particles are transiently present at this stage of morphogenesis. Virions are liberated from infected cells by cell lysis.

The early serologic studies on rotaviruses used a range of techniques and suggested (i) that, irrespective of their species of origin, all rotaviruses contained a common antigen, (ii) that serotypes were species-specific, and (iii) that a limited number of rotavirus serotypes existed. Further experience, however, has shown that none of these postulates is true. Instead, an increasing number of rotaviruses that show no antigenic cross-reactivity with the original rotavirus isolates or with each other has been isolated. These rotavirus isolates, designated antigenically distinct rotaviruses, atypical rotaviruses or pararotaviruses, have now been recovered from human, porcine, chicken, bovine and ovine diarrheal samples (Bohl et al., 1982; Bridger et al., 1982; Dimitrov et al., 1983; Espejo et al., 1984; McNulty et al., 1981; Nicolas et al., 1983; Rodger et al. 1982; Saif et al., 1982; Snodgrass et al., 1984; Hung, et al., 1984). In addition, neutralization experiments with hyperimmune sera as well as cross-protection studies in gnotobiotic piglets, have demonstrated that multiple rotavirus serotypes exist and that strains of animal and human origin are contained within the same group, within the same serotype, or both (Gaul et al., 1982; Sato et al., 1982; Hoshino et al., 1984).

These recent developments emphasize the importance of developing a systematic serologic classification scheme for rotavirus isolates. Pedley et al. (1983) proposed that the rotaviruses be subdivided into groups where a group defines closely related viruses as determined by the presence of a common antigen(s) detected by immunofluorescence, ELISA, immune electron microscopy, complement fixation, or other immunologic assays, and where the viruses studied extensively since 1973 be designated group A viruses. A classification system we use identifies viruses with the following information: group/ species of origin/ place of origin/ strain designation/ year isolated/ subtype (neutralization, subgroup and hemagglutinin) antigens.

The proteins coded for by each of the group A viral genes and the biochemical and biologic properties of the structural proteins (found in virus particles) have been characterized, and this information provides the foundation for a molecular understanding of rotaviral antigens (Figure 1).

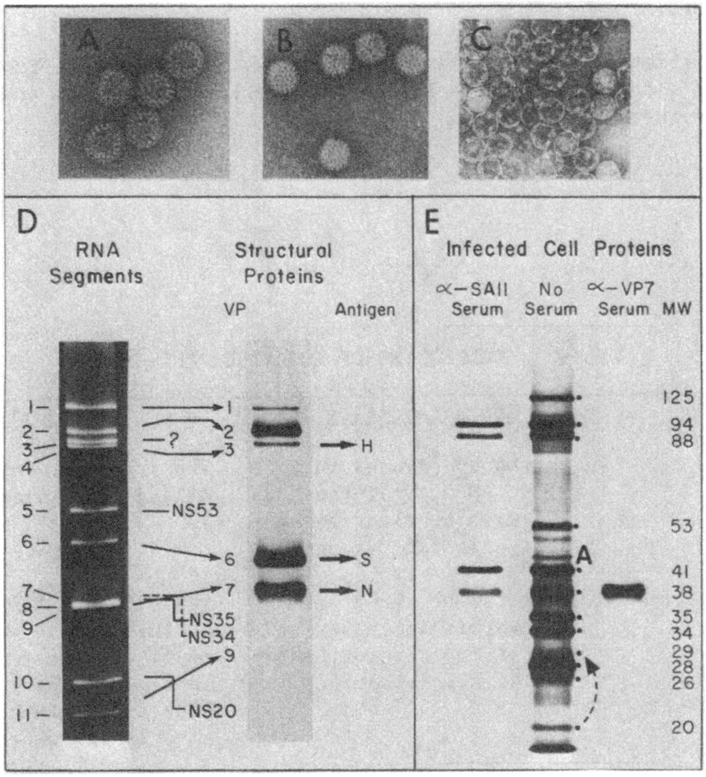

Figure 1. Rotavirus Structure and Antigens
Electron micrographs of SA11 (A) double-capsid, (B) single-capsid and (C) core particles observed after staining with ammonium molybdate. (D) Polyacrylamide gel electrophoresis patterns showing the eleven RNA segments and structural proteins of double-capsid SA11 particles. The viral structural proteins are indicated as VP and the arrows show the genes coding for these proteins and their antigenic specificities (see text). The nonstructural (NS) proteins of molecular weights of 53K, 35K, 34K and 29K are seen only in infected cells (see panel E). Two of the proteins (VP7 and NS29) are glycosylated. (E) The center lane shows the polypeptides found in SA11-infected cells. The viral-specific polypeptides are indicated by a dot and the polypeptides that react with hyperimmune polyclonal guinea pig anti-SA11 serum or polyclonal monospecific anti-VP7 serum are shown. The molecular weights of the polypeptides are indicated (in thousands) on the right of the photograph. Polypeptide nomenclature and processing is reviewed in Estes et al. (1983).

Double-shelled virus particles of purified group A rotaviruses (not exposed to proteolytic enzymes) contain six polypeptides: three (VP1, VP2, and VP6) are associated with the inner capsid and three (VP3, VP7, and VP9) make up the outer capsid of the virus particles. The molecular weights of these polypeptides range from 125,000 (VP1) to 27,000 (VP9). After proteolytic cleavage by exposure of purified virus particles to trypsin, additional polypeptides (VP4*, VP5*, and VP8*) derived from VP2 and VP3 may be found (Espejo et al., 1981; Estes et al., 1981). While one preliminary study suggested that similar polypeptides are present in non-group A viruses (Espejo et al., 1984), this remains to be rigorously established.

Properties of the rotavirus antigens are summarized in Table 1. It should be noted that these antigens are well characterized only for the group A viruses.

Table 1: PROPERTIES OF ROTAVIRUS ANTIGENS

Designation	Characteristics	Comments
Group	Shared by viruses in a group; VP2, VP3, VP6 and VP7 each contain common antigenic determinants.	VP6 is often incorrectly called "the group antigen"
Neutralization[a]	Detected by serum-neutralization assays. VP7 is major serotype specific polypeptide.	VP3, VP6 and VP9 may also be involved in neutralization. The role of these proteins in protection from infection in vivo is unknown.
Subgroup[a]	Detected by ELISA, IAHA or IF with selected post-infection sera or with monoclonal antibodies. Associated with VP6.	Two forms are well-characterized; a third form has been reported. It has been a useful epidemiologic marker.
Hemagglutinin[a]	Detected by HI assays. VP3 is the hemagglutinin.	May be useful to distinguish viruses with the same neutralization subtype. Antibodies to VP6 and VP7 have also been reported to mediate HI, presumably due to steric hindrance.

[a] Based on information available only for group A rotaviruses.

IDENTIFICATION OF THE GROUP (COMMON) ANTIGEN

Recognizing that rotavirus particles can contain nine proteins (including the cleavage products), it was of interest to know if the common antigen was only one protein. We addressed this question by producing polyclonal monospecific antisera (in guinea pigs) to the individual structural polypeptides of SA11. These antisera were used in immunofluorescent (IF) and immunoprecipitation experiments. Monospecific sera made to SA11 VP2, VP6 and VP7 reacted with homologous and heterologous virus strains showing that each of these three polypeptides contain common antigenic determinants (Table 2). Therefore, there does not appear to be a single group antigen, and the cross-reacting antigens detected as "the group antigen" may be located on more than one structural polypeptide. The presence of common determinants on VP7 has been confirmed by monoclonal antibodies to VP7 that cross-react between viruses (Greenberg et al., 1983b; Roseto et al., 1983; Sonza et al., 1984). Monoclonal antibodies to VP2 have not yet been reported.

Table 2: IDENTIFICATION OF ROTAVIRUS GROUP ANTIGENS DETECTED BY IMMUNOFLUORESCENCE OR IMMUNOPRECIPITATION

Antiserum to SA11[a]	Viral Antigens Simian (SA11)	Bovine (NCDV)	Porcine (OSU)	Chicken (CH_2)
DS virus	+	+	+	+
SS virus	+	+	+	+
VP2	+	+	+	+
VP6	+	+	+	+
VP7	+	+	+	+

[a] These antisera were prepared in guinea pigs to purified double-shelled (DS) or single-shelled (SS) particles or to polypeptides purified from denaturing polyacrylamide gels. The monospecificity of the polypeptide antisera has been demonstrated by immunoprecipitation of only the immunizing polypeptide.

Based on immune electron microscopy studies, Bridger (1978) reported that the common antigen is only associated with single-shelled particles. This suggested the common antigen was VP6. These results show that the common determinants on some of the rotavirus proteins (e.g., the outer capsid VP7 and the core proteins) are not detected by immune electron microscopy. In addition, this illustrates that antigen detection depends on both the specificity of the sera and the assay being used. The antigenic determinants on some of the rotavirus proteins may only be accessible following denaturation or solubilization.

ANTIGENS OF VIRUSES WITHIN A GROUP (SUBTYPE ANTIGENS)

Once classified into a group, rotaviruses can be further characterized on the basis of three independent antigenic specificities that include the major neutralization antigen (that functions to identify the serotype), the major inner capsid antigen (the subgroup antigen) and the hemagglutinin. These separate antigens are coded for by independent gene products as shown by genetic studies of reassortants where each antigen segregated with one gene (Kalica et al., 1981). For simplicity of virus characterization, we propose to call these antigens the N, S, and H antigens and to distinguish their individual specificities by the designation of subtypes of each antigen (e.g., N_1, N_2; Graham & Estes, submitted for publication). Using our classification system, SA11, the best characterized serotype 3 virus, would be denoted as A/SI/S. Africa/SA11/58/N_3, S_1, H_{SA11}.

THE NEUTRALIZATION (N) ANTIGEN

Rotaviruses can be distinguished by neutralization (plaque-reduction) assays and genetic studies with virus reassortants have demonstrated that the outer capsid glycoprotein, VP7, is the major neutralization antigen (Kalica et al., 1981; Greenberg et al., 1983b). The structural and biochemical properties of VP7 are summarized in Table 3.

Several of the properties of VP7 are of interest in relation to current efforts to produce subunit or polypeptide rotavirus vaccines by the expression of this protein from cloned genes. One important question is whether glycosylation is required for antigenicity. We addressed this question by studying a nonglycosylated variant of SA11 (Estes et al., 1982). During our studies on the biosynthesis of the SA11 glycoprotein, we demonstrated heterogeneity of the glycoprotein. We isolated a number of glycoprotein variants by subcloning a virus stock that contained glycoproteins of 36,000 (36K) and 38,000 (38K) molecular weight. Individual virus clones containing a glycosylated VP7 that migrated on SDS gels with apparent molecular weights ranging from 38K to 36K, and 35.5K (an unglycosylated variant) were isolated. Further studies with the 35.5K variant showed that glycosylation is not required for viral infectivity, hemagglutination, or antigenicity of particles. That is, hyperimmune antiserum made to glycosylated virus particles or to nonglycosylated particles neutralized either type of virus with equivalent antibody titers (unpublished data). This finding suggests that the synthesis of VP7 from gene 9 clones may not require an expression vector that glycosylates proteins to produce an antigenic product. However, the role of the carbohydrate in the structure or function of VP7 remains unknown. It is possible the carbohydrate is important during the synthesis of VP7, for example, for the correct folding of VP7 or for its interaction with other viral proteins.

Accumulating evidence suggests that conformation is critical for neutralizing specificity of VP7. For example, hyperimmune monospecific antisera produced to gel purified VP7 did not neutralize viral infectivity (unpublished data). In addition, rotavirus proteins that were

Table 3: PROPERTIES OF THE SA11 NEUTRALIZING ANTIGEN[a]

1. Gene	1062 nucleotides. 1 open reading frame (nucleotides 49 --> 1025). Contains two possible in-phase initiation sites, 30 nucleotides apart. Contains two N-terminal hydrophobic regions. 1 carbohydrate acceptor site (asn-ser-thr) at amino acid 69-71. No polyadenylation site at 3' end.
2. Protein	326 (or 297)[b] amino acids 10 prolines, 9 cysteines
Molecular weight	
Predicted:	37,368 (or 33,919)[b]
Observed:	
Precursor	37,000
Precursor lacking signal sequence	35,500
Glycosylated protein	38,000
3. Carbohydrate	N-linked high-mannose oligosaccharides.
4. Biosynthesis	Co-translationally glycosylated. Contains a cleavable signal sequence. Oligosaccharide trimming observed.
5. Miscellaneous	Heterogeneity in polypeptide backbone. Homology with gene of UK bovine virus of a different serotype: 77% nucleotide homology. 86% amino acid homology.

[a] Compiled from Ericson et al. (1982); Estes et al. (1982); Ericson et al. (1983); Both et al. (1983); and Arias et al. (1984).
[b] Predicted if the second AUG is used to initiate translation.

first separated on denaturing gels and then transferred to nitrocellulose filters reacted poorly with neutralizing antibody. This contrasts with the excellent reactivity of these sera with proteins from infected cells solubilized for immunoprecipitation by more gentle procedures (unpublished data). Finally, Bastardo et al. (1981) reported that monospecific antisera produced to VP7 contained better neutralizing activity if the VP7 antigen was not treated with reducing agents. The deduced protein sequence shows that VP7 contains 9 cysteine and 10 proline residues which is consistent with our hypothesis that at least some of the neutralization epitopes of VP7 may be non-linear or conformational.

Comparative analysis of the predicted protein sequence of VP7 from two serotypes provides preliminary information about the potential

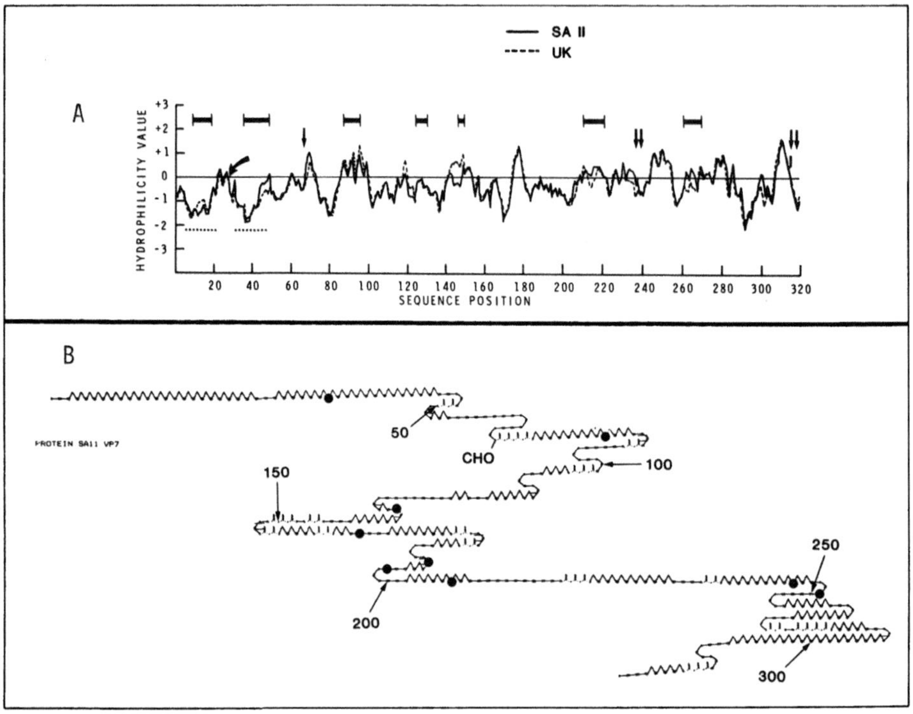

Figure 2. Computer predictions of the structure of the rotavirus neutralization antigen VP7.
(A) Hydrophilicity plots comparing the deduced VP7 amino acid sequence of two rotavirus strains (SA11, ——— and the bovine UK, -----) that belong to different serotypes. The average hydrophilicity for the protein sequences from Both et al. (1983), Arias et al. (1984) and Elleman et al. (1983) was calculated over an average of 6 amino acids according to Hopp and Woods (1983). The hydrophilic and hydrophobic regions are indicated by the "plus" and "minus" values, respectively. Regions where the sequences of these two proteins show divergence are indicated by solid bars. The two hydrophobic regions at the amino terminal that may serve as signal sequence(s) are indicated by dots and the second in-phase initiation AUG is indicated by a curved arrow. The glycosylation site shared by both viruses at residues 68-70 is indicated by a straight arrow. Two additional potential glycosylation sites in the bovine VP7 sequence are indicated by double arrows. (B) Plot of predicted α-helical, β-sheet and turn probabilities of the amino acid sequence of SA11 VP7. The Chou/Fasman rules of predictions were applied according to Corrigan and Huang (1982). Each 50 residues are numbered and the location of the glycosylation site (CHO) and of the 9 cysteine residues (•) are also indicated.

structure and location of the antigenic determinants on this protein. As the sequences of the glycoproteins of viruses representing additional different serotypes become available, clusters of amino acid changes between serotypes may be confirmed. Figure 2 highlights the regions of differences seen by comparing the sequences of the SA11 and UK VP7's, and illustrates the hydrophilic regions of VP7 using the methods of Hopp and Woods (1983) or Kyte and Doolittle (1982). Finally, a computer generated plot of the possible secondary structure of the SA11 VP7 predicted by the Chou-Fasman rules of protein conformation (1978) is shown. Although these analyses have limitations in the absence of X-ray crystallographic data, these approaches have been useful as a guide in predicting the location of some antigenic determinants on viral and bacterial antigens. However, a complete map of the antigenic determinants of VP7 will require more extensive epitope mapping using monoclonal antibodies, sequencing of variants selected under antibody pressure, analysis of chemically modified proteins, and the chemical synthesis and testing of the immunogenicity and antigenicity of overlapping peptides, or a combination of these methods.

Comparison of the hydrophilic regions and of the conformational features of SA11 VP7 (Figure 2) suggests that many of the cysteine residues could be available for disulfide bonding. The predicted conformation of the regions between 128-138 and 210-220 is also highly charged in similar analyses of the VP7 sequence of the serotype-distinct UK virus. If our hypothesis that conformation of this protein is critical for neutralization specificity is correct, then attempts to characterize the determinants by the synthesis and testing of linear synthetic peptides will not be successful. Alternative strategies of epitope mapping using monoclonal antibodies alone, or in combination with recombinant DNA technology, to express this antigen may be required to unravel the precise structure of the rotavirus neutralization antigenic determinants.

Initial studies with monoclonal antibodies have demonstrated the existence of four distinct epitopes on SA11-VP7 (Sonza et al., 1984). One epitope was most critical for neutralization, another was involved to a lesser extent and two epitopes seem to have no role in neutralization. Further work is required to locate the sites of these two neutralization epitopes.

THE SUBGROUP(S) ANTIGEN

The subgroup antigen has an interesting history. In early experiments, researchers attempted to use ELISA assays to distinguish virus serotypes (Yolken et al., 1978). Although animal and human virus strains were able to be differentiated by these assays, viruses characterized as representing different serotypes based on classical neutralization assays were not used as controls. When these isolates and virus reassortants were subsequently compared by both the ELISA and neutralization tests, it was recognized that the ELISA assay did not distinguish virus based on the neutralization antigen but rather on an antigen on VP6. It was then proposed that (i) two forms of subgroup antigen on VP6 existed, (ii) all animal viruses are in subgroup 1, (iii) subgroup 1 viruses could be distinguished by an overall long electropherotype, and (iv) this was a useful marker for epidemiologic studies (Kapikian et al., 1981).

Monoclonal antibodies specific for each of the 2 subgroups have been developed (Greenberg et al., 1983a). However, viruses have been found that cannot be subgrouped, and one additional subgroup specificity has been proposed (Tufvesson, 1983). The characterization of additional isolates has also shown that animal rotaviruses may contain either subgroup specificity and a long electropherotype pattern does not reliably predict subgroup 1 viruses. The subgroup antigen may still be a useful marker for epidemiologic studies. Some questions, however, remain concerning the significance of this antigen. The monoclonal antibodies that "distinguish viruses" do so on the basis of a differential reactivity of the strains and viruses that are not able to be subgrouped react well with both reagents. These results suggest that both epitopes may be present on all viruses and possibly the monoclonals react with overlapping conformational determinants formed by interactions with other viral proteins. If this is correct, and new rotavirus strains evolve by a mechanism of gene reassortment, it is likely the subgroup antigenic reactivity will become more variable and this antigen will become a less useful marker as more virus strains are characterized.

The gene coding for VP6 has been cloned. The protein sequence deduced from the nucleotide sequence shows VP6 contains 397 amino acids, 3 cysteines and it has a proline-rich carboxyl-terminus (Estes et al., 1984). Comparisons of the deduced protein sequences of VP6 from viruses representing two subgroups show minimal (34) amino acid changes. Most of these changes (70%) are clustered in three regions of the protein (Both et al., 1984). Due to this, the use of synthetic peptides may be a feasible approach to map the subgroup epitopes on this protein.

THE HEMAGGLUTININ (H) ANTIGEN

The hemagglutinin or H antigen is located on the outer capsid protein VP3 that is coded for by the fourth gene segment. VP3 has several interesting biologic properties. It can be cleaved by proteolytic enzymes and such cleavage enhances infectivity. In addition, this protein is the hemagglutinin. Although not all rotaviruses are thought to contain a hemagglutinin, recent studies have suggested this may be related to conditions of virus cultivation. For example, the hemagglutinin of human rotavirus strains is only detectable if viruses are grown in the absence of trypsin (Kitaoka et al., 1984). This antigen may also be involved in neutralization in a minor way. That is, some monoclonal antibodies directed against VP3 neutralize viral infectivity (Greenberg et al., 1983b). This independent antigen may become more important if it allows one to distinguish viruses that have the same neutralization subtype. cDNA clones have been produced for this gene but the nucleotide sequence is not yet published.

VACCINES FROM SYNTHESIZED ROTAVIRUS ANTIGENS

Current research efforts in this and other laboratories are focused on determining the feasibility of using expressed protein products as immunogens (and as antigens for serodiagnostic tests). The availability of DNA clones to the rotavirus genes makes this a reasonable approach, especially since the reports of first generation vaccines have been encouraging (Vesikari et al., 1984). Possible novel vaccine strategies, summarized in Table 4, will undoubtedly be evaluated soon. As part of these studies, the relative importance of the three rotavirus antigens in eliciting a protective response in the host must be evaluated. Although VP7 is the major serotype-specific and neutralization antigen, it may not be the protective antigen in vivo, and it may be necessary to administer expressed protein products (VP3, VP6, and VP7) individually and in mixtures as subunit vaccines.

Table 4: PREVENTION OF ROTAVIRUS GASTROENTERITIS

1. Live Attenuated Vaccines

   Formulation?
   - Animal or human viruses
   - Animal/human reassortants

   Problems?
   - Reproducible safety and efficacy?
   - Reversion to virulence?
   - Safe and effective in children in developing countries?

2. Alternate Vaccines

   Subunit vaccines from purified particles

   Neutralization antigen(s) from cloned gene(s) expressed in prokaryotic, eukaryotic, insect or viral vectors

   Vaccination with a live expression vector that colonizes the GI tract and produces the neutralization antigen(s)

   Synthetic peptides

The synthesis of an immunogenic rotavirus protein(s) may be possible using cloned genes, but the successful vaccination of the gastrointestinal tract with nonreplicating antigens will also require overcoming other obstacles. These include: (i) larger amounts of pure antigen may be required; (ii) antigen stability will be unknown; (iii) the induction of intestinal IgA may require oral immunization, and (iv) long lasting immunity is not assured. However, major steps toward overcoming these obstacles are being made and, in the near future, we should know whether modern technology will be able to prevent rotavirus gastroenteritis.

ACKNOWLEDGEMENTS

Research in this laboratory has received partial support by grants from the National Institutes of Health, the Thrasher Research Fund, the Diarrheal Disease Program of the World Health Organization, and the Vicks Health Care Division of Richardson-Vicks, Inc.

REFERENCES

Arias, C. F., Lopez, S., Bell, J. R., and Strauss, J. H., 1984, Primary structure of the neutralization antigen of simian rotavirus SA11 as deduced from cDNA sequence, J. Virol., 50(2):657-661.

Bastardo, J. W., McKimm-Breschkin, J. L., Sonza, S., Mercer, L. D., and Holmes, I. H., 1981, Preparation and characterization of antisera to electrophoretically purified SA11 virus polypeptides, Infect. Immun., 34:641-647.

Bohl, E. H., Saif, L. J., Theil, K. W., Agnes, A. G., and Cross, R. F., 1982, Porcine pararotavirus: detection, differentiation from rotavirus, and pathogenesis in gnotobiotic pigs, J. Clin. Microbiol., 15:312-319.

Both, G. W., Mattick, J. S., and Bellamy, A. R., 1983, The serotype-specific glycoprotein of simian-11 rotavirus: coding assignment and gene sequence, Proc. Natl. Acad. Sci. USA, 80:3091-3095.

Both, G. W., Siegman, L. J., Bellamy, A. R., Ikegami, N., Shatkin, A. J., and Furuichi, Y., 1984, Comparative sequence analysis of rotavirus genomic segment 6 - the gene specifying viral subgroups 1 and 2, J. Virol., 51:97-101.

Bridger, J. C., 1978, Location of type-specific antigens in calf rotaviruses, J. Clin. Microbiol., 8(6):625-628.

Bridger, J. C., Clarke, I. N., and McCrae, M. A., 1982, Characterization of an antigenically distinct porcine rotavirus, Infect. Immun., 35:1058-1062.

Chou, P. Y., and Fasman, G. D., 1978, Prediction of the secondary structure of proteins from their amino acid sequence, Adv. Enzymol., 47:45-148.

Corrigan, A. J., and Huang, P. C., 1982, A basic microcomputer program for plotting the secondary structure of proteins, Computer Programs in Biomedicine, 15:163-168.

Dimitrov, D. H., Estes, M. K., Rangelova, S. M., Shindarov, L. M., Melnick, J. L., and Graham, D. Y., 1983, Detection of antigenically distinct rotaviruses from infants, Infect. Immun., 41:523-526.

Elleman, T. C., Hoyne, P. A., Dyall-Smith, M. L., Holmes, I. H., and Azad, A. A., 1983, Nucleotide sequence of the gene encoding the serotype-specific glycoprotein of the UK bovine rotavirus, Nuc. Acids Res., 11:4689-4701.

Ericson, B. L., Graham, D. Y., Mason, B. B., and Estes, M. K., 1982, Identification, synthesis and modification of simian rotavirus SA11 polypeptides in infected cells, J. Virol., 42:825-839.

Ericson, B. L., Graham, D. Y., Mason, B. B., Hanssen, H. H., and Estes, M. K., 1983, Two types of glycoprotein precursors are produced by the simian rotavirus SA11, Virology, 127:320-332.

Espejo, R. T., Lopez, S., and Arias, C., 1981, Structural polypeptides of simian rotavirus SA11 and the effect of trypsin, J. Virol., 37:156-160.

Espejo, R. T., Puerto, F., Soler, C., and Gonzalez, N., 1984, Characterization of a human pararotavirus, Infect. Immun., 44:112-116.

Estes, M. K., Graham, D. Y., and Mason, B. B., 1981, Proteolytic enhancement of rotavirus infectivity, molecular mechanisms, J. Virol., 39:879-888.

Estes, M. K., Graham, D. Y., Ramig, R. F, and Ericson, B. L., 1982, Heterogeneity in the structural glycoprotein (VP7) of simian rotavirus SA11, Virology, 122:8-14.

Estes, M. K., Mason, B. B., Crawford, S., and Cohen, J., 1984, Cloning and nucleotide sequence of the simian rotavirus gene 6 that codes for the major inner capsid protein, Nuc. Acids Res., 12:1875-1887.

Estes, M. K., Palmer, E. L., and Obijeski, J. F., 1983, Rotaviruses, a review, Current Topics in Microbiology and Immunology, 105:123-184.

Feachem, R. G., Hogan, R. C., and Merson, M. H., 1983, Diarrheal disease control: reviews of potential interventions, Bull. W.H.O., 61:637-640.

Gaul, S. K., Simpson, T. F., Woode, G. N., and Fulton, R. W., 1982, Antigenic relationships among some animal rotaviruses, virus neutralization in vitro and cross-protection in piglets, J. Clin. Microbiol., 16:495-503.

Greenberg, H. B., McAuliffe, V., Valdesuso, J., Wyatt, R., Flores, J., Kalica, A., Hoshino, Y., and Singh, N., 1983a, Serological analysis of the subgroup protein of rotavirus, using monoclonal antibodies, Infect. Immun., 39:91-99.

Greenberg, H. B., Valdesuso, J., VanWyke, K., Midthun, K., Walsh, M., McAuliffe, V., Wyatt, R. G., Kalica, A. R., Flores, J., and Hoshino, Y., 1983b, Production and preliminary characterization of monoclonal antibodies directed at two surface proteins of rhesus rotavirus, J. Virol., 47:267-275.

Holmes, I. H., 1983, Rotaviruses, in: "The Reoviridae," W. K. Joklik, ed., Plenum Press, New York.

Hopp, T. P., and Woods, K. R., 1983, A computer program for predicting protein antigenic determinants, Molecular Immunology, 20:483-489.

Hoshino, Y., Wyatt, R. G., Greenberg, H. B., Flores, J., and Kapikian, A. Z., 1984, Serotype similarity and diversity of rotaviruses of mammalian and avian origin as studied by plaque-reduction neutralization, J. Inf. Dis., 149:694-702.

Hung, T., Chen, G., Wang, C., Yao, H., Fang, Z., Chao, T., Chou, Z., Ye, W., Chang, X., Den, S., Liong, X., and Chang, W., 1984, Waterborne outbreak in adults in China caused by a novel rotavirus, Lancet, i: 1139-1142.

Kalica, A. R., Greenberg, H. B., Wyatt, R. G., Flores, J., Sereno, M. M., Kapikian, A. Z., and Chanock, R. M., 1981, Genes of human (strain WA) and bovine (strain UK) rotaviruses that code for neutralization and subgroup antigens, Virology, 112:385-390.

Kapikian, A. Z., Cline, W. L., Greenberg, H. B., Wyatt, R. G., Kalica, A. R., Banks, C. E., James, H. D., Flores, J., and Chanock, R. M., 1981, Antigenic characterization of human and animal rotaviruses by immune adherence hemagglutination assay (IAHA), evidence for distinctness of IAHA and neutralization antigens, Infect. Immun., 33: 415-425.

Kitaoka, S., Suzuki, H., Numazaki, T., Sato, T., Konno, T., Ebina, T., Ishida, N., Nakagomi, O., and Nakagomi, T., 1984, Hemagglutination by human rotavirus strains, J. Med. Virology, 13:215-222.

Kyte, J., and Doolittle, R. F., 1982, A simple method for displaying the hydropathic character of a protein, J. Mol. Biol., 157:105-132.

McNulty, M. S., Allan, G. M., Todd, D., McFerran, J. B., and McCracken, R. M., 1981, Isolation from chickens of a rotavirus lacking the rotavirus group antigen, J. Gen. Virol., 55:405-413.

Monto, A. S., and Koopman, J. S., 1980, The Tecumseh study. XI. Occurrence of acute enteric illness in the community, Am. J. Epidemiol., 112:323-333.

Nicolas, J. C., Cohen, J., Fortier, B., Lourenco, M. H., and Bricout, F.,1983, Isolation of human pararotavirus, Virology, 124:181-184.

Pedley, S., Bridger, J. C., Brown, J. F., and McCrae, M. A., 1983, Molecular characterization of rotaviruses with distinct group antigens, J. Gen. Virol., 64:2093-2101.

Rodger, S. M., Bishop, R. F., and Holmes, I. H., 1982, Detection of a rotavirus-like agent associated with diarrhea in an infant, J. Clin. Microbiol., 16:724-726.

Roseto, A., Scherrer, R., Cohen, J., Guillemin, M. C., Charpilienne, A., Feynerol, C., and Peries, J., 1983, Isolation and characterization of anti-rotavirus immunoglobulins secreted by cloned hybridoma cell lines, J. Gen. Virol., 64:237-240.

Saif, L. J., Theil, K. W., and Redman, D. R., 1982, Detection and pathogenicity of an enteric bovine rotavirus-like agent, Abst. 63rd Conf. Res. Workers on Animal Dis., p. 98.

Sato, K., Inaba, Y., Miura, Y., Tokuhisa, S., and Matumoto, M., 1982, Antigenic relationships between rotaviruses from different species as studied by neutralization and immunofluorescence, Arch. Virol., 73:45-50.

Snodgrass, D. R., Herring, A. J., Campbell, I., Inglis, J. M., and Hargreaves, F. D., 1984, Comparison of atypical rotaviruses from calves, piglets, lambs and man, J. Gen. Virol., 65:909-914.

Snyder, J. D., and Merson, M. H., 1982, The magnitude of the global problem of acute diarrheal disease: a review of active surveillance data, Bull. W.H.O., 60:605-613.

Sonza, S., Breschkin, A. M., and Holmes, I. H., 1984, The major surface glycoprotein of simian rotavirus SA11 contains distinct epitopes, Virology, 134:318-327.

Tufvesson, B., 1983, Detection of a human rotavirus strain different from types 1 and 2 - a new subgroup? Epidemiology of subgroups in a Swedish and an Ethiopian community, J. Med. Virol., 12:111-117.

Vesikari, T., Isolauri, E., D'hordt, E., Delem, A., Andre, F. E., and Zissis, G., 1984, Protection of infants against rotavirus diarrhea by RIT 4237 attenuated bovine rotavirus strain vaccine, Lancet, i: 977-980.

Yolken, R. H., Barbour, B., Wyatt, R. G., Kalica, A. R., Kapikian, A. Z., and Chanock, R. M., 1978, Enzyme-linked immunosorbent assay for identification of rotaviruses from different animal species, Science, 201:259-261.

MONOCLONAL ANTIBODIES AGAINST BACTERIAL OUTER MEMBRANE ANTIGENS

Robert E.W. Hancock and Lucy M. Mutharia

Department of Microbiology
University of British Columbia
Vancouver, B.C., Canada, V6T 1W5

ABSTRACT

Monoclonal antibodies have proved to be highly specific tools for defining the antigenic epitopes of *Pseudomonas aeruginosa* outer membrane macromolecules. In this article we have highlighted the use of monoclonal antibodies in the study of lipopolysaccharide heterogeneity and in particular have demonstrated that single monoclonal antibodies can recognize epitopes on lipid A which are conserved in all Gram negative bacteria tested. Monoclonal antibodies against *P. aeruginosa* outer membrane proteins have been used to demonstrate the strong conservation of specific antigenic sites in all *P. aeruginosa* strains tested. In the case of one monoclonal antibody, specific for outer membrane lipoprotein H2, the antigenic site recognized by the antibody was also found to be conserved in all group 1 *Pseudomonads*. The implications of these monoclonal antibodies to bacterial taxonomy is discussed. Monoclonal antibodies against two separate conserved surface epitopes on outer membrane protein F were isolated and differentiated according to their reactions with 2 mercaptoethanol-reduced protein F and with proteolytic and cyanogen bromide peptide fragments of protein F. . One of these protein F-specific monoclonal antibodies has been demonstrated to have immunotherapeutic potential.

INTRODUCTION

Monoclonal antibodies provide several advantages over conventional polyclonal sera derived from immunization of animals with antigenic fractions. For example, they have defined specificity for a single epitope (i.e. antigenic site) on one species of macromolecule. This allows one to accurately quantify and identify macromolecules under all circumstances, regardless of the presence of contaminating antigens. In addition, monoclonal antibodies interact with their target antigens in a highly reproducible fashion, whereas polyclonal sera are notorious for their batch to batch variation. Also hybrid cells producing these monoclonal antibodies can be grown in large amounts in tissue culture or in ascites and thus provide potentially inexhaustable supplies of these reagents.

We have applied hybridoma technology to the production of monoclonal antibodies against the outer membrane antigens of Pseudomonas aeruginosa. In this article we will attempt to highlight the potential uses of the monoclonal antibodies in a variety of areas of study.

LIPOPOLYSACCHARIDE HETEROGENEITY

LPS is a major surface component in gram negative bacterial cells (Nikaido and Nakae, 1979). It consists of a tripartite structure with lipid A (adjacent to and inserted into the outer monolayer of the outer membrane) covalently bonded to the rough core which is itself covalently attached to the O antigenic side chain. The lipid A usually consists of diglucosamine phosphate substituted with five or six fatty acids, including around four moles of hydroxy fatty acids. The rough core region contains a variety of saccharides often including octoses, heptoses and hexoses. The most distal (to the membrane) portion of LPS is the O antigen which usually consists of repeated tri-to pentasaccharide units. It is now recognized that the LPS can be quite heterogeneous both in the substitution of various groups (including phosphates, ethanolamines, aminoacids and acetyl groups) at various places along the saccharide chain and in the length of individual O antigenic side chains (reviewed in Darveau and Hancock, 1983). LPS has two properties of major importance to the pathogenesis of gram negative bacterial infections. Firstly the O antigen is usually the immunodominant portion of the bacterial surface and in many bacterial species differences in O antigen composition are the basis of serotyping. Secondly the lipid A, also called endotoxin, confers many interesting properties on the LPS including pyrogenicity, toxicity, mitogenicity, adjuvanticity, tumor necrotising ability, etc. (Bradley, 1979).

We have used monoclonal antibodies to probe the antigenic conservation of different regions of the LPS (Hancock et al., 1982; Mutharia et al., 1984). A monoclonal antibody specific for the O antigen of our wild type P. aeruginosa strain interacted with strains of equivalent serotype (O-5 in the International Antigen Typing Scheme, IATS) and weakly with strains of a related serotype (O-17) (Table 1).

In contrast monoclonal antibodies specific for rough LPS (lacking O side chains) reacted with a limited subset of strains from the 17 serotypes of the IATS (Table 1), demonstrating greater conservation of the rough core. Monoclonal antibodies against lipid A interacted with all of the strains of P. aeruginosa tested as well as many other genera of the Gram negative bacteria (Table 2). Thus, as a general rule it appears that LPS heterogeneity is increased for substituents that are further from the outer membrane surface.

CONSERVATION OF PROTEIN ANTIGENS IN P. AERUGINOSA

The Pseudomonas aeruginosa outer membrane contains a 6-9 species of polypeptides present in high copy number (10^4-10^5 copies per cell) and thus called major outer membrane proteins

Table 1: Reaction of LPS specific monoclonal antibodies with *Pseudomonas aeruginosa* strains

Outer Membranes		Reaction using monoclonal antibodies to LPS: O Antigen type 5 (MA1-8)	Rough Core (MA3-5)	Rough Core (MA3-6)	Lipid A (5E4)
P. aeruginosa	PA01	+	+	+	+
P. aeruginosa	rough	–	+	+	+
P. aeruginosa	serotype 5	+	+	+	+
P. aeruginosa	serotype 17	+[a]	–	–	+
P. aeruginosa	serotypes 7,8,10,14,16	–	+	+	+
P. aeruginosa	serotype 11	–	–	+	+
P. aeruginosa	serotypes 1,2,3,4,6,9 12,13,15	–	–	–	+
P. aeruginosa	clinical strains	+/- (2/34)[b]	+/-(1/20)[b]	+/-(1/20)[b]	+(20/20)[b]

[a]weaker reaction due to a shared determinant between serotype 17 and serotype 5.

[b]postive reactions out of the number of strains tested.

(Mutharia et al., 1982). We have demonstrated that these proteinsare immunogenic in animals and in humans, in that the sera of cystic fibrosis patients with P. aeruginosa pulmonary infections (Hancock et al., 1984), of convalescent patients recovered from P. aeruginosa bacteremia (Lam et al., 1983) and of rats with induced chronic P. aeruginosa lung infections (Lam et al., 1983), all have antibodies that interact with the major outer membrane proteins of our laboratory wild type strain PAO1. The simplest explanation for this data is the outer membrane protein antigens are antigenically conserved in P. aeruginosa. We have now confirmed this suggestion for outer membrane proteins F and H2 using monoclonal antibodies (Table 2).

POTENTIAL USE OF MONOCLONAL ANTIBODIES IN THE TAXONOMY OF BACTERIA

At present the major tools in the taxonomic grouping of bacterial cells are determination of growth characteristics , DNA:DNA hybridization (Palleroni et al., 1973), rRNA homology studies (Palleroni et al., 1973; DeVos and DeLey, 1983) and fatty acid analysis (Moss et al., 1972). Each of these methods is time consuming and in the latter three cases involves expensive reagents or equipment. Thus there is some justification for the development of new taxonomic aids for the identification of groups of gram negative bacteria. We have considered monoclonal antibodies directed against outer membrane antigens as potential taxonomic tools.

To date we have identified three monoclonal antibodies (Table 2) which may have taxanomic value. Monoclonal antibody 5E4 was produced by Centocor, PA and is directed against the lipid A portion of the lipopolysaccharide (LPS) of E. coli strain J5 (Mutharia et al., 1984). This antibody interacts with almost all (97%) Gram negative outer membranes (i.e. lipid A), including those from representatives of 4 Families, 16 species and up to 30 strains of a single species. Unfortunately this is a moderately low affinity antibody (thus giving weak ELISA reactions) and the delineation of bacterial cells into Gram negative and Gram positive is usually more simply accomplished by the Gram stain. However, this or a similar antibody may be of interest in those cases in which Gram staining provides a Gram indefinite phenotype.

A second monoclonal antibody MA1-6, specific for the outer membrane peptidoglycan-associated lipoprotein H2 (Hancock et al., 1982), may have greater taxanomic value. This antibody interacts with all known species of the group ("true") Pseudomonads as classified by Palleroni et al. (1973) and by DeVos and DeLey (1983), but not with those strains of Pseudomonadaceae shown by nucleic acid homology to be distantly related to these strains. The group 1 Pseudomonads (and the related organism Azotobacter vinelandii) share one or more of the following characteristics, DNA homology (Palleroni et al., 1973), rRNA homology (Palleroni et al., 1973; deVos and deLey, 1983), and the presence of the hydroxy fatty acids α-hydroxy decanoic acid, α-hydroxydodecanoic acid (occasionally)and ß-hydroxydodecanoic acid attached to their LPSs (Moss et al., 1972). In contrast they differ substantially in morphology, nutritional requirements and production of fluorescent pigments - all of which are characteristics previously used to differentiate the Pseudomonadaceae.

Table 2: Potential taxonomic value of monoclonal antibodies

Bacterial outer membrane antigen	Reaction using monoclonal antibodies to: Outer membrane proteins			LPS	Relationship to Pseudomonas aeruginosa established by a		
	H2 (MA1-6)	F (MA5-8)	F (MA4-4)	Lipid A (5E4)	rRNA homology	DNA homology	Fatty acid homology
P. aeruginosa (33 strains)	+	+	+	+	+	+	+
P. putida	+	–	+	+	+	+	+
P. syringae	+	–	+	+	+	+	+
P. chloraphis	+	–	–	+	+		
P. aurofaciens	+	–	–	+	+		
P. stutzeri	+	–	–	+	+	+	+
P. fluorescens	+	–	–	+	+	+	+
P. anguilliseptica	+	–	–	+			+
Azotobacter vinelandii	+	–	–	+	+		
P. maltophilia[b]	–	–	–	+	–	–	–
P. acidovorans[b]	–	–	–	+	–	–	–
P. solanacearum[b]	–	–	–	+	–	–	–
P. cepacia[b]	–	–	–	+	–	–	–
Enterobacteriaceae (4 Genera)	–	–	–	+	–	–	–
Vibrionaceae (4 Genera)	–	–	–	+	–	–	–
Rhizobiaceae	ND[c]	ND	ND	+	–	–	–

a. + relationship established; – no relationship; blank – no available data.
b. These strains were established as taxonomically distinct from the *P. fluorescens* group of the Family *Pseudomonadaceae* and fit into 3 distinct rRNA and DNA homology groups.
c. ND – not determined.

A third antibody of interest is MA5-8 specific for porin protein F of Pseudomonas aeruginosa (Mutharia and Hancock, 1983). This antibody cross-reacts with all 33 strains of P. aeruginosa we have examined, but no other Gram negative bacterial strains. Interestingly, other monoclonal antibodies against porin protein F have lesser specificity, since they recognize P. putida and P. syringae strains. Nevertheless, even these antibodies may turn out to be of taxonomic interest.

Thus monoclonal antibodies can distinguish both major subgroupings and more minor subgroupings of bacterial cells. However, due to the potential for removal of given antigens by mutation it seems unlikely that they will become more than an adjunct technique to assist classical taxonomic methods.

CHARACTERIZATION OF TWO ANTIGENIC EPITOPES ON PROTEIN F

In our studies on the cross reactivity of monoclonal antibodies against outer membrane protein F of Pseudomonas aeruginosa we demonstrated monoclonal antibodies with two distinct specificities. One antibody, MA5-8, interacted only with P. aeruginosa strains whereas three other monoclonal antibodies, MA2-10, MA4-4 and MA4-10, cross reacted with P. syringae and P. putida strains. These two antibodies were distinguishable by a variety of other tests (Table 3).

Table 3: Differentiation of two classes of monoclonal antibodies against protein F

	Monoclonal Antibody	
	MA5-8	MA2-10, MA4-4, MA4-10
Reactivity with purified protein F	+	+
Surface labelling of intact P. aeruginosa	+	+
Reactivity with P. putida and P. syringae protein F	-	+
Reactivity with a 31 kD trypsin or 29 kD papain proteolytic fragment of protein F	-	+
Reactivity with cyanogen bromide fragments of protein F	+ (23,28 kD)	-
Antigenic reactivity stable to 2-mercaptoethanol	+	-
Binding to oligomers of protein F on SDS polyacrylamide gels	+	-

Both classes of antibodies recognized a surface epitope as judged by indirect immunofluorescent labelling (Mutharia and Hancock, 1983) and colony immunoblotting (Mutharia and Hancock, submitted) of intact P. aeruginosa cells. The data favour the existence of two separate highly conserved surface epitopes on outer membrane protein F in P. aeruginosa. While these studies have not as yet allowed us to propose a realistic molecular model for protein F, it is anticipated that the isolation and characterization of a greater variety of monoclonal antibodies may permit structural predictions to be made.

USE OF MONOCLONAL ANTIBODIES IN PASSIVE IMMUNOTHERAPY

Monoclonal antibodies are a good candidate for passive vaccines due to their high specificity. We considered the possibility that one of our monoclonal antibodies against protein F would provide protection against a subsequent lethal challenge with P. aeruginosa. The results shown in part in Table 4 strongly suggest that this antibody, which recognizes a conserved surface epitope and can protect mice against P. aeruginosa of a serotype different to the strain used to originally isolate the monoclonal antibody, has good potential as an immunotherapeutic agent.

Table 4. Protective effect of protein F-specific monoclonal antibody MA4-4 against challenge of B6/D2 mice with P. aeruginosa strain M2

Group Protection	Survival	Significance of Difference (Fisher exact test)
Control	1/10	
MA4-4 injected	9/10	$p < 0.01$

Antibody MA4-4 (0.1 mg) of the IgG_{2a} class was injected into the tail vein of 25 Gram mice two hours prior to intraperitoneal challenge with 4×10^6 strain M2 organisms suspended in physiological saline. Survival was recorded after 3 days. No mice died after this time. Results are taken from two separate experiments. A control monoclonal antibody MA1-6 specific for a non surface located epitope of protein H2 did not afford protection.

CONCLUSIONS

This short review has attempted to highlight the potential uses of monoclonal antibodies directed against bacterial surface antigens. It is our feeling that monoclonal antibodies will have a major impact on serology, immunochemistry, taxonomy and immunotherapy. This is emphasized by numerous studies in addition to our own studies reported here. Their continued use should open up many areas of research by providing specific reagents for defining bacterial antigens.

REFERENCES

Bradley, S.G. 1979. Cellular and molecular mechanisms of action of bacterial endotoxins. Annu. Rev. Microbiol. 33:67.

Darveau, R.P., and Hancock, R.E.W. 1983. Procedure for the isolation of bacterial lipopolysaccharides from both smooth and rough *Pseudomonas aeruginosa* and *Salmonella typhimurum* strains. J. Bacteriol. 155:831.

DeVos, P., and J. DeLey. 1983. Intra- and intergenic similarities of *Pseudomonas* and *Xanthromonas* ribosomal ribonucleic acid cistrons. Int. J. Systemat. Bacteriol. 33:487.

Hancock, R.E.W., Mutharia, L.M., Chan, L., Darveau, R.P., Speert, D.P. and Pier, G.B. 1983. *Pseudomonas aeruginosa* isolates from patients with cystic fibrosis: A class of serum sensitive, nontypable strains deficient in lipopolysaccharide O side chains. Infect. Immun. 42, 170.

Hancock, R.E.W., Wieczorec, A.A., Mutharia, L.M., and Poole, K. 1982. Monoclonal antibodies against *Pseudomonas aeruginosa* outer membrane antigens: Isolation and characterization. Infect. Immun. 37:166.

Lam, J.S., Mutharia, L.M., Hancock, R.E.W., Hoiby, N., Lam, K., Belk, L. and Costerton, J.W. 1983. Immunogenicity of *Pseudomonas aeruginosa* outer membrane antigens examined by crossed immunoelectrophoresis. Infect. Immun. 42:88.

Moss, C.W., Samuels, S.B., and Weaver, R.E. 1972. Cellular fatty acid composition of selected *Pseudomonas* species. Appl. Microbiol. 24:596.

Mutharia, L.M., Crockford, G., Bogard, W.C., and Hancock, R.E.W. 1984. Monoclonal antibodies specific for *Escherichia coli* J-5 lipopolysaccharide: Cross reaction with other Gram negative bacterial species. Infect. Immun., in press.

Mutharia, L.M. and Hancock, R.E.W. 1983. Surface localization of *Pseudomonas aeruginosa* outer membrane porin proten F by using monoclonal antibodies. Infect. Immun. 42:1027.

Mutharia, L.M., Nicas, T.I., Hancock, R.E.W. 1982. Outer membrane proteins of *Pseudomonas aeruginosa* serotype strains. J. Infect. Dis. 146:770.

Nikaido, H., and Nakae, T. 1979. The outer membrane of Gram negative bacteria. Adv. Microb. Physiol. 19:163.

Palleroni, N.J., Ballard, R.W., Ralston, E., and Doudoroff, M. 1973. Nucleic acid homologies in the genus *Pseudomonas*. Int. J. Systemat. Bacteriol. 23:333.

GENETIC APPROACHES TO STUDY PSEUDOMONAS AERUGINOSA PROTEIN ANTIGENS *

Thalia I. Nicas and Barbara H. Iglewski

Department of Microbiology and Immunology
Oregon Health Sciences University
Portland, Oregon 97201

ABSTRACT

Pseudomonas aeruginosa produces a large number of extracellular products which may play a role in pathogenesis. We have used genetic techniques to elucidate the relative contribution of these proteins to virulence, and as a method of producing safe toxoids. A mutant has been isolated which produces an immunologically reactive nontoxic form of toxin A, the most toxic extracellular protein produced by *P. aeruginosa*. Although there are difficulties in production of sufficient quantities of this CRM toxoid, these are likely to be solved by further genetic manipulation. Protection studies with toxin A antibody and studies of mutants deficient in toxin A have confirmed that toxin A plays a role in pathogenesis while clearly showing that toxin A alone cannot totally account for the virulence of *P. aeruginosa*. Studies of mutants specifically altered in three other products, exoenzyme S, and the two major proteases of *P. aeruginosa*, elastase and alkaline protease, have clarified the contribution of these products to virulence. Demonstration by genetic studies that exoenzyme S was a major factor in the virulence for one *P. aeruginosa* strain allowed us to correctly predict that antibody to this product would be protective against infection with that strain.

---

*This work was supported in part by grant DAMD 17-83-G-9556 from the United States Army and AI 14671 from the National Institute of Allergy and Infectious Disease. TIN was supported by a fellowship from the Natural Sciences and Engineering Research Council of Canada.

Abbreviations used: ADP, adenosine 5'-diphosphate; NAD, nicotinamide adenine dinucleotide; CRM, cross reactive antigen.

INTRODUCTION

Pseudomonas aeruginosa produces numerous extracellular toxins and enzymes which may contribute to its pathogenicity (Liu, 1974, Woods and Iglewski, 1983). These include two proteins with ADP-ribosyl transferase activity (toxin A and exoenzyme S), several proteases including one that degrades elastin, a hemolysin with phospholipase activity and a leukocin. Recently, genetic techniques have been employed to study the structure and function of these proteins, to produce toxoids, to determine the relative contribution of these proteins to virulence and to elucidate their role in the pathogenesis of P. aeruginosa infections.

STUDIES WITH TOXIN A

Structure and Function of Toxin A

The most toxic extracellular protein produced by P. aeruginosa is toxin A. Its median lethal dose is about 0.2 ug per mouse (Liu, 1974). Toxin A inhibits protein synthesis by catalyzing the transfer of the ADP-ribosyl moiety of NAD to the diphthamide residue of eucaryotic elongation factor 2 (Iglewski and Kabat 1975). Toxin A is released as a 66,583 dalton, toxic proenzyme which is devoid of enzymatic activity (Vasil et al. 1977). Its enzymatic activity can be potentiated by cleavage of the toxin molecule or by mild denaturation and reduction of at least two of its four disulfide bonds (Woods and Iglewski 1983). The enzymatically active fragments and forms of toxin A are non-toxic presumably because they do not bind to the toxin A receptor on the surface of eucaryotic cells. The structural gene encoding toxin A has been cloned in E. coli and its nucleotide sequence determined (Grey et al. 1984). By comparing its deduced primary structure to previous results from chemical cleavage studies, the ADP-ribosyl transferase activity was located on the C- terminus of this protein. This conclusion was confirmed by cloning a restriction fragment encoding a 306 C- terminal polypeptide which had ADP-ribosyl transferase activity.

Development of a Toxin A Vaccine

Toxin A has attracted considerable interest as a potential candidate for protective vaccination against P. aeruginosa infection. It is the most toxic protein produced by this organism and 90% of all clinical isolates produce it. Further, its importance in pathogenesis is indicated by the finding that hypotoxigenic mutants are less virulent than their toxigenic parental strains in a number, but not all, animal models. The principal challenge in the use of toxin A as a vaccine has been the creation of a safe, nontoxic immunogen.

As native toxin A is highly toxic, various procedures for developing a safe toxoid have been attempted. Chemical toxoiding of toxin A has met with mixed success (Cryz et al. 1981, Pavloskis et al. 1981 and Pollack and Prescott, 1982). Formalin toxoiding was shown by Cryz et al. (1981) to effectively reduce toxin A toxicity to both tissue cultures and to mice while retaining immunogenicity. However, toxins prepared with Formalin tended to regain toxicity on storage after the Formalin was removed. Formalin toxoiding in the presence of lysine resulted in stable, nontoxic preparations, but suffered from extensive loss of immunogenicity (Cryz et al. 1981). Toxoid prepared by glutaraldehyde toxoiding was even less effective than Formalin-lysine toxoid in inducing protective antibody

(Pavloskis et al. 1981). Photoaffinity inactivation has been described by Callahan et al. (1984) as an alternative means of toxoiding. Photoaffinity labelled preparations were stable, low in toxicity, and highly immunogenic. In studies so far reported, the toxin A preparation used for photoaffinity labelling, although enzymatically active, was of very low toxicity initially, so the effectiveness of this toxoiding procedure is somewhat difficult to evaluate.

An alternative which avoids many of the drawbacks of chemical toxoiding is the isolation of mutants which produce nontoxic, antigenically crossreacting protein, or CRM. One such toxin A CRM mutant has been isolated by Cryz et al. (1980). This mutant, PAO-PR1, was isolated by screening culture supernatants of mutagenized clones of P. aeruginosa PAO for loss of toxicity to cells growing in tissue culture. Properties of the mutant are summarized in Table 1.

Table 1. Properties of P. aeruginosa PAO-PR1, a mutant producing toxin A CRM-66, and its parent strain PAO.

	PAO	PAO-PR1
Form of toxin A	Wild type	CRM-66
Toxicity of culture supernatant[a]		
MCD for CHO cells	0.05 ng	250 ng
LD_{50} for mice	0.25 ug	>25 ug
Enzymatic activity[b]		
activated	344	0
not activated	21	0
Antigen, ul/ml	1.0	1.2

[a]Summarized from Cryz and Iglewski (1983).

[b]ADP-ribosyl transferase activity expressed as cpm in standard assay (Cryz et al. 1980).

The CRM protein, CRM-66, lacks any appreciable ADP-ribosyl transferase activity and is relatively nontoxic. It is unaltered in apparent molecular weight, and is immunologically indistinguishable from native toxin in immunodiffusion assay and competive binding studies. As the alteration is inherent in the CRM amino acid sequence it cannot revert to a toxic form, a problem with some chemically induced toxoids. The CRM protein possesses the advantage of stability, low toxicity, and high immunogenicity.

The vaccine potential of CRM-66 has not yet been explored. Problems with use of the CRM-66 as a vaccine are the relatively low levels of toxin A antigen produced (a property of its parent strain, PAO), and the difficulty in purification of CRM-66 resulting from the high level of proteolytic activity of derivatives of strain PAO. Low levels of toxin A production have been somewhat ameliorated by development of media which allow optimum production. As production of toxin A appears to be regulated in part by iron levels (Bjorn et al. 1978), low iron media has been adopted. Genetic approaches should be useful in overcoming both the problem of low levels of production and proteolysis. It should be possible to isolate Toxin A hyperproducing mutants deregulated for toxin A production. Mutants deregulated with respect to inhibition of toxin A synthesis by iron have been successfully isolated in the parent strain, PAO (Sokol et al. 1982), but this approach has not yet been successful with the mutant, PAO-PR1. An alternate approach to the development of toxin A hyperproducers is suggested by the work of Hedstrom et al. (Hedstrom, R. C., Funk, C. R., Pavloskis, O. R., and Galloway, D. R., Absrt. Annu. Meet. Am. Soc. Microbiol., 1984, B151, p. 42). These workers have cloned what appears to be a positive regulator of toxin A synthesis. When introduced into PAO or PAO-PR1, the cloned gene appears to result in increased levels of toxin A or CRM-66 production respectively.

Protection Studies With Toxin A Vaccines

Although the problems of developing a toxin A toxoid suitable for active immunization are still to be solved, it has been possible to raise specific antiserum to toxin A in rabbits or horses, and studies using passive immunization have provided some insight into the role of toxin A in pathogenesis of *P. aeruginosa* and the protective capacity of toxin A vaccine. Considerable variation in protection has been reported. Liu and Hsieh (1973) demonstrated passive immunization protected mice against lethality of ip infection with *P. aeruginosa*. Pavloskis et al. (1977) and Snell et al. (1978) showed that passive immunization increased the survival of burned mice infected with toxigenic strains relative to control mice. The degree of protection obtained in these later studies was modest, and depended on the strain used. Other studies have shown very limited or no protection in the burned mouse model (Cryz et al. 1983b), a leukopenic mouse model (Cryz et al. 1983a) and a rat burn model (Walker et al. 1979). Modest but significant protection of mice actively immunized with Formalin-lysine toxoid was reported by Pavloskis et al. (1981), who also showed that survival of immunized mice could be much enhanced with antibiotic treatment. Although the significant protection by both active and passive immunization reported by several groups confirms that toxin A does contribute to virulence, it is clear that the contribution of toxin A alone cannot totally account for the virulence of *P. aeruginosa*, and that other products must be considered in the development of an anti-pseudomonal vaccine.

GENETIC STUDIES OF EXTRACELLULAR PRODUCTS OF *P. AERUGINOSA*

The Genetic Approach to Studying Extracellular Products

The observation that antibody to toxin A is not highly protective suggests that other factors also contribute to virulence. Genetic studies have been useful in confirming that several products are involved in pathogenesis (for a review see Nicas and Iglewski 1984b). *P. aeruginosa* produces a number of extracellular proteins whose biochemical activities suggest a role in pathogenesis. One approach to determining the relative contribution of individual products and identifying which may be protective antigens has been the analysis of mutants in which a given product is specifically altered or deficient. We have used this genetic approach to study the role of four extracellular proteins: toxin A, exoenzyme S, and the two major proteases, elastase and alkaline protease.

Properties of Exoenzyme S, Elastase, and Alkaline Protease

Exoenzyme S, like toxin A, is an enzyme with ADP-ribosyl transferase activity, transferring the ADP-ribosyl moeity of NAD to protein (Iglewski et al. 1978). It differs from toxin A in that it does not modify eucaryotic elongation factor 2 but instead modifies multiple proteins in extracts of eucaryotic cells. The physiological consequences of this activity is not known. The enzymatic activity of exoenzyme S is found in a 49,000 dalton protein. Exoenzyme S producing cells also produce a related enzymatically inactive 53,000 dalton protein. This larger protein cross-reacts immunologically with the enzymatically active form, and proteolytic digestion patterns of the two proteins have shown them to be very similar. Mutants selected for lack of exoenzyme S activity are deficient in both the 53,000 and 49,000 proteins, suggesting that the two proteins are products of the same gene. Exoenzyme S enzyme activity has been shown in about 40% of clinical isolates of P. aeruginosa (Sokol et al. 1981), however, recent studies in our laboratory have shown that the 53,000 dalton form of exoenzyme S is produced by many P. aeruginosa strains in which exoenzyme S activity cannot be detected in vitro. All existing schemes for purification of exoenzyme S have required the use of detergents, and neither 49,000 nor 53,000 dalton exoenzyme S prepared by such protocols have been toxic. Genetic studies, however, have indicated that exoenzyme S may play a major role in pathogenicity in at least some infections, as described below.

Two extracellular proteases, alkaline protease and elastase, are produced by most clinical isolates of P. aeruginosa. These two enzymes are distinct in size, isoelectric point and pH optimum, and are immunologically unrelated (Wretlind and Pavloskis, 1983). Elastase is ca. ten times more active than alkaline protease in degrading casein, and while both proteases are able to attack a broad range of substrates and thus potentially could contribute to tissue damage in infection, elastase has been shown to degrade several biologically important proteins not degraded by alkaline protease. Elastase, but not alkaline protease degrades elastin, a major component of some mammalian tissues, and elastase but probably not alkaline protease has been shown recently to degrade IgG (Doring et al. 1981). Elastase also inactivates several complement components and human alpha-1 proteinase inhibitor (Wretlind and Pavloskis 1983).

Isolation and Characterization of Exoproduct Mutants

To elucidate the role of these four products in pathogenesis, we isolated mutants deficient in specific products. Mutants were isolated from three different parent strains: 388, which produces high levels of exoenzyme S activity and both elastase and alkaline protease but no toxin A; PA103, which produces high levels of toxin A but no elastase and very little total protease; and PAO, a genetically well characterized strain which produces moderate levels of toxin A and both proteases. Both PAO and PA103 produce 53,000 dalton exoenzyme S, but do not show exoenzyme S activity in vitro. Chemical mutagens were used to isolated mutants of strains PAO and PA103. As the use of chemical mutagens always carries the risk of generating multiple or pleiotropic mutations, each mutant was extensively characterized to ensure that the mutation did not affect other extracellular products, growth rate, or other properties. Mutants of strain 388 were generated by transposon insertion, introducing the transposon Tn1 on a temperature sensitive plasmid vector. These mutants were also extensively characterized.

Toxin A mutants were isolated from both PAO and PA103. In both cases, mutagenized clones were screened for production of toxin A antigen by immunoprecipitation of antiserum placed in wells adjacent to isolated colonies (Ohman et al. 1980a and 1980c). A derivative of PAO, PAO-T1, produced no detectable toxin A, while a PA103 derivative, PA103-29, made ca. 0.2% parental levels. A mutant of PAO with an altered elastase was identified by screening for elastolytic activity on nutrient agar. This mutant, PAO-E64, was a CRM mutation producing an altered elastase. The mutant produced parental levels of elastase antigen, and was capable of degrading casein, but was unable to efficiently degrade elastin at temperatures above 30°C. It was, however, still able to degrade elastin at lower temperatures (Ohman et al. 1980b). An alkaline protease deficient mutant of PA103, PA103-AP1, was found by screening for hydrolysis of casein on skim milk agar (Howe and Iglewski 1984). A transposon induced mutant of 388, 388 exs1::Tn1, was isolated by screening clones with transposon insertions for exoenzyme S enzymatic activity. This mutant produced no detectable exoenzyme S activity or antigen.

Effect of Mutational Loss of Exoproducts on Virulence

P. aeruginosa infections in man occur as opportunistic infections, requiring underlying disease or injury before *P. aeruginosa* can establish itself as a pathogen. These disease states are quite varied, including both local and disseminated, acute and chronic infection. Consequently, a number of different animal models are used to assess virulence in *P. aeruginosa*. Studies with mutant strains have shown that products which appear to play a major role in virulence of some kinds of infection may be much less important in others. To illustrate how we have been able to use comparison of mutants in specific exoproducts to elucidate the role of these products in virulence, data for the burned mouse infection model are presented below.

Table 2. Effect of mutational loss of exoproducts on virulence of *P. aeruginosa* in a burned mouse model.

Parent strain (LD_{50})	Mutant	Phenotype	Mutant LD_{50} / Parental LD_{50}
388 (3.0×10^1)	388 exs1::Tn1	exoenzyme S deficient	5,300[a]
	388 leu::Tn1	leucine auxotroph	0.77
PAO (1.3×10^4)	PAO-T1	toxin A deficient	4.3
	PAO-E64	elastase CRM	7.1
PA103 (7.1×10^3)	PA103-29	toxin A deficient	41
	PA103-AP1	alkaline protease deficient	17

[a]Data summarized from Nicas and Iglewski (1984b).

As shown in Table 2, loss of a single exoproduct or activity resulted in significant loss of virulence in the burned mouse infection model. Both of the two toxin A deficient mutants showed a moderate reduction in virulence relative to their respective parental strains. The non-elastolytic mutant PAO-E64 also showed a moderate loss of virulence, as did an alkaline potease deficient mutant of PA103, PA103-AP1. In contrast to the rather small increase in median lethal dose brought about by mutation to deficiency in toxin A or protease, loss of exoenzyme S in strain 388 resulted in a greater than 5,000-fold decrease in virulence. As this mutant was isolated after transposon mutagenesis, the mutant is unlikely to be altered at more than one gene. The presence of the transposon in itself does not appear to affect virulence, as a second transposon-induced mutant auxotrophic for leucine was unaltered in virulence (Nicas and Iglewski 1984a).

The decrease in virulence in the burned mouse model accompanying mutational loss of toxin A or protease activity confirms that these products play a role in pathogenesis, but the finding that the decrease in virulence is rather small for each of these products suggests that other factors are also involved, and that virulence is multifactorial. The large decrease in virulence accompanying exoenzyme S deficiency was an unexpected finding, and perhaps serves to illustrate the usefulness of the genetic approach in identifying products important in pathogenesis. Because of the difficulty of exoenzyme S purification and the lack of toxicity of purified exoenzyme S, no previous evidence existed which supported a role for exoenzyme S in pathogenesis.

Studies with Antibody to Exoenzyme S

The loss of virulence of exoenzyme S deficient mutants suggested that antibody to it may be protective. Antibody was prepared to the enzymatically active 49,000 form of the enzyme, which was purified by extraction from gels after sodium dodecylsulfate polyacrylamide gel electrophoresis. Purified IgG from immunized animals was absorbed with killed cells of the exoenzyme S deficient mutant in order to reduce the possibility that antibody to other components, such as lipopolysaccharide, were present in the immune IgG. The absorbed IgG antibody was then used to passively protect mice from infection with strain 388 in the mouse burn model. Passive protection with anti-exoenzyme S resulted in an increase in LD_{50} from ca. 25 bacteria in the unprotected mice to greater than 10^5 bacteria in the passively immunized animals. The order of magnitude of protection by anti-exoenzyme S closely paralleled the decrease in virulence found in exoenzyme S defficient mutants, for which the LD_{50} in this model was increased to ca. 2×10^5 (Nicas and Iglewski 1984a).

Mutant studies and studies with antibody to exoenzyme S have also been complimentary in elucidating the role of exoenzyme S in infection. In infections of burned mice, the exoenzyme S producing parent strain grows to very high levels (10^9 bacteria per g) at the site of infection, the skin, and within 48 h reaches densities of 10^5 to 10^6 bacteria per ml in the blood and per g in other tissues. In contrast, the exoenzyme S deficient mutant achieves similar levels in the skin, but appears in the blood and other organs only transiently and in relatively low numbers (Table 3). Thus, it would appear that exoenzyme S does not play a role in colonization of the skin, but is involved in dissemination. As might be anticipated anti-exoenzyme S appears to be protective in preventing dissemination of bacteria to blood and other tissues, but does not prevent colonization of the skin.

Table 3. Colonization and dissemination in the burned mouse model of P. aeruginosa 388 and its exoenzyme S deficient mutant, 388 exsl::Tnl, and of 388 in the presence of anti-exoenzyme S.

	Bacteria per g or per ml (48h)[a]		
	SKIN	BLOOD	LIVER
388, no antiserum	$1.9x10^9 \pm 1.3$	$1.0x10^4 \pm 2.3$	$6.0x10^4 \pm 1.2$
388, anti-exo S	$3.1x10^9 \pm 3.4$	$<10^2$	$3.9x10^2 \pm 1.0$
388 exsl::Tnl, no antiserum	$1.0x10^9 \pm 3.1$	$6.7x10^2 \pm 4.0$	$3.2x10^2 \pm 1.9$

[a]Animals were infected with 300 bacteria per mouse. Data shown are geometric means of colony counts from 5 animals, ± standard deviation.

CONCLUSIONS

In this paper we have summarized some of the ways in which genetic approaches may be used to study protein antigens of P. aeruginosa. Isolation of CRM mutants of toxin A is one approach to the development of a safe toxoid suitable for vaccine use. One such mutant, PAO-PR1 is currently being tested, and the genetically modified toxin A CRM produced by this strain promises to be a more effective toxoid than chemically modified toxin A. Genetic manipulation may also prove useful in improving yields of toxin A. Recent studies suggest that the supplementation of genetic techniques with molecular biology may aid the development of strains producing high levels of CRM.

Investigation of mutants deficient in specific exoproducts has proven to be a useful tool in developing an understanding of the role of these products in pathogenesis. A major concept emerging from these studies is that no one product can totally account for the virulence of P. aeruginosa, but rather virulence is multifactorial. In the case of one product, exoenzyme S, genetic studies showed the product to have an unanticipated major role in pathogenesis. From genetic experiments we were able to predict that antibody to exoenzyme would be protective. This has been confirmed in passive protection studies.

The application of current techniques of transposon mutagenesis and molecular cloning to the study of extracellular products of P. aeruginosa promises to expand the potential of genetic studies of these products. One application currently being explored is the use of genetic engineering to obtain larger quantities of antigen. In the future, we can expect to see these techniques used both to clarify the role of these products in virulence, and to create specifically altered CRM products to enhance our understanding of their function and structure.

REFERENCES

Bjorn, M. J., Iglewski, B. H., Ives, S. K., Sadoff, J. C. and Vasil, M. L., 1978, Effect of iron on yields of exotoxin A in cultures of Pseudomonas aeruginosa PA103, Infect. Immun. 19:785.

Callahan, L. T. III, Martinez, D., Marburg, S., Tolman, R. L. and Galloway, D. R., 1984, Toxoids of Pseudomonas aeruginosa exotoxin A: Photoaffinity inactivation of purified toxin and purified toxin derivatives, Infect. Immun., 43:1019.

Cryz, S. J., and Iglewski, B. H., 1983, Toxoids of Pseudomonas aeruginosa toxin A: production by chemical and genetic means, Rev. Infect. Dis., 5:S992.

Cryz, S. J., Freidman, R. L. and Iglewski, B. H., 1980, Isolation and characterization of a Pseudomonas aeruginosa mutant producing a nontoxic, immunologically crossreactive toxin A protein, Proc. Natl. Acad. Sci. U.S.A., 77:7199.

Cryz, S. J., Freidman, R. L., Pavloskis, O. R. and Iglewski, B. H., 1981, Effect of formalin toxoiding on Pseudomonas aeruginosa toxin A: biological, chemical, and immunochemical studies, Infect. Immun., 32:759.

Cryz, S. J., Furer, E., and Germanier, R. 1983a, Passive protection against Pseudomonas aeruginosa infection in an experimental leukopenic mouse model, Infect. Immun., 40:659.

Cryz, S. J., Furer, E., and Germanier, R. 1983b, Protection against Pseudomonas aeruginosa infection in a murine burn wound sepsis model by passive transfer of antitoxin A, antielastase, and antilipopolysaccharide, Infect. Immun., 39:1072.

Gray, G. L., Smith, D. H., Baldridge, J. S., Harkins, R. N., Vasil, M. L., Ellson, Y., and Heynecker, H. L., 1984, Cloning, nucleotide sequencing and expression in E. coli of the exotoxin A structural gene of Pseudomonas aeruginosa, Proc. Nat. Acad. Sci., (in press).

Doring, G., Obernesser, H. J. and Bozenhart, K., 1981, Extracellular toxins of P. aeruginosa II. Effect of two proteases on human immunoglobulins IgG, IgM and secretory IgA, Zbl. Bakt. Hyg. Mikrobiol., 249:89.

Howe, T. R. and Iglewski, B. H., 1984, Isolation and characterization of alkaline protease deficient mutants of Pseudomonas aeruginosa in vitro and in a mouse eye model, Infect. Immun., 43:1058.

Iglewski, B. H. and Kabat D., 1975, NAD-dependent inhibition of protein synthesis by Pseudomonas aeruginosa exotoxin A, Proc. Natl. Acad. Sci. U.S.A., 72:2284.

Iglewski, B. H., Sadoff J. C., Bjorn, M. J. and Maxwell, E. S., 1978, Pseudomonas aeruginosa exoenzyme S: an adenosine diphosphate ribosyltransferase distinct from toxin A, Proc. Natl. Acad. Sci. U.S.A., 75:3211.

Liu, P. V., 1974, Extracellular toxins of Pseudomonas aeruginosa, J. Infect. Dis., 130:S94.

Liu, P. V. and Hsieh, H., 1973. Exotoxins of Pseudomonas aeruginosa III: Characteristics of exotoxin A, J. Infect. Dis., 128:520.

Nicas, T. I. and Iglewski, B. H., 1984a, Isolation and characterization of transposon induced mutants of Pseudomonas aeruginosa deficient in exoenzyme S, Infect. Immun. (in press).

Nicas, T. I. and B. H. Iglewski, 1984b, The contribution of exoproducts to virulence of Pseudomonas aeruginosa, Canadian J. Microbiol., (in press).

Ohman, D. E., Burns, R. P. and Iglewski, B. H., 1980a, Corneal infections of mice with toxin A and elastase mutants of Pseudomonas aeruginosa, J. Infect. Dis., 142:547.

Ohman, D. E., Cryz, S. J. and Iglewski, B. H., 1980b, Isolation and characterization of a Pseudomonas aeruginosa PAO mutant that produces altered elastase, J. Bacteriol., 142:836.
Ohman, D. E., Sadoff, J. C. and Iglewski, B. H., 1980c, Toxin A deficient mutants of Pseudomonas aeruginosa PA103: Isolation and characterization, Infect. Immun., 28:899.
Pavlovskis, O. R., Pollack, M., Callahan, L. T. III and Iglewski, B. H., 1977, Passive protection by antitoxin in experimental Pseudomonas aeruginosa burn infections, Infect. Immun. 18:596.
Pavlovskis, O. R., Edman, D. C., Leppla, S. H., Wretlind, B., Lewis, L. and Martin, K. E., 1981, Protection against experimental Pseudomonas aeruginosa infection in mice by active immunization with toxin A, Infect. Immun. 24:181.
Pollack, M. and Prescott, R. K., 1982, Toxoid from exotoxin A of Pseudomonas aeruginosa: preparation and characterization, J. Infect. Dis., 145:688.
Snell, K., Holder, I. A., Leppla, S. H. and Saelinger, C. B., 1978, Role of exotoxin A and protease as possible virulence factors in experimental infections with Pseudomonas aeruginosa, Infect. Immun., 19:839.
Sokol, P. A., Iglewski, B. H., Hager, T. A., Sadoff, J. C., Cross, A. S., McManus, A., Farber, B. F., and Iglewski, W. F., 1981, Production of exoenzyme S by clinical isolates of Pseudomonas aeruginosa, Infect. Immun., 34:147.
Sokol, P. A., Cox, C. D., and Iglewski, B. H., 1982, Pseudomonas aeruginosa mutants altered in their sensitivity to the effect of iron on toxin A or elastase yields. J. Bacteriol. 151:783.
Vasil, D. E., Kabat, D. and Iglewski, B. H., 1977, Structure activity relationships of an exotoxin of Pseudomonas aeruginosa, Infect. Immun., 16:353.
Walker, H. C., McCleod, G. C., Leppla, S. H., Wretlind, B. and Mason, A. D., 1979, Evaluation of Pseudomonas aeruginosa toxin A in experimental rat burn wound specis, Infect. Immun., 25:828.
Woods, D. E., and Iglewski, B. H., 1983, Toxins of Pseudomonas aeruginosa: new perspectives, Rev. Infect. Dis., 5:S715.
Wretlind, B. and Pavlovskis, O. R., 1983, Pseudomonas aeruginosa proteases and their role in Pseudomonas infections, Rev. Infect. Dis. 5:S998.

RECENT EFFORTS IN THE DEVELOPMENT OF A SPOROZOITE VACCINE AGAINST HUMAN MALARIA

Wayne T. Hockmeyer[1] and John B. Dame[2]

Department of Immunology, Walter Reed Army Institute of Research, Washington, DC 20307[1], and Laboratory of Parasitic Diseases, National Institute of Allergy and Infectious Diseases, National Institutes of Health, Bethesda, MD 20205[2]

ABSTRACT

Nineteen monoclonal antibodies were produced against sporozoites of Plasmodium falciparum. They have been characterized immunochemically, as well as by isotype, CSP reactivity, and stage and species specificity. Five of these monoclonal antibodies were used to screen a genomic library of P. falciparum in the expression vector λgt 11. Aspects of the cloning and sequencing of the gene encoding for the CS protein of P. falciparum are described. The CS gene encodes for a sporozoite surface protein which is thought to be important in eliciting protective immunity in man. The protein has regions that include a signal sequence, charged region, a series of repeating peptides in the middle of the protein, followed by another charged region and then an anchor sequence. There are also two regions of amino acid homology between the CS proteins of P. falciparum and P. knowlesi. Both the repeating region and the conserved sequences could form the basis for a vaccine. The immunogen for such a candidate vaccine could be produced in bacterial fermentation systems or by development of synthetic peptides coupled to a carrier.

INTRODUCTION

Human malaria, a disease of antiquity, was first clinically recognized and described by Hippocrates. Its global impact is well documented by the World Health Organization, which estimates that there are more than 200 million cases per year and 1.2 million deaths. Malaria control and eradication programs reached their zenith by the early 1960's. Political instability, parasite drug resistance and vector resistance to insecticides have been principal factors helping to account for the global resurgence of malaria. With the advent of new techniques in immunochemistry and molecular biology, scientists have been increasingly turning their attention to development of vaccines to help contribute to control of this disease.

Vaccine development is related to the life cycle of the parasite itself and there are principally three main areas of focus. The first is that of the sporozoite (the stage transmitted by Anopheline mosquitoes) which initiates infection in man. Successful development of a sporozoite vaccine has the intrinsic theoretical advantage of completely preventing infection. However, since cross stage protection does not exist in malaria parasites, a single sporozoite escaping the primed immune system could initiate clinical disease. Complementary efforts thus exist to develop vaccines against the asexual erythrocytic stages - namely the schizont or merozoite. This effort is hindered by the antigenic complexity of these stages and the difficulty of sorting out the numerous proteins of these stages to identify successful candidates for a vaccine. The last effort involves the concept of transmission blocking immunity (Miller et al., 1984). Anti-gametocyte antibodies taken in with the blood meal into the mosquitoes midgut can block fertilization, lyse gametes and zygotes or block development of zygotes. Such a vaccine might ultimately be used in malaria control programs along with more traditional methods.

Development of a sporozoite vaccine has been evolving for nearly 20 years - pioneered to a large extent by Ruth Nussensweig, her colleagues at New York University and to a lesser extent others, including ourselves. Though it has been possible to immunize and protect animals and man with irradiated sporozoites, this method is impractical for all but experimental purposes. Sporozoites can be laboriously obtained only from infected mosquitoes and cannot be cultured _in vitro_. Thus adequate quantities for immunization are not obtainable.

Sporozoite immunity is related to development of antibodies to the circumsporozoite (CS) protein (Nussenzweig et al., 1969; Vanderberg et al., 1969). The CS protein covers the surface of the sporozite and was first identified in the rodent malaria _P. berghei_ (Cochrane et al., 1980). Monoclonal antibodies (MAbs) to this protein completely protect recipient animals from sporozoite challenge (Potochjak et al., 1980). Similar CS proteins have also been identified for species infecting monkeys and man (Nardin et al., 1982; Santoro et al., 1982) including _P. falciparum_, the most significant of the human malaria parasites.

The gene encoding for the CS protein of _P. knowlesi_, a monkey malaria, was recently cloned and sequenced (Godson et al., 1983). This gene was cloned first because of the availability of large numbers of _P. knowlesi_ sporozoites from infected mosquitoes for preparation of a cDNA library. Of particular importance was the finding that this gene encodes for a protein with a large repeating sequence - a 12 amino acid peptide repeated 12 times. A 12 residue peptide synthesized from the repeat contains the epitope that finds the protective monoclonal antibodies. This repeating epitope was the major immunogen on the protein as it was shown that MAbs recognizing the repeat blocked access of polyclonal anti-sporozoite sera to a triton X-100 solubilized sporozoite protein in an immuno-radiometric assay (Zavala et al., 1983).

We describe in this manuscript recent work in our laboratories involving the production and characterization of MAbs against sporozoites of *P. falciparum* and aspects of the cloning, sequence and structure of the gene of the CS protein of *P. falciparum*. We compare its structure with that of the previously described CS protein of *P. knowlesi*.

EXPERIMENTAL

Sporozoites

P. falciparum, *P. berghei*, *P. knowlesi*, *P. cynomolgi*, *P. gallinaceum*, and *P. vivax* sporozoites were produced as previously described (Williams et al., 1984).

Monoclonal Antibodies

Monoclonal antibodies (MAbs) were produced in 3-5 mn old female Balb/c mice as described (Williams et al., 1984). Isotype determinations were made by Ouchterlony immunodiffusion assays with commercially prepared heavy and light chain specific goat anti-mouse sera.

Indirect Immunofluoresence (IFA) and Circumsporozoite Precipitin (CSP) Assays

IFA reactions were performed as previously described (Williams et al., 1984). CSP reactions were performed essentially as described by Vanderberg et al. (1969).

Antigen Preparation

P. falciparum sporozoites were isolated from the salivary glands of *An. freeborni* mosquitoes (Bosworth et al., 1975). and preserved as pellets at -80°C in PBS containing 0.2% ovalbumin. The sporozoite extraction buffer consisted of 5.0 ml PBS, 50 ul of a 200 mM solution of phenylmethylsulfonylflouride in 95% ethanol, 12.5 ul of 10 mg/ml solutions of leupeptin and antipain respectively, and 250 ul of a 10% solution of Nonidet P40 (Sigma, St Louis, MO). For antigen extraction, a pellet of sporozoites was resuspended in the volume of freshly prepared extraction buffer required to give $1x10^3$ sporozoites per microliter. The material was incubated at room temperature for 1 hr and vortexed vigorously for 15-30 sec every 10 min. The insoluble material was pelleted by centrifugation at 13,000 x g for 2 min. The supernatant fluid was collected, aliquoted in 10 ul volumes and stored at -80°C. In some experiments 10 mg bovine serum albumin (BSA) or ovalbumin were added to the extraction buffer.

Western Blotting

Western blot analysis was performed using a modification of the method of Towbin et al. (1979). Proteins in the sporozoite extract were separated by SDS polyacrylamide gel electrophoresis (Laemnli et al., 1970) with a 4.5 stacking gel and an 8% to 12% separating gel. The gels were washed twice for 30 min per wash with 200 mls of Towbin's buffer. The proteins were blotted onto a 0.22 micron-nitrocellulose filter at a field strength of 8 volts per centimeter for 14-16 hr at 4°C. Unreacted binding sites on the nitrocellulose were blocked by treatment of the filter with PBS containing 0.05% Tween 20 and 5% BSA for 1 hr at room temperature. The blot was then washed 4 times, at 20 min per wash, with 100 mls of PBS plus 0.05% Tween 20. The filter was reacted with a pool of 5 Mabs (2E6.4, 2F1.1, 4D9.1, 4D11, 5G5.3) diluted with PBS containing 0.05% Tween 20 and 20% fetal calf serum for 90 min at room temperature with continuous rocking. The blots were washed 4 times as before and then treated with sheep anti-serum prepared against whole mouse antibody and labelled with ^{125}I (10 uCi/ug, Amersham, Arlington Heights, ILL.) diluted to $2x10^5$ cpm with the same diluent used for the MAbs. The filters were then washed 4 times as before and dried. Autoradiography was performed using kodak XAR-2 film at -80^oC.

CS Gene of P. falciparum

The cloning and sequencing of the gene encoding for the CS protein of P. falciparum was done as previously described (Dame et al., 1984).

RESULTS

Nineteen MAbs against P. falciparum sporozoites were produced and subsequently characterized as shown in the following table:

Table 1. Characteristics of P. falciparum MAbs.

MAb	Iso-type	CSP Reaction	IFA Reactions Gland Sporozoite	Pattern	Cross Species Reactivity	P.f. 60
1B 2.2	G1	21/25	25/25	rim	neg	+
1G 3.4	G1	23/25	25/25	rim	neg	+
2E 6.4	M	24/22	25/25	rim	neg	+
2E10.2	M	23/25	25/25	rim	P.b	+
2F 1.1	M	25/25	25/25	rim	P.b	+
4D 9.1	G1	15/27	25/25	rim	P.b	+
4D11.6	M	20/29	24/25	rim	neg	+
4F 2.5	M	25/25	25/25	rim	P.b	+
5G 5.3	G1	20/25	25/25	rim	P.b.	+
5H12.7	M	23/25	25/25	rim	P.b.	+
1A 6.4	M	24/25	25/25	rim	neg	+
1H 1.2	G3	3/25	25/25	rim	P.b	-
3G11.1	G1	6/25	24/25	rim	neg	+
4C 7.6	M	8/23	25/25	rim	neg	+
5A 4.1	G3	13/27	25/25	rim	P.b.	+
5B 5.2	M	29/29	25/25	rim	neg	nt
5C 1.1	G1	16/25	25/25	rim	neg	+
6D 6.1	G3	19/25	25/25	uniform	neg	+
6F11.1	M	13/25	25/25	rim	neg	+

The immunoglobulin isotopes are shown column 2, CSP reactivity in column 3, IFA reactivity and pattern of fluorescence in column 4 and 5. Cross species sporozoite reactivity is shown in column 6 (P.b= P. berghei) and column 7 indicates whether the MAbs recognized the P. falciparum M_r60,000 CS protein by Western Blot.

The results of the CSP reactions with ascites fluid diluted 1:10 with PBS are shown in column 3 of Table 1. The majority of the MAbs were strongly CSP reactive. The pattern produced by the MAbs contained in the hybridoma supernatant fluids in the IFA was examined using air dried P. falciparum sporozoites isolated from salivary glands of An. freeborni mosquitoes (Table 1, Column 4,5,). Most of the MAbs produced a uniform fluorescence on the entire sporozoite with a noticably brighter intensity around the outer rim of the parasite. In some instances the fluorescent stained material was also seen just outside of the edge of the parasites, as if some of the antigen had sloughed off or leaked out of the sporozoites. One of the MAbs, 6D6.1, exhibited uniform staining patterns without the brighter rims. The MAbs were also tested for species specificity with air dried sporozoites of avian (P. gallinaceum), murine (P. berghei), simian (P. knowlesi, and P. cynomolgi) and human (P. vivax) malaria parasites. Cross reactions were not seen with any of the heterologous species except P. berghei; four of the MAbs gave consistently positive reactions while four were positive but somewhat variable (Column 6, Table 1). In order to examine the stage specificity of the MAbs, each was tested with both sexual (stage V gametocytes) and asexual (schizonts) blood stage 7G8 parasites. All of the MAbs were specific for sporozoites (data not shown).

Due to the number of antibodies to be examined and the difficulty in obtaining sufficient quantities of metabolically labelled antigen for analysis, the immunoblot technique (Dame et al., 1984) was used in the following experiments, rather than immunoprecipitation. In preliminary studies, a pool of 5 MAbs which had high CSP titers were used to detect antigens in extracts of isolated sporozoites. The pooled MAbs detected 3 polypepetides: one with an apparent molecular weight of 60,000 and two that appeared as a doublet with apparent molecular weights of 53,000 and 51,000 (Figure 1) Dilutions of the extract showed that the technique was sensitive enought to detect the antigens in samples containing as few as 3,333 sporozoites. Ascites fluids containing the MAbs were used at dilutions of $1x10^{-4}$ to $1x10^{-5}$ to eliminate non-specific binding to the filters.

Each of the five MAbs produced the same three-band pattern that the pool produced. Furthermore, analysis of all of the MAbs demonstrated that every CSP positive MAb gave identical results. In contrast, MAbs that did not produce the CSP reaction failed to react with this antigen. Control experiments with ascites prepared in exactly the same manner, but containing MAbs produced against P. vivax, demonstrated that the reaction was not due to non-specific factors in the ascites.

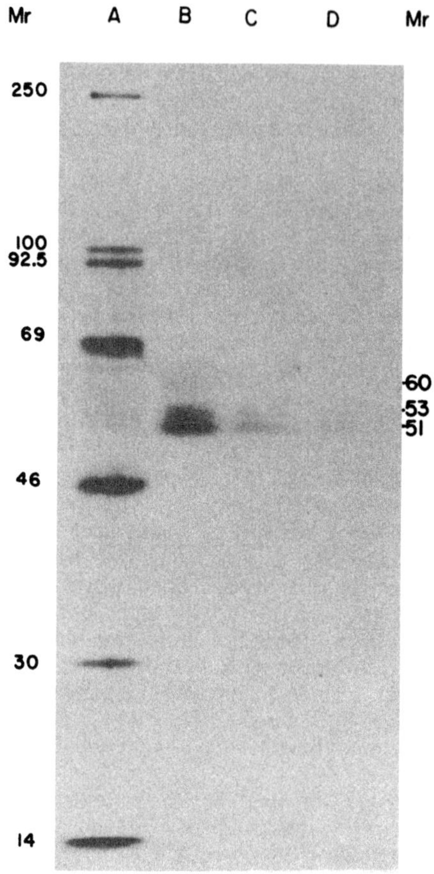

Figure 1. Western blot analysis of P. falciparum sporozoites. Lane A contains molecular weight standards; Lane B, undilute sporozoite antigen (10^4 sporozoites); lane C and D, 3- and 9-fold dilutions of sporozoite antigen, respectively.

The gene encoding the CS protein of P. falciparum was cloned from genomic library in the expression vector λgt 11 (Dame et al., 1984). The library was made from the 7G8 clone of the IMTM 22 isolate of P. falciparum from Brazil. A pool of five MAbs previously described which react with the P. falciparum CS protein were used to screen the library. The clones in λgt 11 were introduced as lysogens in E. coli strain RY1089. Lysates of the IPTG induced bacteria were analyzed for reactivity with the each of the five MAbs by ELISA (Table 2). Seven clones (λmPF 1, 3, 4, 8, 11, 13 and 15) bound all 5 MAbs.

Table 2. Reaction of monoclonal antibodies to the CS protein with lysates of bacteria expressing cloned CS protein gene [a].

Antigen	Monoclonal antibody				
	2E6.4	2F1.1	4D9.1	4D11.6	5G5.3
λmPf1 (2.3 kb)	1.2	1.3	1.7	1.5	1.9
λmPf3 (2.3 kb)	1.3	1.1	1.9	1.4	1.8
λmPf5 (1.3 kb)	1.1	0.9	1.5	1.2	1.4
λmPf8 (1.3 kb)	1.2	0.9	1.6	1.2	1.4
λmPf11 (2.3 kb)	1.1	0.9	1.6	1.3	1.4
λmPf13 (1.3 kb)	0.6	0.5	0.5	0.7	0.6
λmPf15 (1.35 kb)	1.1	1.1	2.0	1.7	2.0
λmgt11, IPTG induced	0.5	0.5	0.4	0.6	0.4
λgt11, not induced	0.5	0.4	0.4	0.6	0.4
RY1089	0.4	0.5	0.3	0.6	0.3

[a] The λmPf clones contain the P. falciparum CS protein gene inserted into the Eco RI site of λgt11. The size of the P. falciparum DNA insert is in parentheses. All bacteria containing recombinant phage were induced with IPTG. See Dame et al.(1984) for details of the ELISA. The data are expressed as the absorbance at 414 nm.

The lysates used for the ELISA were also electrophoresed an SDS-PAGE and electroblotted onto nitrocellulose (Figure 2). The proteins on the nitrocellulose paper were identified by antisporozoite MAbs. The anti-sporozoite MAbs against the protein blots of λmPF 1,3,5,8,11 and 13 bound to 2 doublets at M_r 60,000/57,000 and 53,000/51,000 although the intensity for λmPF13 was reduced. No binding occured to the λgt 11 vector without an insert. MAbs bound to proteins from sporozoites at M_r 60 000, 53,000 and 51,000. Thus, all the sporozoite genes for the CS protein in λgt 11 produced a protein of similar mobility to the unprocessed CS protein synthesized by the sporozoite itself.

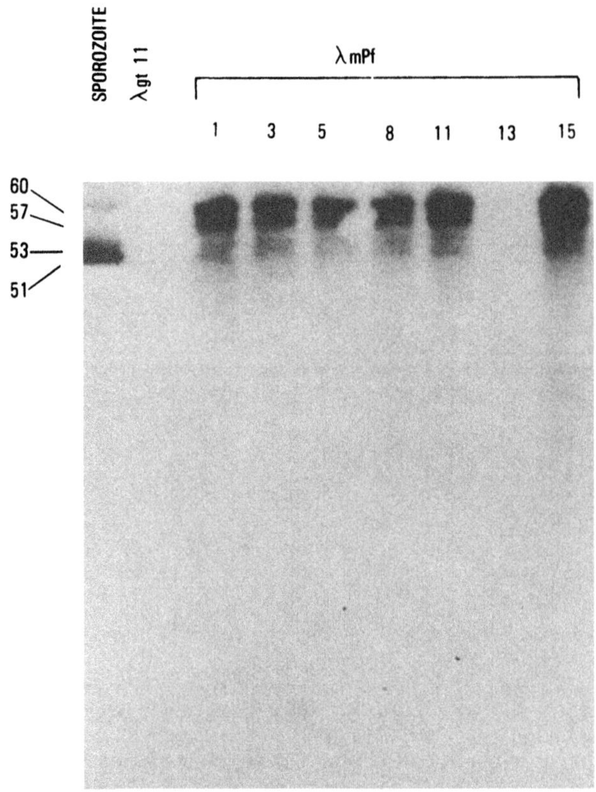

Fig. 2 Western blot analysis of P. falciparum sporozoites, λgt11, and the recombinant products of clones λmPf1, -3 -5, -8,-11, -13, and -15 probed with a pool of five monoclonal antibodies to the CS protein. Details of the analysis are described (Experimental; Dame et al., 1984). Only a small quantity of protein reactive with monoclonal antibodies was produced by λmPf13.

The structure of the protein is in a general sense analogous to that of the CS gene of P. knowlesi (Figure 3). At the amino terminal end of the protein there is a probable signal sequence, a highly charged region, then a repeating region composed of multiply repeated tetrapeptides. Following the repeats, there is another charged region. Lastly, at the carboxyl terminal end, the protein has a sequence of 21 hydrophobic amino acids which represents an anchor sequence for an intergral membrane protein.

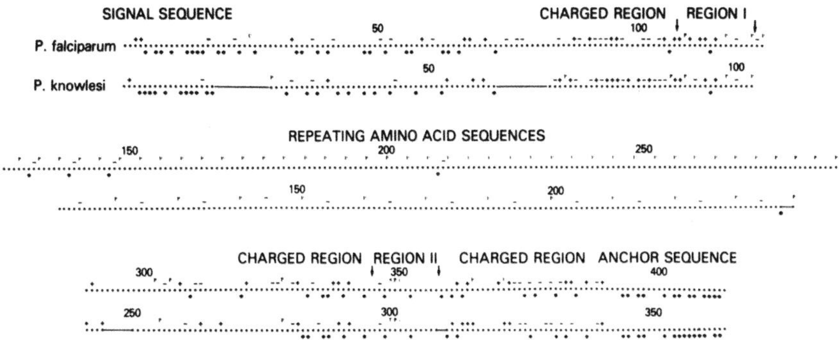

Fig. 3. Circumsporozoite protein structure. The amino acid sequences of the CS proteins of P. falciparum and P. knowlesi (Godson et al., 1983) are displayed together in a symbolic form to highlight the various regions of the two proteins and to illustrate their similar structure. The P. falciparum sequence is divided into three segments, one preceding and one following the repeating amino acid sequence, and it appears in the figure above the analogous parts of the P knowlesi CS protein sequence. Each amino acid is represented by a filled circle. Since the two proteins different in length, spaces (indicated with a solid line) were introduced into the P. knowlesi sequence in a limited number of places to adjust the alignment. Amino acids carrying a charge at physiological pH are indicated above the line (Asp or Glu as (-); Arg or Lys as (+)). The hydrophobic amino acids Ile, Leu, Phe, Trp, Tyr, or Val are indicated below the line as ϕ. Pro and Cys are designated with P and C, respectively. The initial three amino acids of the repeating amino acid sequences of P. knowlesi are included in the pre-repeat part in this comparison since they form part of region I. Also, the Pro was included after the end of the P. knowlesi repeat to illustrate that Pro occurs at this juncture in both proteins. The single amino acid difference in region II is identified as S for Ser in P. falciparum and T for Thr in P. knowlesi.

To prove conclusively that the repeating nucleotide unit of the P. falciparum sporozoite gene was correct, we synthesized peptides of the repeat (Dame et al., 1984). These peptides were then used in a modification of the ELISA to determine if they would inhibit binding of the monoclonal antibody 2F1.1 to λmPf1. The results shown in Figure 4 demonstrate that the 7, 11, and 15 residue peptides significantly inhibit binding of 2F1.1 to λmPf1. Inhibition of binding was evident at $5x10^{-7}$ M with the 15 residue peptide. The 7 residue peptide also inhibited binding of monoclonal antibody 2F1.1 to the sporozoite antigen substituted for λmPf1. Furthermore, the synthetic peptide inhibited binding of the other four monoclonal antibodies to λmPf1. These data indicate that the sequence of the repeating unit is correct. The increased inhibition of binding seen with the 11 and 15 residue peptides may reflect secondary confirmational changes. The data do not suggest that they contain two epitopes since neither could be detected in a double-sided assay with 2F1.1.

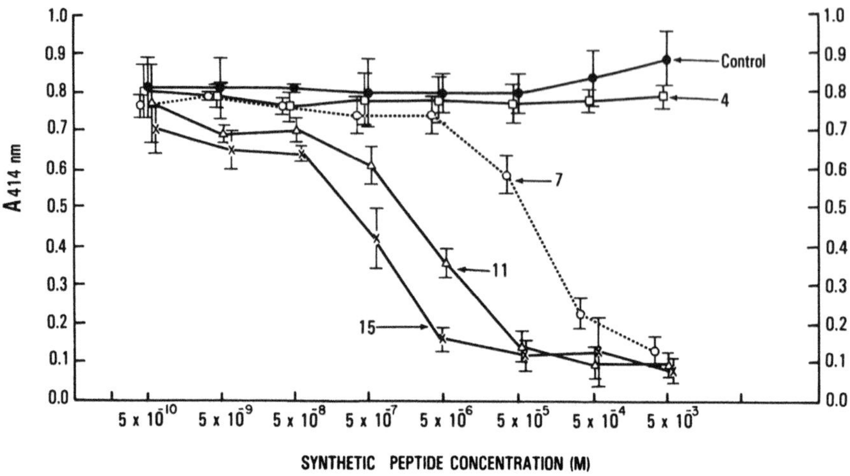

Figure 4. Inhibition of binding of the monoclonal antibody 2F1.1 to CS protein by synthetic peptides of the predominant repeating amino acid sequence. Synthetic peptides (Dame et al., 1984) containing increasing lengths of the predominant repeat sequence were prepared and used to inhibit binding of 2F1.1 to a lysate of mPf1 growing in RY 1089 (Dame et al., 1984). The data are given as the mean ± standard error of three replicates. The synthetic sequences tested were as follows: □, Asn-Pro Asn-Ala; ○, Pro-Asn-Ala-Asn-Pro-Asn-Ala; △, Pro-Asn-Ala-Asn-Pro-Asn-Ala-Asn-Pro-Asn-Ala; x, Pro-Asn-Ala-Asn-Pro-Asn-Ala-Asn-Pro-Asn-Ala-Asn-Pro-Asn-Ala; and ●, an unrelated decapeptide.

DISCUSSION

Five MAbs have been previously described which were produced by Nussenzweig and her colleague against P. falciparum sporozoites (Nardin et al., 1982; Zavala et al., 1983). All of these appeared to be similar if not identical. Because of the importance of falciparum malaria, we produced MAbs to help identify immunogens for possible vaccine development. All but 3 of the 19 MAbs described here (Table 1) produce a strong CSP reaction. MAbs that mediate CSP reactions are of interest because antibodies which neutralize infectivity of sporozoites both in human and animal models, produce this reaction (Cochrane et al., 1980; Potocnjak et al., 1980; Yoshida et al., 1980). Futhermore, MAbs mediating CSP reactions block binding of 70-95% polyclonal immune sera to a solubilized sporozoite antigen (Zavala et al., 1983). Since MAbs were screened by IFA using air dried sporozoites, MAbs recognizing other surface antigens should have been detected. The high percentage (85%) of CSP reactive MAbs reemphasizes the immuno-dominance of the circumsporozoite antigen. The nature of the antigens recognized by MAbs which are IFA positive but do not produce the CSP reaction is not known. They may react with the same protein but in a manner which does not result in precipitation; such MAbs have been described for P. knowlesi (Cochrane et al., 1980). The MAbs described are generally species specific; however, 8 of 19 do react with P. berghei sporozoites. These MAbs may recognize 2 epitopes, one unique to P. falciparum and the other shared by sporozoites of P. falciparum and P. berghei. These epitopes may be different or overlap each other on the antigen. Cochrane et al. (1980) described anti P. knowlesi MAbs that were species specific as well as some which cross reacted with P. falciparum or P. knowlesi. When these MAbs were tested in competitive antigen binding assays, each inhibited binding of the others, suggesting that the epitopes overlapped. We do not know whether this is true for the falciparum MAbs presented here.

The proteins recognized by the MAbs produced against P. falciparum sporozoites has been determind. The MAbs which produce CSP reactions all recognize the same set of 3 proteins by Western Blot. These antigens appear to be identical to those described by Nardin et al. (1982). The multiple band pattern seems to be characteristic of CS proteins. The larger polypeptide represents an intracellular precursor of the smaller surface form of the antigen. The M_rs are slightly lower than those previously reported (Nardin et al , 1982) and likely represent differences in methods rather than any difference in the antigen.

The CS gene of P. falciparum is remarkably analogous to that of P. knowlesi (Figure 3). They are of similar size with calculated molecular weights of 44, 426 and 36,792 for P. falciparum and P. knowlesi, respectively. They have analogous regions that include a signal sequence, charged region, a series of repeating peptides in the middle of the protein, followed by another charged region and then an anchor sequence. There are also two regions of amino acid homology between the proteins. The repeating region of the CS protein of P. falciparum consists of a region of 41 four-amino acid units. Thirty-seven of the repeats in the central region are identical (Asn-Ala-Asn-Pro); four have an alternative sequence (Asn-Val-Asp-Pro).

Analogous CS proteins have been found on all Plasmodium sporozoites to date (Nardin et al., 1982; Santoro et al., 1983). Monoclonal antibodies raised against these proteins neutralize sporozoite infectivity in vitro (Cochrane et al., 1980) or confer protection in vivo (Potocnjak et al., 1980). Such immunity is species specific (Cochrane et al., 1980) and in the case of P. knowlesi, the protective monoclonal antibodies recognize the repeating epitope (Ellis et al., 1983). These MAbs also react in a radio-immunometric assay that requires 2 or more epitopes and led Zavala et al. (1983) to propose that MAb's to CS proteins react with an immunodominant region of repeating epitope. Our own data confirms this hypothesis in that all 5 MAbs (Figure 4) against CS protein of P. falciparum are directed against repeating units (Asn-Ala-Asn-Pro) in the protein. This immunodominant repeating region could thus form the basis for a vaccine. The immunogen for such a candidate vaccine could be produced in bacterial formentation systems or polymerization of a synthetic peptide composed of Asn-Ala-Asn-Pro repeats. In addition to the immunodominant region, we (Dame et al., 1984) described 2 conserved regions (Figure 3) on the basis of their homology to the P. knowlesi CS protein and suggested that they might be involved in reception for liver invasion. If true, the parasites may be unable to vary the sequence in this region. These 2 areas are also tempting targets for sporozoite vaccines.

Though an aura of optimism abounds concerning the development of an efficacious sporozoite vaccine against P. falciparum malaria, much remains to be done. The next several years will be occupied by scientists developing the appropriate immunogen and the means to deliver it to stimulate protective immunity and meet human use requirements.

REFERENCES

Bosworth, A. B., Schneider, I. and Freir, J. E., 1975, Mass Isolation of Anopheles stephensi Salivary Glands Infected with Malaria Sporozoite, J. Parasitol. 61:769.

Cochrane, A. H., Nussenzweig, R. S. and Nardin, E. H. 1980, Immunization Against Sporozoites, In "Malaria", J. P. Krier, ed, Academic Press, New York.

Dame, J. B., Williams, J. L., McCutchan, T. F., Weber, J. L. Wirtz, R. A., Hockmeyer, W. T., Maloy, W. L., Haynes, J. D. Schneider, I., Roberts, D., Sanders, G. S., Reddy, E. P., Diggs, C. L., and Miller, L. H., 1984, Structure of the Gene Encoding the Immunodominant Surface Antigen on the Sporozoite of the Human Malaria Parasite Plasmodium falciparum. Science. 225; 593.

Ellis, J., Ozaki, L. S., Gwadz, R. W., Cochrane, A. H., Nussenzweig, V., Nussenzweig, R. S., and Godson, G. N. 1983, Cloning and Expression in E. coli of the Malarial Sporozoite Surface Antigen Gene from Plasmodium knowlesi. Nature. 302:536.

Lacemnli, U. K., 1970, Cleavage of Structural Proteins During the Assembly of the Head of Bacteriophage T4. Nature. 227:680.

Miller, L. H., David, P. H., and Hadley, T. J., 1984, Perspectives for Malaria Vaccination. Proceedings of the Royal Society Part B. (In press).

Nardin, E. H., Nussenzweig, V., Nussenzweig, R. S., Collins, W. E., Harinasuta, K. T., Tapchaisri, P., and Chomcharn, Y., 1982, Circumsporozoite Proteins of Human Malaria Parasites Plasmodium falciparum and Plasmodium Vivax. J. Exp. Med. 156:20.

Nussenzweig, R., Vanderberg, J., and Most, H., 1969, Protective Immunity Produced by the Injection of X-Irradiated Sporozoites of Plasmodium berghei. IV. Dose Response, Specificity and Humoral Immunity. Military Medicine Special Issue. Sept: 1176.

Potochjak, P., Yoshida, N., Nussenzweig, R. S., and Nussenzweig, V., 1980, Monovalent Fragments (Fab) of Monoclonal Antibodies to a Sporozoite Surface Antigen (Pb44) Protect Mice Against Malarial Infection. J. Exp. Med. 151:1504.

Santoro, F., Cochrane, A. H., Nussenzweig, V., Nardin, E. H., Nussenzweig, R. S, Gwadz, R. W., and Ferreira, A., 1983, Structural Similarities Among the Protective Antigens of Sporozoites from Different Species of Malarial Parasites. J. Biol. Chem. 258:3341.

Towbin, H., Staehelin, T., and Gordon, J., 1979, Electrophoretic Transfer of Proteins from Polyacrylamide Gels to Nitrocellulose Sheets: Procedures and Applications. Proc. Natl. Acad. Sci. 76:4350.

Vanderberg, J., Nussenzweig, R., and Most H., 1969, Protective Immunity Produced by the Injection of X-Irradiated Sprozoites of Plasmodium berghei. In vitro Effects of Immune Serum on Sporozoites. Miliary Medicine Special Issue. Sept.:1183.

Williams, J. L., Hockmeyer, W. T., Wirtz, R. W., Burkot, T. R., . Schneider, I., Weber, J. L., and Esser, K. M., 1984, Characteristics of *Plasmodium falciparum* Sporozoite Monoclonal Antibodies; Western Blot Detection and Chemical and Physical Characteristics of the Circumsporozite Protein. *J. Immunol.* (Submitted).

Yoshida, N., Nussenzweig, R. S., 1980, Hybridoma Produces Protective Antibodies Directed Against the Sporozoite Stage of Malaria Parasite. *Science.* 207:71.

Zavala, F., Cochrane, A. H., Nardin, E. H., Nussenzweig, R. S., and Nussenzweig, V., 1983, Circumsporozoite Proteins of Malaria Parasites Contain a Single Immunodominant Region With Two or More Identical Epitopes. *J. Exp. Med.* 157:1947.

THE PRIMARY SEQUENCE AND ANTIGENIC STRUCTURE OF GONOCOCCAL PILIN: APPROACHES TOWARDS A GONOCOCCAL VACCINE

Jonathan B. Rothbard and Gary K. Schoolnik

Departments of Medicine and Medical Microbiology

Stanford University Medical School
Stanford, CA. 94305

The complete amino acid sequence of pilin from strain MS11 and the partial sequence of R10 pilin are reported. The protein consists of 159 amino acids and appears to be divided into constant and variable domains. The region of conserved sequence appears to be involved in the polymerization of the molecule and composing the ligand binding site. The carboxyl terminal portion of the molecule contains a thirty amino acid disulfide loop that significantly varies in sequence between the two strains and is postulated to be site of the serological variation in the molecule. The antigenic structure of the molecule was investigated by assaying the binding of antisera engendered by intact pili from strains MS11 and R10 and their two major cyanogen bromide generated fragments, CNBr-2 (residues 9-92) and CNBr-3 (residues 93-159), to synthetic peptides corresponding to the amino acid sequence of MS11 pilin. Two independent, strain specific determinants were found within the disulfide loop. A weakly immunogenic, common determinant was located between residues 48-60. Antisera raised against either MS11 or R10 CNBr-2 bind a separate peptide, residues 69-84. This region is only immunogenic as a fragment, not in the pilus filament. Antisera generated to each of the peptides were assayed for their ability to crossreact with intact pili from both homologous and heterologous strains. Of the peptides corresponding to regions of conserved sequence, antisera directed at residues 69-84 was the most efficient in binding pili from all strains tested in both solid phase assays and immunoblots. Sera against the two peptides shown to contain strain specific epitopes crossreacted with MS11 equally well, but differed in their ability to bind pili from heterologous strains. Anti-121-134 was strain specific, whereas anti-135-151 bound all pilin tested. Sera against 41-50 and 69-84 successfully inhibited intact gonococci from binding to a human endometrial cell line. These peptides might be important components of an effective vaccine for the prevention of gonorrhea.

INTRODUCTION

A pathogen's ability to evade the host immune response by antigenic variation provides an intriguing model system to study the molecular details of the antigenicity and immunogenicity of proteins. It also presents an important obstacle to the development of efficacious vaccines for a variety of diseases (Cross, 1978; Wiley et al., 1981; Krystal et al., 1983; Colman et al., 1983; Dale et al., 1980; Barbour et al., 1982). In only a few systems are the molecular details of the sequence variation of the membrane proteins and the genetic mechanism for their variation partially understood (Borst and Cross, 1982; Webster et al., 1982).

Efforts at developing a successful vaccine against gonorrhea have concentrated on blocking the organism's binding to eucaryotic cell surfaces. Gonococcal adhesion to mucous membranes is mediated in part by pili, filamentous polymers which extend out from the bacterial surface (Swanson, 1973). These organelles are composed of a single repeating protein subunit, pilin. Pili prepared from different gonococcal strains are antigenically heterogeneous and attempts to use them as components of a vaccine have resulted only in protection against the homologous strain (Brinton et al., 1978).

To investigate the structural basis of these serological properties we have determined the complete amino acid sequence of pilin isolated from the transparent colonial variant (Tr) of strain MS11 and the partial sequence of R10 (Tr) pilin. The comparison of these two sequences and the serological properties of antisera directed against the two principal cyanogen bromide fragments of pilin (CNBr-2, residues 9-92 and CNBr-3, residues 93-159) have enabled us to construct an antigenic model of the protein. The immunodominant antigenically variable region is within CNBr-3, while CNBr-2 contains a conserved receptor binding region and possibly an immunorecessive epitope common to all gonococcal pilin.

Peptides corresponding to both variable and constant regions of MS11 pilin were synthesized in order to (i.) more precisely determine the antigenic structure of pilin, (ii.) understand the molecular details of the antigenic variation in the protein, and (iii.) identify regions of the protein which might elicit antisera that would inhibit the binding of all gonococcal strains to eucaryotic cells. Each peptide was tested for its ability to be bound by polyclonal sera directed against pili isolated from homologous and heterologous gonococcal strains. In addition, antisera directed against CNBr-2 and CNBr-3 were tested for their ability to bind the peptides. An antigenic determinant common to all strains of pili and two separate strain specific epitopes were defined. We also have demonstrated that antisera against CNBr-2 recognizes a different common determinant than that seen by sera against the intact pilus filaments.

Antisera generated against the synthetic peptides were examined for their ability to crossreact with intact pili from homologous and heterologous strains by solid phase binding assays and immunoblots.

Finally, sera against 41-50 and 69-84 separately were shown to inhibit intact bacteria from binding human endometrial carcinoma cells. These results demonstrate that the immune response to peptides corresponding to small regions of gonococcal pilin dramatically differ from that elicited by the intact protein. Consequently by using peptides as immunogens we have succeeded in generating antibodies directed at the conserved residues implicated in receptor binding that are normally poorly immunogenic in the intact protein. These peptides might be important components of an effective vaccine for the prevention of gonorrhea.

EXPERIMENTAL

Gonococcal strains and growth medium. The isolation and maintenance of gonococcal strains MS11 and R10 have been described (Schoolnik et al., 1984). Strain 1896 was isolated in Seattle from a patient with a disseminated gonococcal infection, whereas R16, F62, and 2686 were provided by Cornell University School of Medicine, New York and D.S. Kellogg, Jr. (center for Disease Control, Atlanta, GA.) from patients with uncomplicated gonorrhea. The four colonial variants (piliated/opaque (O) or transparent (Tr) and nonpiliated/opaque or transparent) of each strain were independently propagated on solid typing medium by selected daily passage of single colonies.

Purification of pili. Pili were purified from MS11 (Tr) and R10 (Tr) strains of *N. gonorrhoeae* as described previously (Schoolnik et al., 1984).

Cyanogen bromide cleavage. Chemical cleavage at methionine residues with cyanogen bromide was performed as described (Gross and Witkop, 1962).

Arginine-specific hydrolysis. Pili were citraconylated by a modification of the previously documented method (Atassi and Habeeb, 1972) in 0.05 M sodium phosphate buffer, pH 9.0.

Iodosobenzoic acid cleavage of CNBr-3. 5 mg of lyophilized CNBr-3 were dissolved in 0.75 ml of 80% (vol/vol) acetic acid-3 M guanidine-HCl and mixed with 0.25 ml of 80% acetic acid containing 12 mg of *o*-iodosobenzoic acid (Pierce Chemical Co.) as described (Mahoney and Hermodson, 1979).

Isolation of peptides. Cyanogen bromide fragments were separated by reverse-phase HPLC. They were eluted from a Whatman Partisil-10 ODS-2 column in a linear gradient of acetonitrile in water containing 0.05% (vol/vol) trifluoroacetic acid (TFA). Tryptic peptides of citraconylated pili and the iodosobenzoic acid fragments of CNBr-3 were resolved by HPLC SEC in which one TSK G3000SW and two TSK G2000SW columns (Varian Associates, Inc., Walnut Creek, CA) were connected in series and the peptides eluted in 0.1 M ammonium bicarbonate butter, pH 8.1.

Sequence analysis. Automated Edman degradation was performed with a Beckman 890C sequencer (Beckman Instruments, Palo Alto, CA), using a modified Quadral program of Beckman Instruments in combination with polybrene. Thiazolinone derivatives of amino acids were converted to Pth-derivatives with aqueous 1.0 N HCl at 80 degrees C for 10 min. Pth-amino acids were identified by HPLC with a Hewlett Packard 1084B HPLC (Hewlett Packard, Co., Cupertino, CA), using Dupont Zorbax ODS columns and confirmed by gas chromatography and/or thin-layer chromatography.

Pilus peptide receptor binding assay. The receptor binding properties of ^{125}I-labeled peptide TC-2 was determined with human endocervical and buccal epithelial cells and with cultivated HeLa cells.

^{125}I-labeled TC-2 was prepared by the chloramine-T procedure (Greenwood et al., 1963) with carrier-free Na ^{125}I (New England Nuclear, Boston, MA).

Selection of the peptides synthesized. Peptides believed to contain strain specific epitopes were chosen according to differences in sequence between MS11 and R10 pilin. All known sequence differences between these two proteins occur between residues 93 and the carboxyl terminus, residue 159. Particular importance was given to regions where amino acid substitutions significantly affected either the charge, the hydrophobicity/hydrophilicity, or the predicted secondary structure of the region. Regions of sequence homology between residues 8 and 92 were synthesized in an attempt to define a common epitope. We avoided regions which were very hydrophobic, but did not simply choose areas of great hydrophilicity. The association of the hydrophilicity of a region with antigenicity neglects the importance of hydrophobic interactions when an antibody binds an antigen (Hopp, 1981). Consequently regions which had both hydrophobic and hydrophilic character were preferred.

The possible conformations that the peptide can adopt when conjugated to a carrier protein also were considered. Only if a peptide adopts the identical secondary structure that it exists in the intact protein will it both bind sera against the intact protein and elicit antibodies which crossreact with that protein. Of the three common secondary structures present in proteins, β-pleated sheets, α-helices, and reverse or β-turns, we believe reverse turns have the highest probability of being present in an isolated oligopeptide less than twenty residues in length. Therefore a peptide containing a region with high turn probability will be more likely to be identical to the coresponding region in the intact protein than regions predicted to be sheets or helices.

The β-turns were emphasized further by conjugating the peptide in an orientation that resulted in the β-turn being distal from the linkage to the carrier protein.

Synthesis and characterization of peptides. Peptides were synthesized by solid phase techniques (Erickson and Merrifield, 1976) on a Beckman model 990B peptide synthesizer using commerically available amino acid polystyrene resins and t-Boc-protected amino acids (Peninsula Laboratories, Belmont, CA). All couplings wre performed using a 2.5 molar excess of t-Boc amino acid and dicyclohexylcarbodiimide over the number of milliequivalents of amino acid on the resin. In the case of Asn and Gln, a 2.5 molar excess of the amino acid, DCC, and N-hydroxy triazole was used.

All couplings were greater than 99% complete as determined by the reaction of the resin with ninhydrin (Kaiser et al., 1970). The peptides were deprotected and removed from the resin simultaneously by treatment with anhydrous hydrogen fluoride in the presence of anisole, dimethyl sulfide, and indole. The peptides were separated from the various organic side products by extraction with ether and isolated from the resin by extraction with 5% acetic acid and subsequent lyophilization. The purity of the crude product was determined by high pressure liquid chromatography on a C-18 reverse phase column (Merck, Darmstadt, Germany) and by amino acid analysis. The peptides used in these experiments were not further purified since they all contained greater than 90% of the desired product.

Conjugation of the peptides to carrier proteins. The peptides were conjugated to bovine serum albumin using succinimidyl 4-(N-maleimido-methyl) cyclohexane-1-carboxylate (SMCC) (Pierce, Rockford, IL) as described (Yoshitake et al., 1979). They were conjugated to thyroglobulin for immunization using a slightly different crosslinker, m-maleimidobenzoyl N-hydroxysuccinimide ester (MBS) (Pierce). The conjugates used in this study all contained between 15 and 25 peptides per molecule of BSA and between 25 and 35 peptides per 100 kilodaltons of thyroglobulin.

The determination of the crossreactivity of sera against intact pili and CNBr-2 and CNBr-3 of pilin. A two-step competitive ELISA was used to determine the crosreactivity of sera directed against MS11 and R10 pili and their cyanogen bromide generated fragments, CNBr-2 and CNBr-3 (Engvall, 1980).

Binding of polyclonal sera to peptide-protein conjugates. Rabbit polyclonal antibodies against either whole pili or cyanogen bromide generated fragments of pilin, CNBr-2 and CNBr-3, were assayed for their ability to bind peptide-BSA conjugates by using solid phase binding assays.

Absorption of antisera with peptide-carrier-sepharose conjugate. The peptide-BSA conjugates were reacted with cyanogen bromide activated sepharose (Pharmacia, Piscataway, NJ.) as described (Porath, 1974). Anti-R10 and anti-MS11 pilus sera (1 ml) were absorbed with 100µl of either peptide 121-134-BSA or 135-151-BSA sepharose for two hours at room temperature. The mixture was spun and the absorbed sera was separated and assayed for the complete removal of the specific anti-peptide antibody by a solid phase binding assay.

Blocking bacterial adherence to endometrial cells. The human endometrial carcinoma cell line (ENCA-4) was grown on coverslips in DMEM with 15% fetal calf serum. Piliated, transparent colonies of gonococcal strain F62 were harvested from agar plates, incubated with dilutions of either preimmune or anti-peptide sera for two hours, and added to the washed cover slips. The mixture was incubated for one hour, washed, fixed, and treated with Geisma stain. The number of bacteria bound per cell was visually counted.

RESULTS

Amino acid sequence analysis. The proposed sequence of gonococcal pilin prepared from strain MS11 is shown in Figure 1. A unique sequence was derived consisting of 159 amino acids in a single polypeptide chain. The strategy used for the sequence analysis of the molecule entailed the selection of chemical and enzymatic cleavages resulting in large overlapping peptides (Figure 2). HPLC of the first phenylthiohydantoin amino acid comigrated with synthetic Pth-N-methyl phenylalanine.

Isolation and alignment of cyanogen bromide peptides. Cyanogen bromide cleavage resulted in three fragments that were separated by reverse phase HPLC. Each was subjected to automated Edman degradation. CNBr-1 was sequenced in its entirety; the sequence of CNBr-2 was determined through residue 49 and was identical to the result from sequencing the intact protein. CNBr-3 peptides isolated from MS11 and R10 pilin were sequenced through residues 143 and 139, respectively. When the first 46 residues of both CNBr-3 sequences were

MePhe-Thr-Leu-Ile-Glu-Leu-Met-Ile-Val-Ile-Ala-Ile-Val-Gly-Ile-Leu-Ala-Ala-Val-Ala- 20

Leu-Pro-Ala-Tyr-Gln-Asp-Tyr-Thr-Ala-Arg-Ala-Gln-Val-Ser-Glu-Ala-Ile-Leu-Leu-Ala- 40

Glu-Gly-Gln-Lys-Ser-Ala-Val-Thr-Glu-Tyr-Tyr-Leu-Asn-His-Gly-Lys-Trp-Pro-Glu-Asn- 60

Asn-Thr-Ser-Ala-Gly-Val-Ala-Ser-Pro-Pro-Ser-Asp-Ile-Lys-Gly-Lys-Tyr-Val-Lys-Glu- 80

Val-Glu-Val-Lys-Asn-Gly-Val-Val-Thr-Ala-Thr-Met-Leu-Ser-Ser-Gly-Val-Asn-Asn-Glu- 100

Ile-Lys-Gly-Lys-Lys-Leu-Ser-Leu-Trp-Ala-Arg-Arg-Glu-Asn-Gly-Ser-Val-Lys-Trp-Phe- 120

Cys-Gly-Gln-Pro-Val-Thr-Arg-Thr-Asp-Asp-Asp-Thr-Val-Ala-Asp-Ala-Lys-Asp-Gly-Lys- 140

Glu-Ile-Asp-Thr-Lys-His-Leu-Pro-Ser-Thr-Cys-Arg-Asp-Lys-Ala-Ser-Asp-Ala-Lys (COOH)

Figure 1. Covalent structure of MS11 gonococcal pilin. MEPhe, N-methyl phenylalanine.

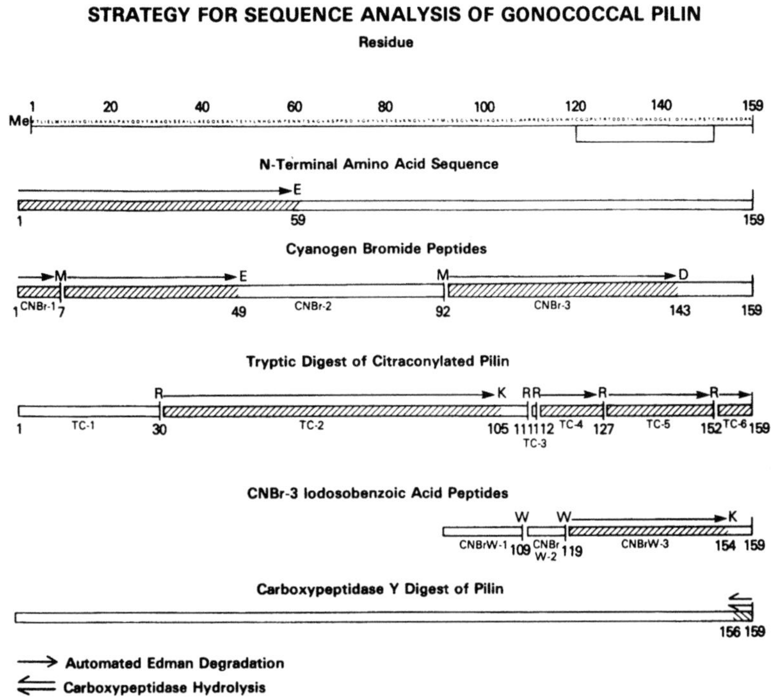

Figure 2. Peptides prepared from gonococcal pilin. Shaded areas indicate regions sequenced.

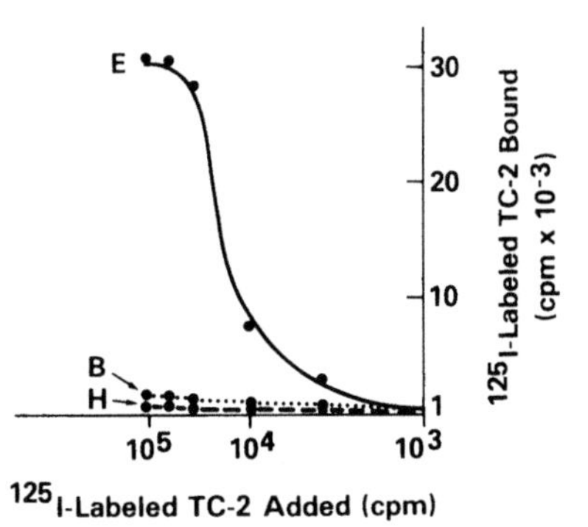

Figure 3. Adhesion of ^{125}I labeled TC-2 to either human endocervical (E), buccal (B), or Hela (H) cells.

compared, 31 positions (67%) were found to be homologous. CNBr-1 and CNBr-2 were aligned by reference to the amino terminal sequence.

Identification of a receptor-binding tryptic peptide. CNBr-2 was proposed to contain residues involved in ligand binding (Schoolnik et al, 1983). TC-2 (residues 31-111), a soluble, monomeric tryptic peptide that contains the last 62 amino acids of CNBr-2, was labeled with ^{125}I, and incubated with endocervical, buccal mucosal, or Hela cells. After one hour bound peptide was separated from unattached material by centrifugation. The fragment specifically bound only the endocervical cells (Figure 3). The relationship between counts added and counts bound was linear and saturable.

Shared antigenicity of R10 and MS11 pili. Previous studies (Brinton et al., 1978) demonstrated that antisera against gonococcal pili are directed principally at strain specific epitopes. However, the sera also contains antibodies that crossreact with pili of other strains. This crossreactivity could be due to either differential binding at strain specific epitopes as a result of amino acid sequence variations or due to the presence of a weakly immunogenic determinant common to all gonococcal pili. Polyclonal sera against the R10 and MS11 pili used in this study initially were tested in a two step inhibition solid phase binding assay to determine their specificity (Engvall, 1980). As seen in Figure 4, the anti-whole pili sera is quite strain specific. The shared antigenicity detected by both sera is only about five per cent. Whether the crossreactivity of the sera is due to epitopes common to all pili or due to differential binding at strain specific determinants was partly answered by examining the antigenic relatedness detected by the sera against two cyanogen bromide generated fragments of pilin, CNBr-2, residues 8-92 and CNBr-3, residues 93-159. A marked difference is apparent in the shared antigenicity detected by the sera directed against CNBr-2 (Figure 4C and 4D) from that against CNBr-3 (Figure 4E and 4F). In the former case the crossreactivity was almost complete, whereas antisera against the carboxyl terminal fragment displayed high strain specificity. These experiments indicate that a weakly immunogenic, common epitope(s) exist between residues 9 and 92 and that the strain specific, immunodominant determinant(s) is located between residues 93-159. The precise location of each of these epitopes was determined by examining the antigenicity of synthetic peptides corresponding to regions of the MS11 sequence.

Location of strain-specific antigenic determinants. To locate more precisely the strain-specific epitopes within CNBr-3 of gonococcal pilin, four peptides were synthesized corresponding to residues 95-107, 107-121, 121-134, and 135-151 (Table 1). Peptides 121-134 and 135-151 compose a disulfide loop while 95-107 and 107-121 correspond to the region of the amino terminal side of this loop. The reduced sulfhydryl of the cysteine at either the carboxyl or amino terminus of each peptide was used to conjugate it to BSA using the heterobifunctional crosslinker, SMCC. Microtiter plates were coated with the BSA-peptide conjugates and antisera against MS11 and R10 pili were tested for their ability to bind them.

The homologous anti-MS11 serum binds the two peptides corresponding to the disulfide loop, but not the peptides outside this region (Figure 5A). Anti-R10 serum does not bind all four of these conjugates (Figure 5C). Additional proof that these peptides contain an epitope specific for MS11 pili is provided by binding experiments with antisera against CNBr-3 of MS11 and R10. Anti-MS11 CNBr-3 serum bound the two peptides comprising the disulfide loop even at dilutions as high as 1:6400 (Figure 5B), whereas antiserum to the same fragment from R10 pilin binds very weakly and only at high serum concentrations (Figure 5D).

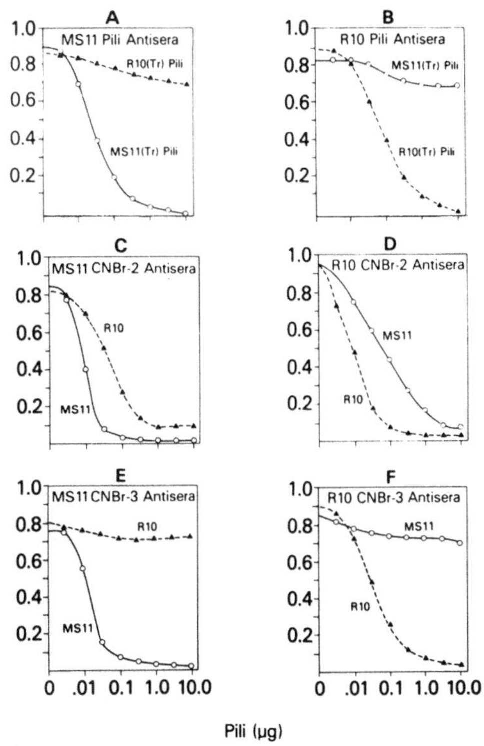

Figure 4. Antisera against intact MS11 pili (A), MS11 CNBr-2 (C), and MS11 CNBr-3 (E) were assayed in a two step competitive ELISA for their ability to bind wells coated with MS11 pili in the presence of increasing amounts of either soluble MS11 or R10 pili. Similarly, antisera against intact R10 pili (B), R10 CNBr-2 (D), and R10 CNBr-3 (F) were tested for their ability to bind wells coated with R10 pili in the presence of increasing amounts of soluble MS11 and R10 pili.

Table 1. Amino Acid Sequences of the Eight Synthetic Oligopeptides and Their Corresponding Location in MS11 Pilin.

L P A Y Q D Y T A R A Q V S E G C*	21 – 35
E G Q K S A V T E Y G C	41 – 50
T E Y Y L N H G K W P E N G C	48 – 60
P P S D I K G K Y V K E V E V K G C	69 – 84
C S G V N N E I K G K K L S	95 – 107
S L W A R R E N G S V K W F C	107 – 121
C G Q P V T R T D D D T V A	121 – 134
D A K D G K E I D T K H L P S T C	135 – 151

*Underlined residues do not exist in the primary structure of MS11 pilin. Additional glycine and cysteine residues were added as spacers and as a mode of attachment to carrier molecules, respectively.

The observation that the two peptides which are bound both by polyclonal sera against whole MS11 and against CNBr-3 are juxtaposed in sequence creates the possibility that either there is a single epitope whose component amino acids are shared by two peptides or that there are two independent epitopes, one within each peptide. This ambiguity was resolved easily by absorbing the MS11 pilus antiserum with one peptide conjugate and then studying the absorbed serum's ability to bind the second peptide conjugate. If there is a single shared determinant the absorption should reduce the amount of of antibodies binding to the second peptide. If there are two epitopes the absorption should not affect the serum's ability to bind the other peptide. Absorption with peptide 121-134 has no affect on the serum's ability to bind 135-151 and reciprocally, absorption with peptide 135-151 does not affect the binding to 121-134 (Table 2). Each peptide contains a strain specific epitope.

Identification of a common determinant in gonococcal pili. Even though the immunodominant epitopes of gonococcal pilin are strain

Table 2. Reaction of Anti-MS11 Pili Sera with Peptides Corresponding to Residues 121-134 and 135-151 Before and After Absorption*.

	Polyclonal Sera Absorbed w/ Sepharose			Polyclonal Sera Absorbed w/ 121-134-BSA-Sepharose			Polyclonal Sera Absorbed w/ 135-151-BSA Sepharose		
Dilution	121-134-BSA	135-151-BSA	BSA	121-134-BSA	135-151-BSA	BSA	121-134-BSA	135-151-BSA	BSA
	(cpm)	(cpm)	(cpm)	(cpm)	(cpm)	(cpm)	(cpm)	(cpm)	(cpm)
1:100	9483	10801	428	507	8769	428	9408	563	549
1:500	2229	2856	145	193	2062	103	2352	186	135
1:1000	1245	1450	89	150	1101	78	1496	110	75
1:5000	234	266	47	99	231	45	321	61	64
1:10000	152	249	37	69	155	26	167	78	88

* Wells were coated with 10 µg/ml of either 121-134-BSA, 135-151-BSA, or BSA, treated with either absorbed or unabsorbed sera, washed, and then exposed to 50,000 cpm of ^{125}I protein A.

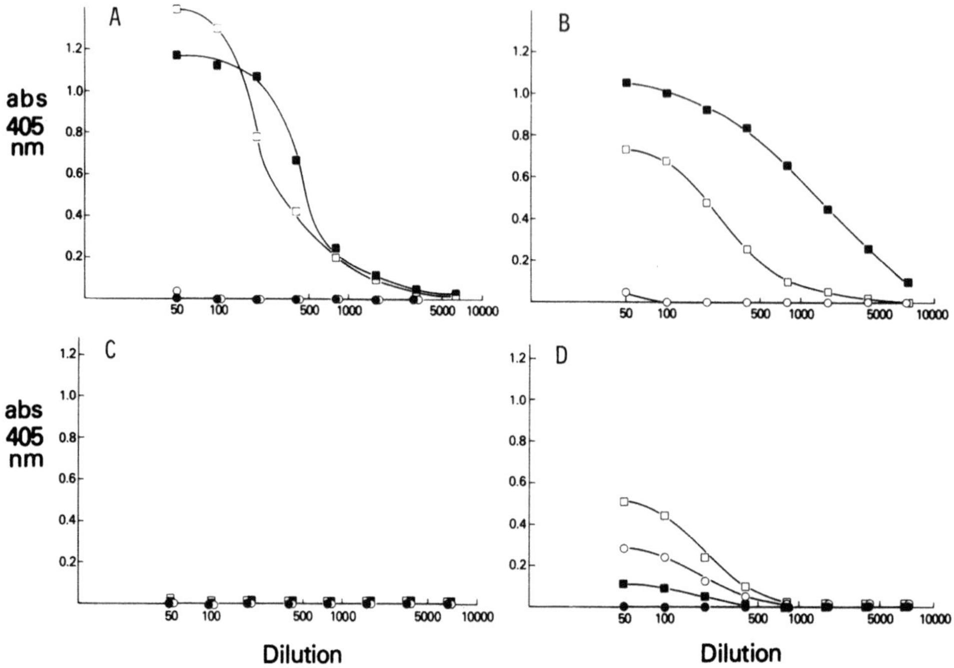

Figure 5. Reaction of anti-gonococcal pili sera with peptide-BSA conjugates corresponding to regions of sequence variation between strains. Anti-MS11 pili (A), anti-MS11 CNBr-3 (B), anti-R10 (C), and anti-R10 CNBr-3 (D) antisera were assayed for their ability to bind peptides corresponding to residues 95-107 ( ), 107-121 ( ), 121-134 ( ), and 135-151 ( ) of the MS11 pilin sequence.

specific, a common determinant(s) also has been demonstrated (Figure 5) to exist within CNBr-2. Peptides chosen to span most of this region of conserved sequence between MS11 and R10 pilin were synthesized as possible epitopes (Table 1). Glycine and cysteine were added to the carboxyl terminus of each peptide to allow them to be conjugated to BSA as previously described. Antisera against both MS11 and R10 pili and against CNBr-2 of each pilin were tested for their ability to bind these conjugates. Anti-whole pili, either MS11 or R10, bound only the region corresponding to residues 48-60 (Figure 6A and 6C). As expected for a common, immunorecessive determinant, both sera bind equally well, but significantly less than the sera bind to the strain specific epitopes. Neither sera bound peptides corresponding to residues 21-35, 41-50, or 69-84.

A different binding pattern is evident when antisera against the CNBr-2 of either MS11 or R10 is used (Figure 6B and 6D). Anti-MS11 CNBr-2 binds the peptide corresponding to residues 69-84 more

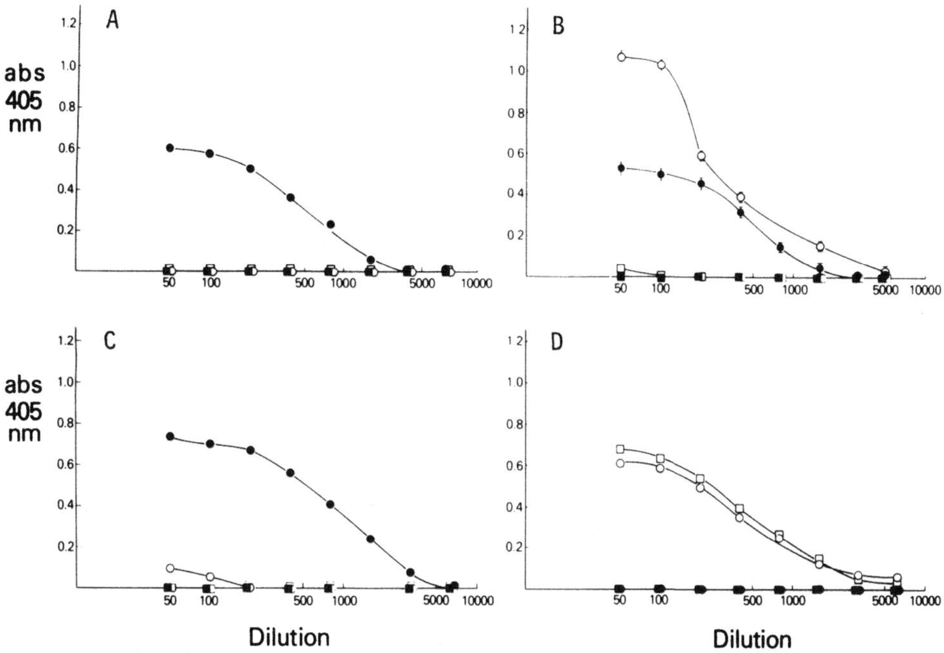

Figure 6. Reaction of anti-gonococcal pili sera with peptide-BSA conjugates coresponding to regions of conserved sequence between strains. Anti-MS11 pili (A), anti-MS11 CNBr-2 (B), anti-R10 pili (C), and anti-R10 CNBr-2 (D) antisera were assayed for their ability to bind peptides corresponding to residues 21-35 ( ), 41-50 ( ), 48-60 ( ), and 69-84 ( ).

effectively than 48-60. Anti-R10 CNBr-2 does not bind 48-60 at all. Instead this serum binds regions corresponding to 69-84 and 41-50. These experiments indicate that an epitope common to both R10 and MS11 pili exists between residues 48 and 60 and that the immunogenicity of this region is determined by interactions present in intact pili, but absent from CNBr-2. Another common determinant, corresponding to residues 69-84, evidently is exposed when CNBr-2 is used as an immunogen.

A solid phase radio-binding assay was performed with the MS11 and R10 anti-whole pili sera to more precisely compare the relative amount of antibodies bound by the common and each of the strain specific determinants. Table 3 is a compilation of the results of this experiment. At a 1:50 dilution of anti-MS11 pili sera approximately three times as many antibodies are bound by each of the strain specific epitopes as by the common determinant. Consequently less than 15% of the antigenicity of whole pili resides in the common determinant, consistent with the data shown in Figure 4.

Generation and analysis of anti-peptide sera. Two rabbits were immunized with each of the seven peptides conjugated to thyroglobulin. The resultant antisera were screened for their ability to bind the homologous peptide conjugated to bovine serum albumin as well as their ability to crossreact with intact MS11 and R10 pili (Figure 8). Though not identical in their immunogenicity, all of the peptides elicit good anti-peptide responses. However, they substantially differed in their ability to generate antibodies which crossreacted with the intact pilus proteins. Not surprisingly, the sera engendered by immunizing with peptides from regions of sequence identity between MS11 and R10 pili crossreacted equally well with the two proteins (Figure 7A-D). However, residues 69-84 are far more effective in generating crossreactive antibodies than 21-35, 41-50, and even 48-60, a natural epitope of gonococcal pilin. The three peptides from variable portions of the molecule 107-121, 121-134, and 135-151, also are roughly equal in their ability to elicit a crossreactive response MS11 pili. However, they

Table 3. Relative Binding of Polyclonal Rabbit Sera Against Intact MS11 Pili to Each of the Peptide-BSA Conjugates*.

	BSA (cpm)	21-35 (cpm)	41-50 (cpm)	48-60 (cpm)	69-84 (cpm)
Anti MS-11 1:50	612 ±12	656 ±21	810 ±28	6519 ±112	649 ±15
relative %	-	-	-	14.5	-

	95-107 (cpm)	107-121 (cpm)	121-134 (cpm)	135-151 (cpm)
ANTI MS-11 1:50	635±26	1824±41	17251±127	18020±95
relative %	-	1.5	40.7	42.6

*Wells were coated with 10)g/ml of the various peptide-BSA conjugates, washed, treated with a 1:50 dilution of anti-MS11 sera, washed, exposed to 50,000 cpm of ^{125}I protein A, washed, and the plates were cut and counted.

differ in their ability to bind R10 pili. Antisera to 135-151 bound R10 almost as well as MS11 (Figure 7G), whereas antisera to 121-134 was specific for MS11 (Figure 7F). Antisera against 107-121 was intermediate in its specificity, binding R10 only at low dilutions of antisera (Figure 7E). Peptides 121-134 and 135-151 each contain a strain specific epitope, yet when used as immunogens, 121-134 elicits a strain specific response whereas 135-151 generates antibodies which crossreact with heterologous pili.

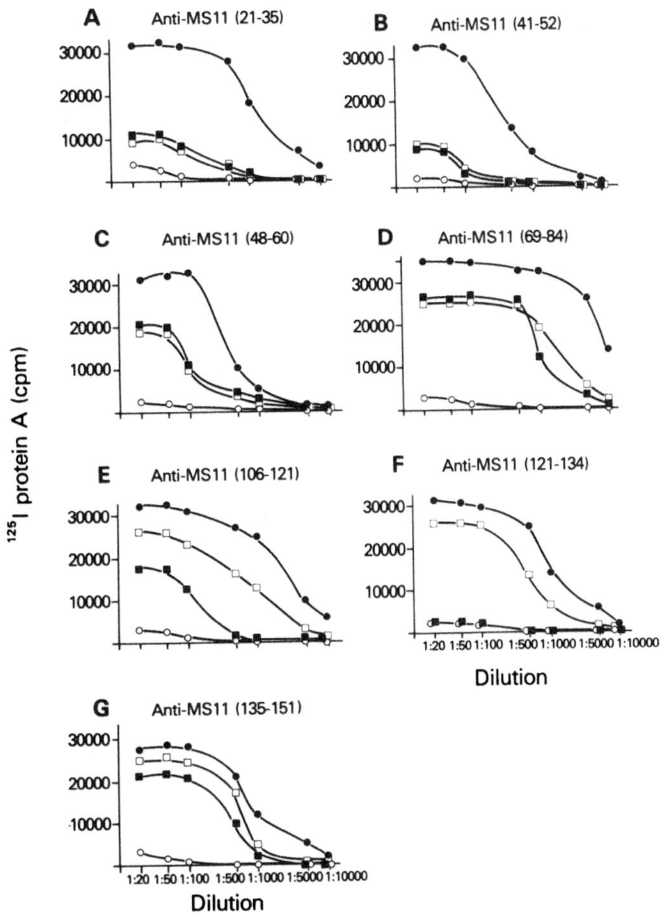

Figure 7. The crossreaction of anti-peptide sera with either homologous (MS11) or heterologous (R10) pili on solid phase. Antisera generated to peptide thyroglobulin conjugates corresponding to residues 21-35 (A), 41-50 (B), 48-60 (C), 69-84 (D), 106-121 (E), 121-134 (F), and 135-151 (G) of MS11 pilin were allowed to bind either the homologous peptide conjugate (●), intact MS11 pili (□), intact R10 pili (■), or BSA (○).

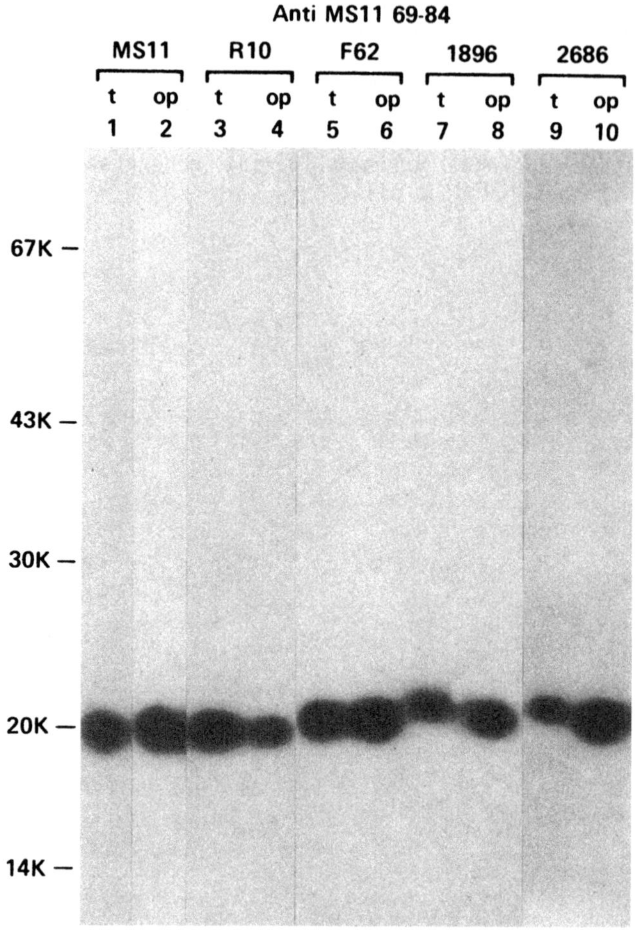

Figure 8. Autoradiograph of an immunoblot displaying the binding specificity of anti-69-84 sera for homologous and heterologous pilin. Intact tranparent (t) or opaque (op) gonococcal colonial variants from strains MS11 (lanes 1,2), R10 (lanes 3,4), F62 (lanes 5,6), 1896 (lanes 7,8), 2686 (lanes 9,10) were probed with a 1:400 dilution of anti-69-84 thyroglobulin conjugate.

Crossreaction of antipeptide sera with heterologous strains of pilin using immunoblots. The crossreactivity of the seven peptide antisera with intact pilin from MS11, R10, and three additional gonococcal strains was examined using immunoblots. The opaque and transparent colonies of each strain were selected from agar plates, boiled in sample buffer, run on acrylamide gels, transferred to nitrocellulose, and probed with each of the seven antisera. All of the sera bound MS11 pilin to varying extents. The strongest signal was seen with antisera to 69-84, 121-134, and 135-151. Antisera against 69-84 bound pilin from both opaque and transparent colonies of strains MS11, R10, F62, 1896, and 2686 (Figure 8). Sera from rabbits immunized with 135-151 also bound all strains tested (data not shown). In contrast, sera against 121-134 bound MS11, but not R10 pilin (data not shown). This sera crossreacts only with heterologous pilin from strain F62 (data not shown).

Inhibition of bacterial adhesion. Antisera to each peptide were screened for their ability to block the binding of viable gonococci derived from piliated, transparent colonial variants to human endometrial carcinoma cells grown as monolayers on coverslips. Endometrial cells were used because (i.) gonococcal endometritis is a well described phenomenon (Novak and Woodruff, 1979), (ii.) they more closely resemble the epithelial cell in natural infection, and (iii.) they are more readily bound by gonococci than either Hela or buccal cells. Strain F62 was employed in these assays because it displayed the greatest capacity to bind this cell line, has caused urethritis in male volunteers (Kellogg et al., 1968), and elaborates pili that possess regions of sequence homology and heterogeniety with MS11. The bacteria were incubated with each of the seven, heat-inactivated peptide antisera and then mixed with the endometrial cells on cover slips. The number of bacteria adhering to the cells were counted and compared with experiments that used identical dilutions of preimmune serum. Only antisera to peptides 41-50 and 69-84 efficiently prevented bacterial attachment to the eucaryotic surface (Table 4). A 1:10 dilution of either sera was most effective, but a 1:50 dilution of anti 41-50 and a 1:100 dilution of anti-69-84 still inhibited approximately 90% of the binding. In contrast, antisera to the other two peptides of the constant region (residues 21-35 and 48-60) did not block attachment. Interestingly, antisera against the two immunodominant regions of intact pili, residues 121-134 and 135-151, previously shown to crossreact with F62 pilin on immunoblots, did not inhibit the bacteria from binding the endometrial cells.

DISCUSSION

Defining the antigenic structure of gonoccocal pili is important for both the design of an efficacious vaccine and for providing molecular details which will lead to a general understanding of how pathogens create antigenic variation to avoid the host immune system.

The amino acid sequence of MS11 gonococcal pilin has been determined by automated Edman degradation of the intact protein and cyanogen bromide, citraconylated tryptic, and iodosobenzoic acid peptides. The relative hydrophilicity of the molecule (Levitt, 1976) and the relative probabilities of the molecules secondary structure are shown in Figure 9. The hydrophobic amino terminus of the molecule appears to be involved in the formation of pilin dimers. The intact protein, or any fragment containing the first thirty amino acids forms

Table 4. Inhibition of Gonocooccal Binding to Endometrial Cells by Synthetic Peptide Antisera.

	Peptide Antisera													
	Constant Region								Variable Region					
Serum Dilution	21-35		41-50		48-60		69-84		107-121		121-134		135-151	
	Pre*	Post+	Pre	Post	Pre	Post	Pre	Post	Pre	Post	Pre	Post	Pre	Post
1:10	47±14‡	12±10	32±8	3±1	48±14	56±15	26±9	1±1	42±13	48±19	61±16	72±10	80±13	73±22
1:25	41±7	56±9	65±15	9±4	N.D.§		48±10	8±3	60±12	72±11	63±14	58±9	74±18	99±23
1:50	N.D.		83±9	9±5	N.D.		170±21	10±4	N.D.		49±11	58±18	76±12	52±11
1:100	N.D.		54±10	49±12	N.D.		94±11	11±2	N.D.		N.D.		89±11	89±26
1:200	N.D.		38±8	45±8	N.D.		41±9	46±7	N.D.		N.D.		N.D.	

*Preimmune sera
+Immune serum against peptide
‡Average number of bacteria per cell ± standard deviation
§Not determined

high molecular weight aggregates under physiological conditions. This region is highly conserved not only among pilin from different strains of gonococci, but also in pilin isolated from P. aeruginosa (Sastri et al., 1983), N. meningitidis (Hermodson et al., 1978), and M. nonliquifaciens (Froholm and Sletten, 1977). Interestingly, residues 2-21 of gonococcal pilin displays partial homology with residues 202-221 of both muscle and cytoplasmic actin (Figure 10). The homology becomes much more substantial when the two areas are arranged in a helical conformation and modeled in a two dimensional helical wheel. Eight of the ten identical residues are clustered on a hydrophobic face. This topography might favor subunit dimerization in both actin and pilin due to hydrophobic packing between helices.

The conserved region of sequence also appears to contain all the residues necessary for the ligand binding site. Either TC-2, residues 31-111, or a tryptic fragment of CNBr-2, residues 31-92 can bind endometrial cells, but not either human buccal or Hela cells.

The experiments described in this paper have defined partially the natural epitopes of gonococcal pili by using synthetic peptides corresponding to linear sequences of the protein. Two strain specific determinants are located within the disulfide loop near the carboxyl terminus and a single epitope common to both strains R10 and MS11 is present between residues 48-60. We recognize that these areas might not constitute the complete antigenic structure of pili because if there are any determinants composed of nonlinear amino acids, created by the juxtaposition of residues due to either the folding of the pilin

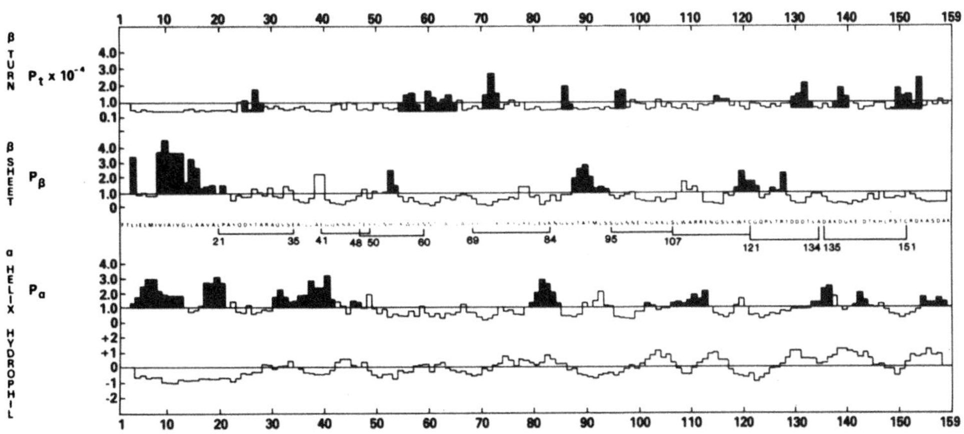

Figure 9. Predictive analysis of the secondary structure (Chou and Fasman, 1978) and the local average hydrophilicity (Levitt, 1976) of MS11 gonococcal pilin.

molecule or its polymerization, they would not be detected by these reagents. However, an analysis of the regions containing the linear epitopes indicates that they can account for many of the interesting features of the antigenicity of gonococcal pili.

The location of the three peptides containing the antigenic determinants (48-60, 121-134, and 135-151) as well as the five other areas synthesized within the MS11 pilin sequence are shown in Figure 9. Table 5 compares the sequences of R10 and MS11 pilin in the eight regions synthesized as well as the quantitative predictions for β-turns (Chou and Fasman 1978). All known sequence variation between R10 and MS11 pilin occurs in the carboxyl terminal cyanogen bromide generated fragment (CNBr-3, residues 93-159). The most striking feature of the variations in sequence between R10 and MS11 pilin is that most of the differences occur within or adjacent to predicted β-turns and yet none of the substitutions affect the predicted secondary structure. Turns are likely regions for advantageous mutations to occur within microbial proteins exposed to host immunity. A number of investigators have postulated that turns exist as a means of keeping a protein globular by reversing the direction of a sheet or helix back into the center of a protein which results in the turn being present on the surface of a protein (Kuntz, 1972, Levitt and Chothia, 1976, Rose, 1978, Rose and Roy, 1980). Thus, unlike sheets and helices, the amino acid side chains in turns interact with solvent and not other regions of the protein.

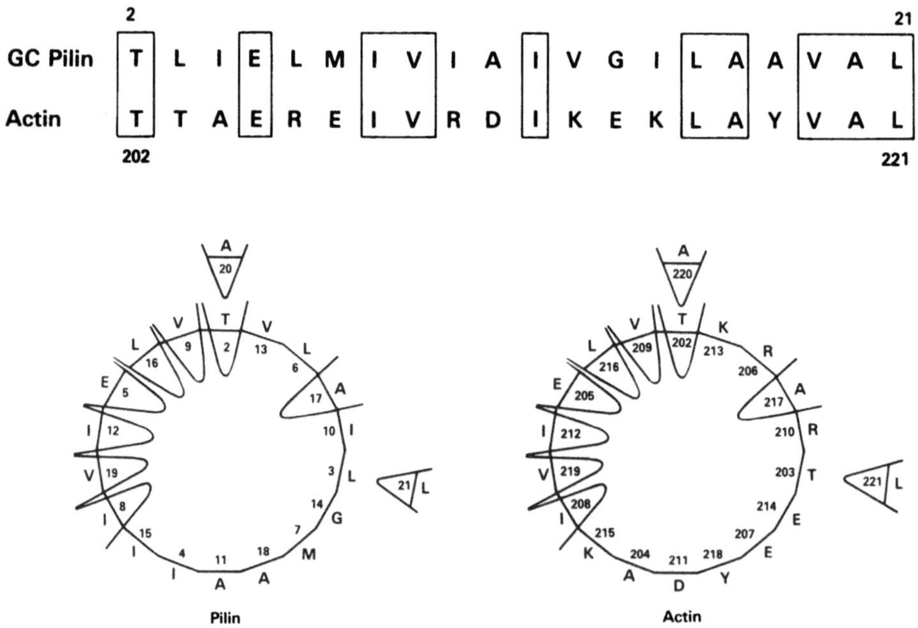

Figure 10. Comparison of the gonococcal pilin and the actin sequences. Identical amino acids are enclosed in boxes.

Table 5. The Comparison of MS11 and R10 Pilin Sequences in the Regions Synthesized and the Predicted Location of β-Turns.

		sequences	Turn probability*
MS 11	21	L P A Y Q D Y T A R A Q V S E 35	1.79×10^{-4}
R 10		- - - - - - - - - - - - - - - θ	
MS 11	41	E G Q K S A V T E Y 50	0.76×10^{-4}
R 10		- - - - - - - - - -	
MS 11	48	T E Y Y L N H G K W P E N 60	1.43×10^{-4}, 1.62×10^{-4}ψ
R 10		- - - - - - - - - - - - -	
MS 11	69	P P S D I K G K Y V K E V E V K 84	3.10×10^{-4}
R 10		[ ]¢	
MS 11	95	S G V N N⁺ E I K G⁻ K K L S 107	0.89×10^{-4}
R 10		- N - - K - - - D - - - -	0.84×10^{-4}
MS 11	107	S L W A R R E N G S V K W F C 121	0.94×10^{-4}
R 10		- - - - K - - A - - - - - - -	0.86×10^{-4}
MS 11	121	C G Q P V T R T D D D⁺² T V⁺ A 134	2.29×10^{-4}
R 10		- - - - - - - - A - K - N R T	2.75×10^{-4}
MS 11	135	D⁺ A K⁻² D⁺ G K E I D T K H L P S T C 151	1.88×10^{-4}, 1.81×10^{-4}
R 10		V - D A [ ]	

* Probability of the underlined four amino acids to be present in a reverse turn based on the algorithm derived by Chou and Fasman. Regions whose probability is greater than 0.7×10^{-4} are assigned as turns by this algorithm. Relative charge differences between the two sequences are shown above the indicated position of the MS11 peptide.

θ--- indicates sequence identity between MS11 and R10 pilin.

ψTwo overlapping turns are predicted in this region, HGKW and WPEN.

Consequently as long as the substitution of an amino acid maintains the turn there is no need to mutate a second amino acid to conserve the correct tertiary structure of the protein, as might be the case if the mutation occurs in a helix or sheet.

The proposed antigenic structure of the molecule supports this theory. Both of the peptides containing strain specific determinants (121-134 and 135-151) have a predicted turn around which the sequence variations are localized (Table 5). Each peptide also contains a turn within which there exists a position where one strain has a lysine whereas the other contains an aspartic acid. Such a variation in sequence involving a double charge change could explain the total absence of crossreactivity of the complementary antibody combining sites. Another feature of the strain specific determinants is that they are localized within a disulfide loop, a region of the protein known not to be involved in either pilin polymerization or receptor binding. Single disulfide loops also are found in Moraxella bovis (Schoolnik, unpublished results), Pseudomonas aeruginosa (Sastri et al., 1983), and common and Gal-Gal binding Escherichia coli pili (Baga et al., 1984, Svanborg-Eden, 1983). We believe that the location of the principal epitopes within the disulfide loop is part of a sophisticated molecular design that enables the organism to segregate its principal antigenic determinants in a region distinct from the residues involved in receptor binding and subunit polymerization. By using a disulfide loop, we propose that the molecule has evolved a domain which (i.) prominently orients the region on the surface of the molecule, perhaps even projecting out from the polymeric structure, resulting in its increased antigenicity and immunogenicity, (ii.) allows amino acid substitutions, additions, and deletions within the antigenic determinants, thereby maximizing antigenic variation, without disrupting areas critical for the biological role of the protein and (iii.) permits an antibody to bind pili and yet not interfere with gonococcal adhesion to the mucosal cell surface (vide infra). The highly variable antigenic determinants of influenza virus neuraminidase recently have been shown to be designed in a similar fashion. They are located in β-turns, in a region distinct from the active site, which enables antibodies to bind to the epitopes without inhibiting the enzyme (Colman et al., 1983).

In addition to making the regions within the disulfide loop immunodominant, the design of pilin also has resulted in the conserved regions involved in receptor binding being immunorecessive. Our experiments have shown that the small, crossreactive antibody fraction of anti-pili sera is directed at an epitope located between residues 48-60. The poor immunogenicity of this region compared to the strain specific determinants explains why pili from a single gonococcal strain is not suitable for use as a vaccine. In contrast, CNBr-2 (residues 9-92) elicits antiserum that crossreacts well with heterologous gonococcal pili. The principal epitope of antibodies generated against this fragment is present between residues 69-84. This is the first indication that peptides from pilin can elicit a different population of antibodies than when the intact protein is used as an immunogen.

The same peptides used to map the antigenic determinants of gonococcal pilin also were used as immunogens. Experiments described in this paper provide evidence that synthetic peptides corresponding to weakly immunogenic regions of the protein can elicit antibodies that not only crossreact with the intact protein, but also inhibit heterologous gonococcal strains from binding eucaryotic cells. The relative immunogenicity of these peptides can be correlated with the antigenic structure of pilin to allow us to begin to contrast the general features of the immunogenicity of peptides and proteins.

The seven peptides used in these experiments as immunogens all evoked relatively equal anti-peptide responses. The regions containing the highly immunogenic, strain-specific epitopes of the intact protein (121-134 and 135-151) failed to generate any greater anti-peptide response than the other five peptides. These results are comparable to the responses generated by the approximately seventy other peptides we have used as haptens in related studies. All peptides which are sufficiently hydrophilic to be soluble in aqueous buffers have been found to generate an approximately equal anti-peptide response. However, antisera elicited by each of the seven peptide analogues of gonococcal pilin substantially differ in their ability to crossreact with the intact protein when assayed either by solid phase binding, immunoblotting, or immune precipitation. The variation in the peptides' ability to generate a crossreactive response is best explained by the number and type of B-cell clones the peptides stimulate. When a synthetic peptide is used as an immunogen, the possible conformation that it can adopt will determine which clones in the B-cell repertoire are stimulated to produce antibodies. Only if the peptide adopts a secondary structure identical to that it assumes in the intact protein will it elicit antibodies that crossreact with that protein. We believe that the peptides that generate the highest anti-pili response (69-84, 121-134, and 135-151) have the highest probability of existing in the identical conformation they exist in pilin or have the lowest energy requirements for being induced to adopt the correct conformation by a complementary antibody combining site.

The regions of pilin that were synthesized were chosen based on their likelihood to adopt stuctures similar to their predicted conformation in the intact protein. Each peptide has a predicted reverse turn located distal to its site of linkage to the carrier protein. We believe that of the three known secondary structures, α-helix, β-sheet, and reverse or β-turns, reverse turns have the highest probability to exist in a peptide less than twenty amino acids in length. This conclusion is based on the supposition that the forces involved in sheet and helix formation are primarily found in the tertiary structure of the protein. Such interactions as inter-strand hydrogen bonding in sheets and hydrophobic bonds found in helix packing and sheet-sheet interactions will be absent in a linear peptide. The structural forces responsible for reverse turns are much simpler. Turns invariably are composed of hydrophilic residues prominantly exposed on the surface of globular proteins, often with the first and fourth amino acids associated with a hydrogen bond (Richardson, 1981). Usually they are flanked by hydrophobic residues whose association is thought to be the driving force for turn formation (Rose and Roy, 1980). Consequently, relatively short, linear peptides can contain all the residues involved in turn formation.

Additional evidence that the immune response against peptides is distinct from that against the intact protein is the specificity of the antibodies generated against residues 135-151. In the intact protein this region elicits antibodies that are specific for MS11 pili. In contrast, the isolated peptide generates antiserum that binds not only MS11, but pili from all other strains tested. Comparison of the sequences of pili from different gonococcal strains indicates that residues 135-145 are variable, but residues 146-151 are constant. Evidently, only the variable region is immunogenic when the intact protein is used as an immunogen, whereas both the variable and conserved regions are immunogenic when the peptide is used. We assume that the different responses are due to events involved in antigen presentation. Free antigen does not stimulate B-cells directly. Only when antigen is associated with immune response gene products and the T-cell receptor is it capable of stimulating a humoral response (McDevitt and Benacerraf, 1969). These components of the immune system must associate with pili and may prevent portions of its surface from being seen by the B-cell repertoire.

Such a model can explain why a region such as 69-84 which is very immunogenic as a free peptide or when it is included in a cyanogen bromide generated fragment fails to elicit a response when it is part of intact pili. Steric inaccessibility cannot explain this phenomenon because antisera against CNBr-2 or residues 69-84 both bind pili avidly, proving that this region is antigenic in the intact protein.

Reverse turns also were attractive to emphasize in this study because of their presence in the ligand binding sites of the few carbohydrate binding proteins whose three dimensional structure has been determined (Colman et al., 1983; Phillips, 1967; Becker et al., 1976; Hardman and Ainsworth, 1976; Newcomer, 1981). Based on the mechanism of adhesion of other gram negative bacteria (Salit and Gotschlich, 1977; Ofek et al., 1977; Gaastra and deGraaf, 1982), pilin is most likely a lectin and binds a carbohydrate moeity on the eucaryotic cell surface. We felt that an antibody generated against a predicted turn in the conserved region of gonococcal pilin might block pili from binding eucaryotic membranes. These ideas appear to be correct because antibodies elicited by residues 41-50 and 69-84 from the constant portion of MS11 pilin inhibit strain F62 from attaching to endometrial cells in vitro. In contrast, antibodies to two peptides distal to the receptor binding segment, 121-134 which shares an epitope with F62 (data not shown) and 135-151 which elicits high titer, crossreacting antibodies, do not block attachment. Thus we propose that regions 41-50 and 69-84 are involved with receptor binding, although neither allosteric interactions nor nonspecific inhibition of attachment by the crosslinking of pili by divalent antibody has been ruled out.

These results are consistent with earlier observations (Tramont et al., 1981) that an intact pilus vaccine elicits receptor blocking antibodies directed at a common epitope, since they could be absorbed with heterologous pili. In contrast, others have recently reported that strain specific, but not crossreacting monoclonal antibodies block attachment, although the location of their epitopes was not determined (Virgi and Heckels, 1984). Taken as a group, these studies and the data presented here suggest that both linear and nonlinear strain specific and common epitopes probably exist and that representatives of each may elicit receptor blocking antibodies depending on their spatial proximity to the residues involved in ligand binding in the folded molecule. These considerations not withstanding, by eliciting sequence specific, functionally defined antibodies, the synthetic peptides described in this report have extended our previous observations on the structure-function analysis of this organelle. The peptides that elicit crossreacting, receptor blocking antibodies are promising candidate immunogens for the prevention of gonnorrhea.

ACKNOWLEDGEMENT

We thank Rosemary Fernandez and Irene McClain for excellent assistence with these expeiments, Dr. Nelson Teng for the ENCA-4 cell line, and Robin Martin for preparation of the manuscript. This research was supported by a grant from Institut Merieux, in part from a grant from the Veterans Administration, and N.I.H. grant CA-36422. G.K.S. is a recipient of a Hartford Foundation Fellowship.

REFERENCES

Atassi, M., and Habeeb, A. 1972. Reaction of proteins with citraconyl anhydride. Methods Enzymol. 25B:546.

Baga, M.S., Normark, S., Hardy, J., O'Hanley, P., Lark, D., Olsson, O., Schoolnik, G., and Falkow, S. 1984. Nucleotide sequence of pap A gene encoding the pap pilus subunit of uropathogenic E. Coli. J. Bacteriol. 157:330.

Barbour, A.G., Tessier, S.L., and Stoenner, H.G. 1982. Variable major proteins of Borrelia hermsii. J. Exp. Med. 156:1312.

Becker, J.W., Reeke, G.N.Jr., Cunningham, B.A., and Edelman, G.M. 1976. New evidence on the location of the saccharide-binding site of concanavalin A. Nature (Lond.) 259:406.

Borst, P. and Cross, G.A.M. 1982. Molecular basis for trypanosome antigenic variation. Cell 29:291.

Brinton, C.C.Jr., Bryan, J., Dillon, J.A., Guerina, N., Jacobson, L.J., Kraus, S., Labik, A., Lee, S., Levene, A., Lim, S., McMichael, J., Polen, S., Rodgers, K., To, A.C., and To, S.C. 1978. in Immunobiology of Neisseria gonorrhoeae. Brooks, G., Holmes,K. Sawyer, W.D., and Young, F.E. editors. American Society for Microbiology, Washington, p. 155.

Chothia, C., Levitt, M., and Richardson, D. 1977. Structure of proteins: packing of α-helices and β-pleated sheets. Proc. Natl. Acad. Sci., U.S.A. 74:4130.

Chou, P.Y., and Fasman, G.D. 1978. Empirical Predictions of protein conformation. Ann. Rev. Biochem. 47:251.

Colman, P.M., Varghese, J.N., and Laver, W.G. 1983. Structure of the catalytic and antigenic sites in influenza virus neuraminidase. Nature (Lond.) 303:41.

Cross, G.A.M. 1978. Antigenic variation in trypanosomes. Proc. R. Soc. Lond. B. 202:55.
Dale, J., Ofek, I., and Beachey, E.H. 1980. Heterogeneity of type specific and cross reactive antigenic determinants within a single M protein of group A streptococci. J. Exp. Med. 151:1026.
Engvall, E. 1980. Enzyme immunoassay methods. Methods in Enzymology. 70:419
Erickson, B.W., and Merrifield, R.B. 1976. Solid phase peptide synthesis. In The Proteins, vol. II. Neurath, H., editor. Academic Press, N.Y. p. 255.
Froholm, L.O. and Sletten, K. 1977. Purification and N-terminal sequence of fimbrial protein from Moraxella nonliquefaciens. FEBS Lett. 73:29.
Gaastra, W. and deGraaf, F.K. 1982. Host specific fimbrial adhesins of noninvasive enterotoxigenic Escherichia coli strains. Microbiol. Rev. 46:129.
Greenwood, F.C., Hunter, W.M., and Glover, J.S. 1963. The preparation of ^{125}I-labeled human growth hormone of high specific radioactivity. Biochem. J. 89:114.
Gross, E. and Witkop, B. 1962. Nonenzymatic cleavage of peptide bonds: the methionine residue in bovine pancreatic ribonuclease. J. Biol. Chem. 237:1856.
Hardman, K.D. and Ainsworth, C.F. 1976. Structure of the Concanavalin A-methyl α-D-mannopyranoside complex. Biochemistry 15:1120.
Hermodson, M.A., Chen, K., and Buchanan, T.M. 1978. Neisseria pili proteins: amino terminal amino acid sequences and identification of an unusual amino acid. Biochemistry 17:442.
Hopp, T.P. and Woods, K.R. 1981. Prediction of protein antigenic determinants from amino acid sequences. Proc. Natl. Acad. Sci., U.S.A. 78:3824.
Jaenicke, R. editor 1980. Protein folding, Proceedings of the 28th conference of the German Biochemical Society. Amsterdam: Elsevier/North Holland Biomed. Press. 587pp.
Kaiser, E., Colescott, R.L., Bossinger, C.D., and Cook, P.I. 1970. Color test for the detection of free terminal amino groups in solid phase synthesis of peptides. Anal. Biochem. 34:595.
Kuntz, I.D. 1972. Tertiary structure in carboxypeptidase. J. Am. Chem. Soc. 94:8568.
Levitt, M. 1976. A simplified representation of protein conformations for rapid simulation of protein folding. J. Mol. Biol. 104:59.
Levitt, M. and Chothia, C. 1976. Structural patterns in globular proteins. Nature (Lond.) 261:552.
Mahoney, W. and Hermodson, M.A. 1979. High-yield cleavage of tryptophanyl peptide bonds by o-iodosobenzoic acid. Biochemistry. 18:3810.
Newcomer, M.E., Gilliland, G.L., and Quiocho, F.A. 1981. L-arabinose binding protein-sugar complex. J. Biol. Chem. 256;13213.
Ofek, I. Mirelman, D., and Sharon, N. 1977. Adherence of Escherichia coli to human mucosl cells mediated by mannose receptors. Nature (Lond.). 265:623.
Phillips, D.C. 1967. The hen egg white lysozyme molecule. Proc. Natl. Acad. Sci. U.S.A. 57:484.
Porath, J. 1974. General methods and coupling procedures in affinity

chromatography. Methods Enzymol. 34:13.
Richardson, J.S. 1981. The anatomy and taxonomy of protein structure. Advances in Protein Chem. 34:167.
Rose, G.D. 1978. Prediction of chain turns in globular proteins on a hydrophobic basis. Nature (Lond.). 272:586.
Rose, G.D. and Roy, S. 1980. Hydrophobic basis of packing in globular proteins. Proc. Natl. Acad. Sci. U.S.A. 77:4643.
Salit, I.E. and Gotschlich, E.C. 1977. Hemagglutination by purified Type I Escherichia coli pili. J. Exp. Med. 146:1169.
Sastry, P.A., Pearlstone, J.R., Smillie, L.B., and Paranchych, W. 1983. Amino acid sequence of pilin isolated from Pseudomonas aeruginosa. FEBS Lett. 151:253.
Schoolnik, G.K., Tai, J.Y., and Gotschlich, E.C. 1983. A pilus peptide vaccine for the prevention of gonorrhea. Prog. Allergy. 33:314.
Schoolnik, G.K., Fernandez, R., Tai, J.Y., Rothbard, J.B., and Gotschlich, E.C. Gonococcal pili: primary structure and receptor binding domain. J.Exp.Med. 159:1351.
Svanborg-Eden, C., Gotschlich, E.C., Korhonen, T.K., Leffler, H., and Schoolnik, G.K. 1983. Aspects on the structure and function of pili on uropathic E. Coli. Prog. Allergy. 33:189.
Swanson, J. 1973. Studies of gonococcal infection, IV. Pili: their role in attachment of gonococci to tissue culture cells. 1973. J. Exp. Med. 137:571.
Tramont, E.C., Sadoff, J.C., Boslego, J.W., Ciak, J., McChesney, D., Brinton, C.C., Wood, S. and Takafuji, E. 1981. Gonococcal pilus vaccine. J. Clin. Invest. 68:881.
Virgi, M. and Heckels, J.E. 1984. The role of common and type-specific pilus antigenic domains in adhesion and virulence of gonococci for human epithelial cells. J. General Micro. 130:1089.
Wiley, D.C., Wilson, I.A., and Skehel, J.J. 1981. Structural identification of the antibody combining sites of Hong Kong influenza haemagglutinin and their involvement in antigenic variation. Nature (Lond.). 289:373.
Yoshitake, S., Yamada, Y., Ishikawa, E., and Masseyeff, R. 1979. Conjugation of glucose oxidase from Aspergillus niger and rabbit antibodies using N-hydroxysuccinimide ester of N-(4-carboxycyclohexylmethyl)maleimide. Eur. J. Biochem. 101:395.

INDEX